Weather Analysis and Forecasting

Weather Analysis and Forecasting

Applying Satellite Water Vapor Imagery and Potential Vorticity Analysis

Second Edition

Christo G. Georgiev

*National Institute of Meteorology
and Hydrology
Bulgarian Academy of Sciences*

Patrick Santurette

*Forecasting Operations Department
Météo-France*

Karine Maynard

*Forecast Laboratory
Météo-France*

AMSTERDAM • BOSTON • HEIDELBERG • LONDON
NEW YORK • OXFORD • PARIS • SAN DIEGO
SAN FRANCISCO • SINGAPORE • SYDNEY • TOKYO
Academic Press is an imprint of Elsevier

Academic Press is an imprint of Elsevier
125 London Wall, London EC2Y 5AS, UK
525 B Street, Suite 1800, San Diego, CA 92101-4495, USA
50 Hampshire Street, 5th Floor, Cambridge, MA 02139, USA
The Boulevard, Langford Lane, Kidlington, Oxford OX5 1GB, UK

Notices
Knowledge and best practice in this field are constantly changing. As new research and experience broaden our
understanding, changes in research methods, professional practices, or medical treatment may become
necessary.

Practitioners and researchers must always rely on their own experience and knowledge in evaluating and using
any information, methods, compounds, or experiments described herein. In using such information or methods
they should be mindful of their own safety and the safety of others, including parties for whom they have a
professional responsibility.

To the fullest extent of the law, neither the Publisher nor the authors, contributors, or editors, assume any
liability for any injury and/or damage to persons or property as a matter of products liability, negligence or
otherwise, or from any use or operation of any methods, products, instructions, or ideas contained in the
material herein.

British Library Cataloguing-in-Publication Data
A catalogue record for this book is available from the British Library

Library of Congress Cataloging-in-Publication Data
A catalog record for this book is available from the Library of Congress

ISBN: 978-0-12-800194-3

For information on all Academic Press publications
visit our website at https://www.elsevier.com/

Working together
to grow libraries in
developing countries

www.elsevier.com • www.bookaid.org

Publisher: Candice Janco
Acquisition Editor: Sara Scott
Editorial Project Manager: Tasha Frank
Production Project Manager: Vijayaraj Purushothaman
Designer: Mark Rogers

Typeset by TNQ Books and Journals

The cover illustration is made by original images produced for the purposes of this book by the authors using
data available in Météo-France. These are satellite water vapor images overlaid by meteorological fields for
analysis of cases considered in the material: Top left in Chapter 3, Section 3.5.3.1; Top right in Chapter 3,
Section 3.6; Bottom in Chapter 4, Section 4.3.2.

Contents

PART 2 PRACTICAL USE OF WATER VAPOR IMAGERY AND THERMODYNAMIC FIELDS

Preface

The main purpose of this book is to provide weather forecasters and operational meteorologists with a practical guide for interpreting satellite water vapor imagery in combination with meteorological fields from numerical models to enable an optimal analysis of atmospheric thermodynamics.

The developments in dynamic meteorology have proved the use of potential vorticity fields as an efficient approach in operational meteorology for more than three decades. This guide illustrates the potential vorticity concept and the current techniques for interpreting imagery in water vapor channels to understand thermodynamic characteristics and evolution of the synoptic situation. The book focuses on numerous examples showing superimpositions between operational numerical model fields and satellite images in the context of understanding the relevant atmospheric processes over the midlatitude, subtropical, and tropical areas. Brief explanations are included, where appropriate, to show the role of the imagery in a forecasting environment. Conceived as a practical training manual for weather forecasters, the book will be of interest and value to university students as well.

Acknowledgments

This manual has been developed in the framework of cooperation between Météo-France and the National Institute of Meteorology and Hydrology of Bulgaria. The calculations by the RTTOV radiative transfer model for illustration the material in Chapter 2 and Section A.4 of Appendix A were performed by Fabienne Dupont (Météo-France). The authors are very grateful to Fabienne Dupont also for computer and software support provided during the studies on the work. Sincere thanks are conveyed to Dr. Johannes Schmetz (EUMETSAT) for the useful discussions on the interpretation of radiances in water vapor channels and to Pascal Brunel (MétéoFrance) for the helpful suggestions on the use of the RTTOV model output. The calculations by the LBLRTM radiative transfer model were kindly performed by Stephen Tjemkes (EUMETSAT) for developing the material presented in Section A.3 of Appendix A. Special thanks are also due to the anonymous reviewers and to the developmental editor for their generous contributions of time and insight. Météo-France funded the joint work of P. Santurette and C. Georgiev on the manual from 2005 to 2013. In 2015, Elsevier provided a grant to defray travel expenses, allowing the authors to meet in Météo-France for final developments of the manuscript. Thanks also to Kevin Eagan, the copyeditor who worked on this title.

Introduction

Meteorological satellites set out around the Earth allow a complete surveillance of the atmosphere. Satellite imagery gives a global and consistent view of the organization of atmospheric features in a great variety of scales over large areas. These data are now operationally available in various spectral channels of three generations of geostationary satellite systems, and the operational updating rate of the images has increased with a cycle of repetition of 30, 15, or even 5 min. Such tools help the human forecaster in the early recognition of high-impact weather phenomena. In operational weather forecasting, satellite imagery is used in combination with other meteorological data, especially with relevant numerical parameter fields. This combined use (see EUMeTrain/EUMETSAT, 2012; COMET, 2016) is a major requirement for an optimal detection of ongoing physical processes as well as to overcome the problem of an excessive amount of material in the forecasting environment.

Water vapor (WV) channels provide meteorologists with valuable information about the moisture and dynamical properties of the troposphere. Sequences of WV images of the geostationary satellites may be used in detecting thermodynamic characteristics of the troposphere and understanding how the significant factors for the development of high-impact weather systems interact. Understanding of important large-scale atmospheric processes calls for diagnosing imagery jointly with meteorological fields representing atmospheric circulation at middle and upper levels. Such dynamical fields include absolute vorticity or potential vorticity (PV) owing to their close relationship with WV channel imagery.

Circulation and vorticity have been recognized as helpful quantities since the beginning of the 20th century, and on this basis, PV theory was first developed by Rossby and Ertel in the late 1930s. Although PV was introduced as a dynamic atmospheric parameter in the early 1940s, its application was limited, mainly because of the complexity involved in calculating PV fields. With the advent of modern computer technology and its application to meteorology, various computer-generated PV fields have begun to appear since 1964. Hoskins et al. (1985) acknowledged the analysis of isentropic PV maps as a crucial diagnostic tool for understanding dynamical processes in the atmosphere. As a consequence, there has been enormously increased interest in using PV for diagnosing atmospheric behavior, especially of cyclogenesis, for research and operational forecasting purposes (eg, Mansfield, 1996; Hoskins, 1997; Molinari et al., 1998; Agustí-Panareda et al., 2004).

During the last two decades, powerful techniques were developed to invert the PV fields into winds and temperatures that can then be used to correct the numeric analysis and to rerun the numerical weather prediction (NWP) models (Demirtas and Thorpe, 1999; Hello and Arbogast, 2004; Arbogast et al., 2008). These PV inversion methods are used for applying changes in the NWP initial conditions in sensitivity studies about the impact of the selected PV anomalies on the structure of the initial atmospheric fields and on the subsequent dynamical evolution of the simulated circulation systems. This approach has been shown as a powerful tool toward the understanding of important atmospheric aspects related to baroclinic and barotropic development of cyclogenesis and convection, and it has been widely used in numerical studies of severe weather cases (Huo et al., 1999; Romero, 2001; Homar et al., 2002, 2003).

Overlaying PV fields onto satellite WV channel images shows a close relationship in the circulation systems of extratropical cyclones. The relationship facilitates image interpretation and helps to validate NWP output. A mismatch between the vorticity fields and the imagery can indicate a model analysis or forecasting error. The relationship has also been applied to the adjustment of initial fields in NWP models (eg, Pankiewicz et al., 1999; Swarbrick, 2001; Santurette and Georgiev, 2005; Argence et al., 2009; Arbogast et al., 2012). Based on the work of Santurette and Georgiev (2005), training materials have been produced by COMET (2015) on the examination of the relationship between model data and WV imagery to help assess the validity of the NWP model's forecasts. This COMET lesson presents diagrams to help understand how to apply the concept and provides an experience applying the assessment procedures and forecast modifications to four different cases over the North Atlantic and Europe.

Part I presents the fundamentals essential for understanding the specific material presented in Part II. Chapter 1 provides knowledge on basic points of atmospheric dynamics. Chapter 2 describes the information content of radiances measured by satellites in WV channels and illustrates the approach for interpreting imagery gray shades. In response to the broad international interest about the application of WV imagery from the current geostationary satellites, Part II of the book includes analyses of atmospheric processes over Europe, North America, Eastern and Western Atlantic, Asia and Indian Ocean, Southern and Northern Pacific. Processes of interaction between midlatitude and tropical circulation systems are also considered. However, the considerations are focused especially on the European region for two main reasons:

1. This region is rich in observational data that allows much easier verification/illustration of the applied concepts and conclusions regarding the association of WV imagery structures to corresponding circulation patterns as well as the NWP model performance.
2. The European Meteosat Second Generation (MSG) geostationary satellites provide imagery in two WV channels, and the book is aimed to offer guidance for application of 7.3 μm images, in addition to those of the broadly used 6.2 μm (for some satellites 6.3, 6.7, or 6.8 μm; see Appendix A) WV channel.

The three chapters of Part II are devoted to various operational applications. WV images are matched with various fields to provide operational forecasters with knowledge about the relationship between the patterns of main thermodynamical fields and the satellite images. The focus is put on the use of PV fields jointly to 6.2 μm channel imagery for diagnosis of synoptic processes, which are driven by the upper-level dynamics. Imagery in the 7.3 μm channel is considered as a tool for observation of structures in the mid-level moisture field and their relation to the thermodynamic features, which are present from the low to the middle troposphere. Chapter 3 illustrates the dynamical insight offered by WV images for interpreting the evolution of significant synoptic-scale circulation patterns. Chapter 4 is dedicated to deep convection, which is a new topic as regards the first edition of Santurette and Georgiev (2005). This chapter discusses important aspects of convective development: the dynamical forcing and inhibition, temperature, and humidity structures of convective environments associated with mid- and upper-level features that can be identified in WV imagery and PV fields. Chapter 5 is focused on the problem of validating NWP fields from analyses and early forecasts. A methodology is presented for helping to improve operational forecasts by comparing PV fields, satellite WV imagery, and pseudo WV images, which are synthetic products of the numerical model.

Although much of the material in Chapters 2, 3, and 5 has appeared elsewhere (Santurette and Georgiev, 2005, 2007; Georgiev and Santurette, 2009; Arbogast et al., 2012), it has been necessary to integrate it here to enable a better understanding of the new material discussed in this edition.

Chapters 3, 4, and 5 conclude with summaries, which let the reader refer easily to any of the specific interpretation problems discussed in the book.

FUNDAMENTALS

A DYNAMICAL VIEW OF SYNOPTIC DEVELOPMENT

CHAPTER OUTLINE

1.1 VORTICITY AND POTENTIAL VORTICITY

Some meteorological parameters are more effective than others for studying the appearance and evolution of dynamical structures at synoptic scale. The conservative parameters—those that remain unchanged when one follows a particle of fluid in motion—are best suited to detect and monitor the structures that play various key roles in a meteorological scenario. With the assumption of adiabatic motions, the potential temperature θ and wet-bulb potential temperature θ_w are thermodynamic tracers for the air particles. They allow us to compare the thermal properties of air particles without taking into account the effects due to thermal advection and pressure changes. However, they only represent a few of the important properties that determine the evolution of the atmosphere. To better understand the observed phenomena, dynamical properties must also be taken into account.

In midlatitudes, at synoptic scale, the important dynamical properties are those related to the rotation of air particles. This rotation is linked both to the motion of Earth and to the rotation component of the wind. The rotation of fluid particles is described by the variable *vorticity*. Vorticity is a measure of the local rotation or spin of the atmosphere: It is the key variable of synoptic dynamics. As illustrated in Fig. 1.1, the vorticity vector gives the direction of the spin axis, and its magnitude is proportional to the local angular velocity about this axis. The fluid particles turn around their vorticity vector, and the *absolute vorticity* is equal to the relative spin around a local cylinder plus the rotation of the coordinate system.

Weather Analysis and Forecasting. http://dx.doi.org/10.1016/B978-0-12-800194-3.00001-7
Copyright © 2016 Elsevier Inc. All rights reserved.

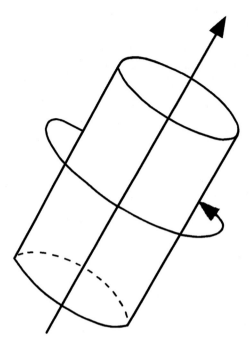

FIGURE 1.1

A vorticity vector and the local rotation in the atmosphere indicated by the circulation around a cylinder of air oriented along the vorticity vector.

Adapted from Hoskins, B., 1997. A potential vorticity view of synoptic development. Meteorol. Appl. 4, 325–334.

To interpret a process in terms of quasi-geostrophic theory, only the vertical component of the vorticity equation is explicitly considered. The vertical component of absolute vorticity is $\zeta = f + \xi$, where f is the Coriolis parameter and the relative vorticity is given by

$$\xi = \frac{\partial v}{\partial x} - \frac{\partial u}{\partial y}.$$

It is also supposed that, at synoptic scale, Earth's rotation dominates (ie, $\zeta \cong f$), in which case the relative vorticity equation contains only stretching and shrinking of this basic rotation (Hoskins, 1997). Two examples are presented in Fig. 1.2. Along the zero vertical motion at the ground, we can make the following observations:

- Tropospheric ascent implies stretching and creation of absolute vorticity greater than f, that is, cyclonic relative vorticity, in the lower troposphere.
- Similarly, tropospheric descent implies shrinking and creation of relative anticyclonic vorticity in the lower troposphere.
- If the initial relative vorticity is zero, the two situations in Fig. 1.2A and B correspond to cyclonic and anticyclonic surface development.

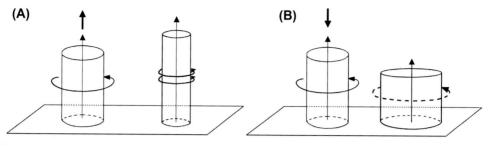

FIGURE 1.2

Tropospheric (A) ascent and (B) descent that leads to respectively (A) stretching and (B) shrinking of vorticity associated with (A) increase and (B) decrease of vorticity and circulation.

Adapted from Hoskins, B., 1997. A potential vorticity view of synoptic development. Meteorol. Appl. 4, 325–334.

Consistent with this discussion, synoptic development can be viewed in terms of vertical velocity (derived in the framework of the quasi-geostrophic theory) associated with the evolution of vorticity in the middle and upper troposphere. Pedder (1997) shows that such a quasi-geostrophic approach can be used for the purposes of subjective analysis to diagnose the vertical circulation associated with a large-scale distribution of pressure and temperature.

Together with quasi-geostrophic theory, the so-called potential vorticity (PV) thinking has proven to be quite useful to study synoptic development in midlatitudes (for theoretical background and references see Hoskins et al., 1985, which contains an exhaustive review of the use of PV). A simple isentropic coordinate version of PV is given by the expression

$$PV = \sigma^{-1}\zeta_{a\theta}, \tag{1.1}$$

where

$$\sigma = -g^{-1}\partial p/\partial \theta > 0 \tag{1.2}$$

is the air mass density in $xy\theta$ space, θ is the potential temperature, p is the pressure, g is the acceleration due to the gravity, and

$$\zeta_{a\theta} = f + \zeta_\theta \tag{1.3}$$

is the absolute isentropic vorticity.

Eq. [1.1] says that PV is a product of the absolute vorticity and the static stability. The units commonly used for the presentation of PV are 10^{-6} m^2 s^{-1} K kg^{-1}, termed the PV-unit (PVU).

PV combining temperature and wind distribution into a single quantity is a powerful parameter for applied meteorology. The dynamic meteorology has shown the usefulness of the PV thinking to identify and understand the synoptic development in midlatitudes. Three properties underlie the use of PV to represent the dynamical processes in the atmosphere:

1. The familiar Lagrangian *conservation principle* for PV, which states that if one neglects the contributions from diabatic and turbulent mixing processes, then the PV of an air parcel is conserved along its three-dimensional trajectory of motion.

2. The second is the *principle of invertibility* of the PV distribution, which holds whether or not diabatic and frictional processes are important. Given the PV everywhere and suitable boundary conditions, then Eq. [1.1] can be solved to obtain, diagnostically, geopotential heights, wind fields, vertical velocities, θ, and so on under a suitable balance condition, depending on access to sufficient information about diabatic and frictional processes.
3. Together with the two principles, another property of PV that allows its use as a concept to describe and understand atmospheric dynamics is the specific *climatological distribution of PV*.

1.2 THE CONCEPT OF POTENTIAL VORTICITY THINKING
1.2.1 THE CONSERVATION PRINCIPLE

The conservation of PV enables us to identify and follow significant features in space and time. In Fig. 1.3, we consider a small vorticity tube whose lower section is at a potential temperature θ and whose upper section is at the potential temperature $\theta + d\theta$. In a dry atmosphere moving adiabatically, this small cylindrical element with a constant mass necessarily moves between these constant potential temperature surfaces (iso-θ), with each particle preserving its potential temperature. Since the vorticity tube follows the two iso-θ surfaces, the quantity $d\theta$ remains constant. At the same time, the PV should be preserved for the fluid element during evolution of the tube. Thus, when h increases (decreasing the θ gradient), the vorticity also increases, and conversely, when h decreases, the vorticity decreases. The stretching/shrinking effect on the vorticity tube bounded by the two isentropic surfaces therefore coincides with the variation of the θ gradient.

Therefore, the conservation of PV in the atmosphere induces changes by the stretching/shrinking effect. The transport of a maximum of PV affects the synoptic flow and, as a consequence, produces vertical motion. From an operational point of view, PV thereby provides a very powerful and succinct

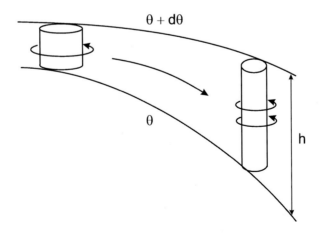

FIGURE 1.3

Conservation of the potential vorticity during the descent of a vorticity tube along two iso-θ surfaces.

view of atmospheric dynamics. Superimposing various PV fields onto a satellite image is a natural diagnostic tool, well suited to making dynamical processes directly visible to the human eye. In particular, a joint interpretation of upper-level PV fields and water vapor imagery provides valuable information because PV structures and water vapor features are well correlated.

1.2.2 THE INVERTIBILITY PRINCIPLE

The conserved nature of the PV parameter and the invertibility of PV enable us to build up the flow and temperature structure associated with a given PV anomaly. The approach described in Section 1.2.1 suggests a method to assess numerical model behavior by making meaningful comparisons between an atmosphere simulated by a model and reality, that is, between numerical weather prediction (NWP) output and satellite imagery. In cases of significant disagreement, the invertibility of PV is the principle that allows us to use local PV modifications to adjust initial conditions of operational numerical models. Thus modifying PV in a local area in the direction given by the observations, mainly by the satellite imagery, can lead to improvement in the model initial state, with all other variables (temperature, winds, etc.) being retrieved via PV inversion. Errors in the forecast track and depth of a cyclone may be reduced by calculating a new forecast from this new initial state (see Section 5.5).

1.2.3 CLIMATOLOGICAL DISTRIBUTION OF POTENTIAL VORTICITY

The climatological PV distribution in the atmosphere is remarkable. It shows that on average in the low levels of the atmosphere PV is uniform (see Fig. 1.4):

1. In high and midlatitudes the PV ranges on average approximately from 0.4 to 0.7 PVU in the troposphere and reaches 1 PVU around 400 hPa. Then it increases rapidly with height and takes on values much higher than 2 PVU in the stratosphere, becoming rapidly greater than 3 PVU in the low stratosphere, owing to the strong increase of static stability.
2. In tropic areas (between 10° Lat. and ∼30° Lat.), PV ranges from 0.3 to 0.5 PVU and reaches 0.5 PVU near 300 hPa.

This discontinuity of the PV near the middle of the atmosphere together with its conservation property allows us to define the 1.5-PVU surface for midlatitudes and the 0.7 PVU surface for tropics (excluding the equatorial area, within 10° latitude) as a tropopause in the view of the PV concept. This new tropopause is called dynamical tropopause, separating the troposphere, with weak and quasi-uniform PV, from the stratosphere, with its strong PV. Defining the tropopause in terms of the PV concept is more efficient in practical meteorology than the classical definition (based on the lapse rate change), because taking into account not only the temperature field but also the motion field through the vorticity is what is crucial in dynamic meteorology. This dynamical tropopause is justified by Fig. 1.5 that highlights its following features:

1. In midlatitudes the correlation between the area where the PV is on average comprised between 1.5 and 2.0 PVU and the area where the mean mixing ratio of ozone is between 150 and 200 ppm (value of ozone giving a good diagnostic of the mean tropopause).
2. In tropics the region of ozone concentration descending below 150 ppm value, well correlated with the area where the PV is on average between 0.5 and 1 PVU (see Figs. 1.4 and 1.5).

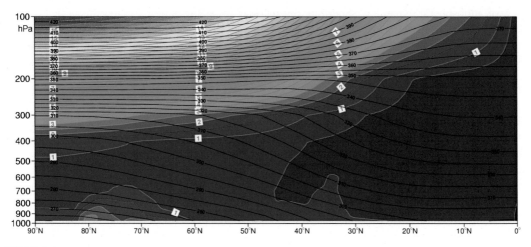

FIGURE 1.4

Zonal year average (vertical cross-section) of the potential vorticity (PV; *color areas*, every 0.5 PVU) and of the potential temperature (*black lines* in K, interval 5K) in the Northern Hemisphere. The contour of PV value = 1.5 PVU (the so-called dynamical tropopause) is given in *red*. This chart used data from 44 years of the European Center for Medium Range Weather Forecasting reanalysis, 1958–2001.

From Malardel (2008).

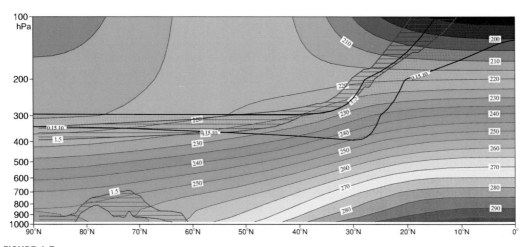

FIGURE 1.5

Comparison in the Northern Hemisphere between the thermal diagnosis, the ozone diagnosis, and the potential vorticity (PV) diagnosis of the tropopause. The temperature field is a zonal and a seasonal average, December-January-February (in *color and red lines*, every 10K). The zone of PV with values comprised between 1.5 and 2.0 PVU (for the same period) is represented in *magenta dashed area*; the contour of PV = 1.5 PVU at the bottom of the *dashed area* is the dynamical tropopause for midlatitudes. Also represented are values of mean mixing ratio of ozone between 150 and 200 ppm (*dotted black zone*, 1 ppm = 10^{-9} kg/kg). This chart used data from 44 years of the European Center for Medium Range Weather Forecasting reanalysis, 1958–2001.

From Malardel (2008).

We can notice (as Highwood and Hoskins, 1998), that a PV definition of the tropopause is useful also for the tropics but not close to the equator (the equatorial area, within 10° latitude) where PV surfaces become almost vertical.

1.2.4 POSITIVE POTENTIAL VORTICITY ANOMALIES AND THEIR REMOTE INFLUENCE

We can now think about the PV distribution itself rather than the behavior of a cylinder between isentropic surfaces. The results must be the same as those depicted in Fig. 1.2 by considering a coherent PV structure at upper levels, referred to as "PV anomalies" (Hoskins et al., 1985). A positive PV anomaly is defined as a coherent region of high values of cyclonic PV (positive PV values in the Northern Hemisphere and negative PV values in the Southern Hemisphere).

The concept of coherent structure is used to associate many of the anomalies of interest to meteorology to such features, which are localized and keep their coherence in time (Plu et al., 2008). The time scales of air particles traveling inside such a coherent structure are shorter than the typical time scales of the evolution of such a structure. Therefore, such a coherent positive PV anomaly undergoes a time evolution that may be interpreted by dynamical diagnoses.

Fig. 1.6 schematically shows the effect of a positive PV anomaly (ie, a region with an isolated maximum of cyclonic PV) surrounded by an atmosphere originally at rest with uniform PV.

The PV anomaly modifies the temperature field and induces a cyclonic circulation.

- At the center of the anomaly the static stability increases (the iso-θ surfaces become closer together); therefore the stability decreases above and beneath the PV anomaly.
- To conserve PV, the absolute vorticity increases above and beneath the anomaly to compensate for this decrease in stability.

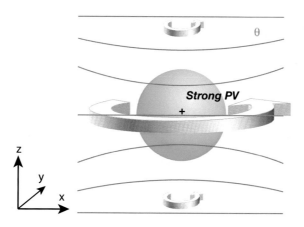

FIGURE 1.6

A schematic cross-section, showing an idealized model of the modification of the troposphere associated with an upper-level cyclonic potential vorticity anomaly, which is referred to as a dynamical tropopause anomaly.

Hence two effects are associated with a cyclonic PV anomaly introduced into an atmosphere with a uniform PV distribution: a decrease of static stability and an increase of vorticity above and below the anomaly. Thus, an upper-level PV anomaly induces a cyclonic circulation that weakens toward the ground. The circulation induced by a PV anomaly will penetrate a vertical distance, whose scale H, referred to as the Rossby penetration height, is given by the equation (Hoskins et al., 1985)

$$H = f \frac{L}{N},$$ [1.4]

where f is the Coriolis parameter, L is the horizontal scale, and N is the Brunt-Väisälä frequency, which is a measure of static stability. H is the scale in physical xyz space, measuring the vertical penetration of the induced modification above and below the location of the anomaly. There is an obvious scale effect, whereby small-scale features have a relatively weak effect on the velocity field and large-scale features have a relatively strong effect.

1.3 OPERATIONAL USE OF POTENTIAL VORTICITY FIELDS TO MONITOR SYNOPTIC DEVELOPMENT

1.3.1 UPPER-LEVEL DYNAMICS, DYNAMICAL TROPOPAUSE, AND DYNAMICAL TROPOPAUSE ANOMALY

The properties of PV allow its use as a tracer in upper-level dynamics, which is crucial in midlatitude synoptic developments. Upper-level disturbances can be considered as upper-level PV anomalies (strong positive PV in the Northern Hemisphere and strong negative PV in the Southern Hemisphere) penetrating into the upper troposphere. The anomaly's influence on the surrounding air is depicted in Fig. 1.7. The troposphere below the PV anomaly is modified as seen in Fig. 1.6. In particular, the iso-θ contours are attracted toward the anomaly. If the synoptic flow is a "zonal" wind increasing with height, such an anomaly moving in this baroclinic environment produces vertical motion: The

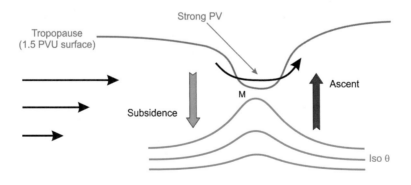

FIGURE 1.7

A schematic cross-section, showing an idealized model of the modification of the troposphere associated with an upper-level positive potential vorticity anomaly, which is referred to as a dynamical tropopause anomaly.

deformation of the iso-θ imposes ascending motion ahead (to the east) of the anomaly and subsiding motion behind (most often to the west of) the anomaly.

It has been considered good synoptic practice to use maps at upper levels to supplement the surface map. The assumptions of balance in the atmosphere and uniform tropospheric PV tell us that only one more level is really needed and that this level is at the tropopause, such as the dynamical tropopause (Santurette and Joly, 2002). Thus, a good way to practically apply the PV concept in an operational forecasting environment is to use maps of the height of the dynamical tropopause (the 1.5 or 2 PVU surface). As this surface represents the transition between the low values of PV in the troposphere and the high values in the stratosphere, a drop of the dynamical tropopause signals an air intrusion characterized by a maximum of PV (as shown in Fig. 1.7).

An upper-level positive PV anomaly advected downward to the mid-troposphere corresponds to an area where the 1.5 PVU surface moved downward to the mid-troposphere or low troposphere; this generally happens when the low tropopause area moves in a baroclinic environment interacting with a jet (which increases vorticity and produces clear vertical motion; see Section 1.3.3); we call such a low tropopause area a *dynamical tropopause anomaly*.

The dynamical tropopause anomaly is the active region of the dynamical tropopause and is characterized by two features:

1. A region of low tropopause height with a minimum of geopotential (or trough, marked "M" on Fig. 1.7).
2. A bordering area where the tropopause is very tilted and thus marked by a pronounced gradient of the 1.5 PVU surface geopotential; this strong gradient zone is associated with a maximum of horizontal wind.

Dynamically meaningful information can be illustrated on a single PV chart. Inspecting the maps of the height of the dynamical tropopause (the 1.5 PVU surface) is a powerful way to practically monitor the upper-level dynamics. An upper-level trough—including a short wave trough—on a constant pressure surface can be interpreted as an area of upper-level PV maximum that corresponds to a trough or an area of low 1.5 PVU heights. Actually upper-level troughs are easier to see and to follow as troughs or minimum of height of the 1.5 PVU surface than as troughs on any isobaric surface, partially due to the conservative property of PV. Also, due to the presence of strong static stability just above the tropopause, the dynamical tropopause is a specific surface, restricting the free atmosphere similar to the ground being the other restricting surface to a free atmospheric flow. For that reason, the horizontal perturbations of the atmospheric flow are more pronounced on the dynamical tropopause than they are on any mid–upper-level isobaric surface. Jet evolutions and especially jet streaks represent the synoptic perturbations of the atmospheric flow, and it has been noted that the dynamical tropopause is the level that best reveals these upper-troposphere disturbances (see Section 1.3.2).

The real tropopause is never perfectly flat, as is the case for isobaric surfaces and isotherms. The minima or troughs of the dynamical tropopause are not all systematically linked to any evident synoptic vertical motion: Some of these tropopause deformations (or minimum height of the tropopause) are not very well pronounced in a quasi-barotropic environment and are surrounded by a weak slope of the tropopause, that is, by a weak geopotential gradient of the 1.5 PVU surface. For that reason, these areas can be considered as *latent tropopause anomalies*, where the balance of the atmosphere is hardly disturbed, and the synoptic vertical motion is not present or very weak. Nevertheless, two key points need to be considered:

- The latent tropopause anomalies are important to follow, because their evolution, under the influence of the large-scale circulation, can lead to dynamical tropopause anomalies that induce strong synoptic-scale vertical motions. For that reason, the latent tropopause anomalies are precursors of actual dynamical tropopause anomalies, and it is therefore essential to locate them.
- The latent tropopause anomalies are regions of dry and cold air aloft that are favorable for increasing convective instability.

1.3.2 JET STREAM AND JET STREAKS

The large- and synoptic-scale circulation is well characterized at the upper levels where the evolutions of the planetary jet stream reflect the variations of the circulation pattern. According to the theory of planetary waves and their propagation, the jet stream moves in latitude, its intensity varies, and there are more or less great changes in the horizontal wind. Areas of large changes in the jet stream horizontal wind are associated with jet streaks—regions of isolated maximum wind embedded within the jet stream—that are perturbations of the large-scale jet stream. There is a link between vertical motions and strong horizontal wind variations in the confluence/diffluence areas respectively associated with entrance/exit regions of the jet streaks. Its influence for the development of severe weather phenomena has been discussed in many papers, and conceptual models of ageostrophic circulations and vertical motions associated with jet streaks have been developed by numerous authors (Shapiro, 1981; Uccellini et al., 1979; Bluestein, 1993; Carrol, 1997a). Right entrances and left exits of jet streaks are associated with pronounced upper-level divergence producing ascending motion in the troposphere that is a positive factor for development of many bad weather systems. So jet streaks are considered as important tracers of the synoptic perturbations at the midlatitudes. However, jet streaks are actually difficult to detect and to follow in time mainly because they are not a conservative structure. Jet streaks are not simply advected with the wind; because of the strong variability of wind, the jet streaks are subject to rapid changes in intensity with time. The accelerations associated with curved flow as well as along-flow thermal advections can modify significantly the simple jet streak conceptual model (Shapiro, 1982). It is impossible for forecasters in operational environments to concretely detect and understand in real time the jet streak evolutions and consequently the areas of forced vertical motions by studying the wind field alone and the jet structures. The need to synthesize more effectively the dynamics information is therefore imperative. Such a synthesis has to take into account atmosphere properties as well as the human spirit. A good solution is to use conservative parameters; indeed they can give efficient tracers to specify the dynamic zones of the motion field at any moment. The PV concept offers a more efficient method to identify and follow the source of upper-level perturbation-generating ascending motion.

It has been noted that the dynamical tropopause is the level which best reveals the jet streaks. A jet streak formation can be explained using the PV concept: a jet streak can be seen as the result of an interaction between the large-scale jet and a tropopause anomaly. Fig. 1.8 depicts the process of jet streak formation in the view of the PV concept.

When a low tropopause area (a latent tropopause anomaly) moves close to the large-scale jet stream (Fig. 1.8A), this low tropopause anomaly interacts with the jet, producing important changes in the

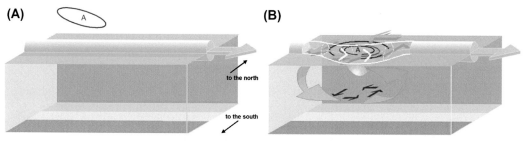

FIGURE 1.8

Schematic representation of a jet streak formation by the mechanism of an interaction between a tropopause anomaly—that is, an upper-level potential vorticity maximum—and the jet stream. The jet stream is represented as the *light blue tube ended by an arrow*. The bottom of the parallelepipeds represents the temperature field near the surface (warm temperature in *orange*, cold temperature in *green*). (A) schematizes the atmospheric situation before the interaction between the jet stream and the tropopause anomaly (noted "A," *red ellipse*). (B) represents the situation when the tropopause anomaly interacts with the jet stream: this induces deformation of the jet and increasing of the wind (*white lines and white arrows*), increasing of vorticity (*blue arrows*) and appearance of vertical motion *(ochre and green large arrows)*. The ochre cone denotes the tropopause going down (see text).

atmosphere. As a consequence of this interaction, additional wind coming from the anomaly is added directly to the jet stream so that a local maximum of wind appears in the jet stream along the tropopause anomaly (Fig. 1.8B). Folding of the tropopause in turn tends to increase at the edge of the anomaly as a response to the increasing cyclonic wind shear; this magnifies locally the wind maxima. In this manner, a maximum wind area—a jet streak—is created at the edge of the zone where the tropopause is most folded. This wind maximum is characterized by an upstream wind confluence (acceleration) area and a downstream wind diffluence (deceleration) area. These perturbations of the wind speed—acceleration and deceleration—produce convergence and divergence zones respectively that imply vertical motion: subsiding motion under the acceleration zone and ascending motion under the deceleration zone (Fig. 1.8B).

The method proposed in this guide to monitor upper-level dynamics is based on the dynamical ideas born from the continuing research efforts on the synoptic scale, mainly the use of PV concepts. First we know the existence of a balance in the atmosphere, namely, the atmosphere evolves close to thermal-wind balance; so we can say that atmospheric synoptic conditions can be summarized by the characteristics of the situation at two levels: one upper level in high troposphere and one low level representing the conditions near the surface (only one upper level is needed in order to supplement surface conditions). Considering the PV properties, it is most relevant for operational purposes to use the *dynamical tropopause*, that is, the 1.5 PVU surface, as a reference upper level (Santurette and Joly, 2002). Studying areas of low dynamical tropopause heights and related strong horizontal gradient zones of the 1.5 PVU surface heights, in addition to the examination of the jet stream evolution, provides meaningful information about the synoptic dynamics, and this is a very efficient approach to identify and follow in time the features responsible for forcing ascending motion. The dynamical tropopause anomaly is a significant upper-level dynamic feature, which may be used for upper-level diagnosis in operational forecasting purposes.

1.3.3 SYNOPTIC DEVELOPMENT AS SEEN BY POTENTIAL VORTICITY CONCEPTS

As shown in Section 1.3.2, a perturbation of the jet stream caused by an interaction between this jet stream and a tropopause anomaly results in the appearance of a jet streak and in a more pronounced tropopause anomaly, that produces related synoptic development as follows:

- A deep tropopause anomaly interacting with a strong jet will lead to a very intense disturbance and then will be accompanied by intense vertical motion and strong vorticity in the troposphere.
- A very pronounced sloping of the tropopause can even appear on the cold side of the jet streak, leading to a "tropopause fold" (the minimum height of the tropopause going down under the jet streak core).

As seen from Eq. [1.4], the cyclonic circulation induced by an upper-level PV maximum may reach the surface and influence more or less strongly the low-level circulation. Fig. 1.9 illustrates the process of cyclogenesis by baroclinic interaction using PV concepts: An area of upper-level PV maximum (a tropopause anomaly, area of low dynamical tropopause) moving close to the jet and approaching the baroclinic zone interacts with these structures; the tropopause exhibits a dynamical tropopause anomaly and significant changes occur in the wind and temperature fields as follows:

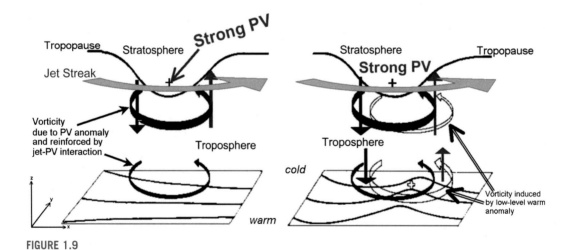

FIGURE 1.9

A schematic picture of the mechanism of cyclogenesis associated with the interaction between an upper-level positive potential vorticity (PV) anomaly (referred to here as a dynamical tropopause anomaly, indicated by a solid "+" sign), a jet, and a low-level baroclinic region. The solid cyclonic arrow indicates the circulation induced by the anomaly, and potential temperature contours are shown at the lower boundary by thin lines. A low-level PV anomaly (the open "+" sign in the right panel), can also induce a cyclonic circulation, indicated by the open cyclonic arrow in the right panel, that acts to reinforce the circulation pattern induced by the upper-level PV anomaly.

Adapted from Hoskins, B.J., McIntyre, M.E., Robertson, A.W., 1985. On the use and significance of isentropic potential vorticity maps. Q. J. R. Meteorol. Soc. 111, 877–946.

- The low tropopause anomaly increases the cyclonic vorticity on the polar side of the jet; the tropopause anomaly deepens and the cyclonic vorticity affects the troposphere up to the low levels (*black arrows*).
- The cyclonic circulation associated with the vorticity (*solid cyclonic arrow*) modifies the structure of the baroclinic zone, producing a wave on the thermal gradient zone. Consequently, thermal advections take place with a low-level warm anomaly (see potential temperature contours at the lower boundary by *thin lines*) appearing slightly east of the upper-level vorticity anomaly.
- This baroclinic wave and the associated thermal wave in turn induce a cyclonic circulation (*open cyclonic arrow*) as shown by the white arrows in Fig. 1.9 (*right panel*) that reinforces the circulation pattern induced by the upper-level anomaly. The resulting vertical motion amplification then leads to further cyclogenesis, and so forth.

If static stability is sufficiently low, the surface cyclonic circulation associated with the warm anomaly may reach the level of the upper PV anomaly, with the following consequences:

- The combined induced upper-level circulation will advect parts of the anomaly southward, reinforcing the upper-level anomaly, and slowing down its easterly movement.
- The circulation induced by the upper-level PV anomaly becomes stronger, further strengthening the warm anomaly.

Thus the tight coupling between upper and lower levels leads to further reinforcement of the upper-level anomaly, to stronger low-level warm advection (increasing the moisture supply as well as the surface θ anomaly), and, hence, to intensified surface development (see Sections 3.3.2 and 3.5.3).

As regards to synoptic/subsynoptic-scale developments, both theory and synoptic experience suggest that the moist processes can greatly enhance the surface development. If condensation occurs in the rising air (for instance, when there is a sufficient supply of moisture from a warm, moist low-level air stream), then the effective Rossby height scale H given by the expression [1.4] will be increased owing to the reduced static stability. If such a moist-air effect is included, the impact of the static stability becomes even more pronounced, and vertical penetration increases significantly.

From Eq. [1.4] we also see that when the static stability becomes very low, the horizontal length scale can also be small. In this case, the intensification of the anomalies becomes stronger and more rapid and tends to take place on a smaller scale.

1.3.4 ANALYSIS OF A REAL-ATMOSPHERE STRUCTURE

Figs. 1.10 and 1.11 show a jet streak evolution resulting from an interaction between a jet and a tropopause anomaly to illustrate various features considered in the section above in a real situation over the northeast Atlantic coast of the United States. A dynamical tropopause anomaly coming from the north (Fig. 1.10A, *red arrow*) approaches the jet stream branch in the leading part of a trough and begins to interact with it (Fig. 1.10B) as the wind in the jet increases (Fig. 1.10B, *blue arrow*). Then the interaction between the anomaly and the jet takes an effect that the wind and, accordingly, the horizontal gradient of the tropopause height continue to strengthen (Fig. 1.10C, *blue arrow*). At the same time, the tropopause anomaly deepens (Fig. 1.10C and D, *red arrow*). Finally the initial large jet is deformed and an isolated maximum of wind separate from this initial jet—a jet streak—appears (Fig. 1.10E, *blue arrow*), clearly associated with the tropopause anomaly.

(A) **(B)**

(C) **(D)**

(E)

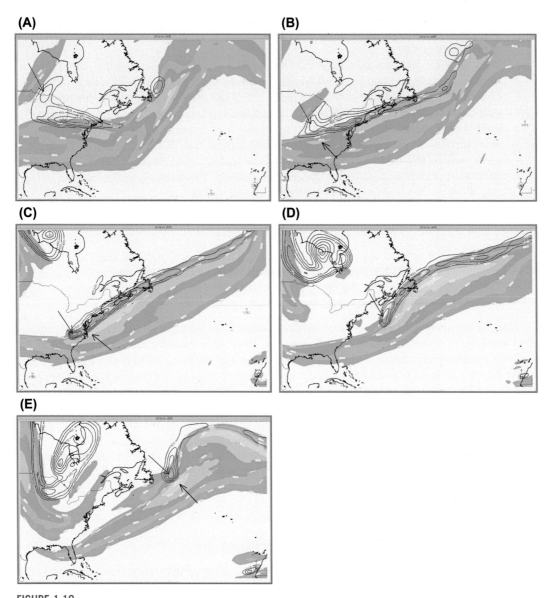

FIGURE 1.10

Isotachs of the wind at 300 hPa greater than 80 kt (*color*, every 20 kt) and geopotential of the 1.5 PVU surface (*brown*, every 75 dam, only ≤600 dam). ARPEGE model analysis (A) 23 January 2014 1200 UTC, (B) 24 January 2014 0000 UTC, (C) 24 January 2014 1200 UTC, (D) 25 January 2014 0000 UTC, (E) 25 January 2014 1200 UTC.

(A)   **(B)**

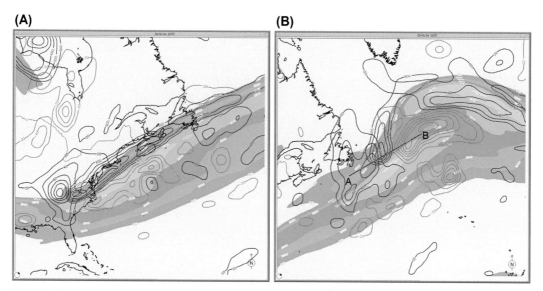

FIGURE 1.11

Isotachs of the wind at 300 hPa greater than 80 kt (*color*, every 20 kt) and vertical velocity at 600 hPa (ascending motion in *orange*, descending motion in *blue*, every 20 10^{-2} Pa/s). ARPEGE model analysis (A) 24 January 2014 1200 UTC, (B) 25 January 2014 1200 UTC. The *black line* in (B) is the axis of the cross-sections presented in Fig. 1.12.

Fig. 1.11 shows that vertical motion develops around the dynamical tropopause anomaly interacting with the jet as described in Sections 1.3.1 and 1.3.2 (see Fig. 1.8 and 1.9), with ascending motion (*orange contours*) just ahead of the anomaly and descending motion (*blue contours*) just at the back of it. It is important to notice that at the time of Fig. 1.11A, the pattern of vertical motion is not easy to understand only by considering the jet structures. The classical pattern of the distribution of ascending and descending motion associated with left entrance and left exit of a maximum of wind (see Carroll, 1997) is not respected at this moment. By using PV concepts, considering tropopause anomalies as sources of the wind perturbations that generate synoptic vertical motion, it is much easier to detect and to understand the vertical motion development closely related to a dynamical tropopause anomaly.

Fig. 1.12 presents vertical cross-sections through the tropopause anomaly interacting with the jet streak along the southwest–northeast axis indicated in Fig. 1.11B. The vertical distribution of PV (*brown*), potential temperature (*green*), wind component transverse to the plan of the cross-section (*black*), and vertical velocity (ascending in *orange*, descending in *blue*) in Fig. 1.12 depicts various features of the real situation that are considered schematically in the section above.

- There are wind maxima, seen in Fig. 1.12A, on both sides of the area of low tropopause, illustrating the strong vorticity associated with the maximum of PV moving downward. Such a wind field structure appears also down to low levels.
- The θ surfaces (see the green contours in Fig. 1.12A and B) curve upward toward the tropopause anomaly. The vertical motion sets up around the tropopause anomaly, with ascending motion forward and subsidence rearward (Fig. 1.12B, downward and upward arrows referring to the large-scale circulation).

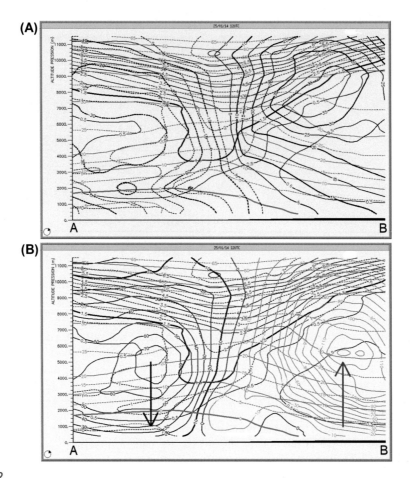

FIGURE 1.12

Vertical cross-section along the southwest–northeast (noted SW NE) axis marked in *black line* in Fig. 1.11B, from ARPEGE model analysis on 25 January 2014 at 1200 UTC. Potential vorticity contours are in *brown* (intervals of 0.5 PVU, 1.5 PVU contour solid); iso-θ surfaces (°C) are in *green*. Also shown in (A) the wind component transverse to the plane of the cross-section (in *black*, kt: *solid lines* indicate the wind in the cross-section plane, and the *dashed lines* indicate the wind out of the cross-section plane). Also shown in (B) the ascending (in *orange*) and the descending (in *blue*) motion (10^{-2} Pa/s).

The aim of Fig. 1.13 is to show an example of the added value of the PV concepts compared to the classical approach using the geopotential of isobaric surfaces. The most dynamical part of the trough crossing the northeast of USA is easier to detect by following the areas of low values of the geopotential of the 1.5 PVU surface (*red contours* in Fig. 1.12) than by looking at the structures of the geopotential of 500-hPa isobaric surface (*brown contours*) generally used in operational forecasting. Thanks to the properties of PV, mainly to its conservation, the dynamical tropopause anomalies (equally maximums of upper-level PV) are nearly conservative that allow easy detection of the

(A)   **(B)**

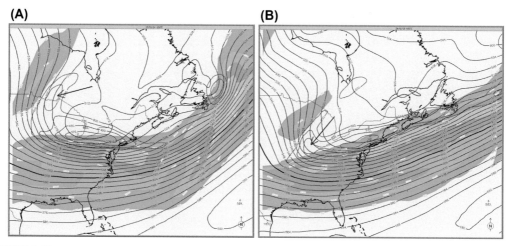

FIGURE 1.13

Isotachs of the wind at 300 hPa greater than 80 kt (*color*, as indicated in the *white labels*, in kt: >80 kt in *blue*, >100 kt in *light blue*, >120 kt in *green*, >140 kt in *yellow-green*) with superimposition of the 500 hPa isobaric surface heights (*brown*, every 4 dam) and of the geopotential of the 1.5 PVU surface (*red*, every 75 dam, only ≤600 dam), at (A) 1200 UTC on 23 January 2014, and (B) 0000 UTC on 24 January 2014.

dynamic zones of the circulation. On the contrary, the isobaric surfaces are constantly changing in order to adapt to wind and temperature changes. Therefore, the advection of tropopause anomalies seen in the field of 1.5 PVU surface heights can be considered efficiently as a precursor of synoptic developments.

This example illustrates that analyzing troughs or minimums of the geopotential height of 1.5 PVU surface can lead to important insight into the critical aspects of the upper-level perturbations, which are more distinctly pronounced on the dynamical tropopause than they are on any mid- or upper-level isobaric surface. This is partially due to the nature of PV and its properties. Thus, the upper-level diagnosis by considering the behavior of the dynamical tropopause is an efficient approach to reveal the upper-troposphere disturbances that are the causes of a majority of severe weather system developments.

THE INTERPRETATION PROBLEM OF SATELLITE WATER VAPOR IMAGERY

CHAPTER OUTLINE

2.1 INFORMATION CONTENT OF 6.2 AND 7.3 μm CHANNELS

Instruments of meteorological satellites measure infrared (IR) radiation in several wavelength ranges in which the radiation is significantly absorbed and reradiated by water in its gaseous state, liquid, or ice crystal form. Such wavelength ranges are referred to as water vapor (WV) "bands", and the wavelengths where they are centered are referred to as WV channels. Due to their specific sensitivity to the moisture and temperature profiles in the path of radiation, the WV channels can provide information for a wide range of atmospheric processes. The ability to infer such information depends on the knowledge on how to interpret the radiances in the great variety of situations that calls for a corresponding knowledge on the effects, which influence the radiation along its path through the real atmosphere to the satellite. The radiation transfer theory applicable to the IR wavelengths in the WV absorption band as well as the associated approach for quantitative interpretation of the radiances measured by the satellite are presented in Appendix A.

Weather Analysis and Forecasting. http://dx.doi.org/10.1016/B978-0-12-800194-3.00002-9

This section is focused on the information that can be inferred by specific image gray shades on the WV images about the vertical distribution of humidity. For that purpose, the response of the two WV channels of Meteosat Second Generation (MSG) to various cases of vertical moisture distribution are considered. Such a material provides the user of satellite imagery with a basic knowledge, which is needed to relate specific patterns and features of gray shades on the images and their changes with time to corresponding atmospheric circulation systems and processes. The considerations are focused especially on the European region because up to 2015 only the European geostationary MSG satellites have provided imagery in two WV channels. Thus, the book provides guidance for application of 7.3 μm images, in addition to those of the broadly used WV channel 6.2 μm (for some satellites 6.3, 6.7, or 6.8 μm; see Appendix A).

2.1.1 ORIGIN OF THE RADIATION, RADIANCE, BRIGHTNESS TEMPERATURE, AND IMAGE GRAY SHADES

Within the WV channels, as they are parts of the IR band, the radiation is emitted by solid objects such as cloud elements, precipitation, and the surface of the Earth. For that reason, the channels in the WV absorption band are sensitive to the profiles of both the temperature and the humidity. Since the intensity of IR radiation is closely correlated to the temperature of the object or substance from which the radiation is emitted, by measuring the intensity of the radiation coming from below, the satellite instrument can detect the temperatures of cloud tops, land, and sea surface. The radiation in IR window channels will reach the satellite after no or very little absorption. For the WV channels, if WV exists between the object and the satellite, it will absorb some portion of the upcoming radiation and re-radiate. Since temperature varies considerably along the vertical path to the satellite, this re-radiation by the WV will occur at a different energy level. The WV usually radiates at a lower energy level, since it is commonly cooler than the Earth's surface or the cloud tops radiating from bellow. The concept is illustrated in Fig. 2.1.

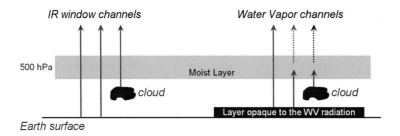

FIGURE 2.1

Idealized scheme of radiation transfer in the atmosphere for infrared window channels and water vapor channels.

The MSG satellites operate by means of the Spinning Enhanced Visible and InfraRed Imager (SEVIRI), which provides observational data in two WV channels centered at 6.2 and 7.3 μm wavelengths. Radiation intensity measured at the satellite in these IR channels, also referred to as radiance, may be converted to a brightness temperature or to an image gray shade. The following analytic relation between the equivalent brightness temperatures (T_b) and the SEVIRI radiances (R) is adopted (EUMETSAT, 2012):

$$T_b = \frac{C_2 \nu_c}{\alpha \log\left[C_1 \nu_c^3 / R + 1\right]} - \frac{\beta}{\alpha} \qquad [2.1]$$

With: $C_1 = 1.19104 \; 10^{-5}$ mW m^{-2} sr^{-1}(cm^{-1})$^{-4}$; $C_2 = 1.43877$ K(cm^{-1})$^{-1}$; ν_c = central wavenumber of the channel; α, β coefficients, which are specific for each IR SEVIRI channel and may vary with the MSG satellites (EUMETSAT, 2012).

From the radiances in different WV channels, usually the 6.2 μm (for some satellites 6.3, 6.7, or 6.8 μm; see Appendix A) are used in an image format, since the WV absorption in these wavelengths is stronger and these channels are more sensitive to variations in humidity than is the 7.3 μm channel (Weldon and Holmes, 1991; Santurette and Georgiev, 2005). In some studies, the information content of 7.3 μm WV images and how to use them to better understand atmospheric situations also has been considered (Santurette and Georgiev, 2007; Georgiev and Santurette, 2009). In this chapter, images in 6.2 and 7.3 μm channels are compared with parameters derived by the operational numerical weather prediction (NWP) model ARPEGE in cases when the model well simulates the thermodynamic conditions of the troposphere. Thus, the material demonstrates important aspects of the WV imagery interpretation and helps the reader to understand how to infer information about the moisture conditions in the troposphere. Since the emphasis here is on image analysis and interpretation, we refer you to Appendix A for a discussion on radiative transfer theory and its implications to elucidate some radiation effects for the WV channels of MSG 6.2 and 7.3 μm.

The radiation in a specific channel exhibits a specific ability to pass through the atmosphere, depending on the density of the absorbing atmospheric substances. For operational applications, WV radiances measured by the satellite are usually displayed in image gray shades. The common convention is that lighter shades on the imagery indicate colder brightness temperatures or lower energy measurements. On the contrary, an area of darker gray shades indicates radiation arriving from a warmer source. Fig. 2.2 shows WV images in 7.3 and 6.2 μm channels. If the temperature decreases monotonically with height (in most of the cases) the result of an increase in humidity is a decrease in the measured radiance and the image gray shade will become lighter.

Since the WV absorption is stronger at 6.2 μm, the radiances in this channel presented in an image better represent moisture distribution in the mid- to upper troposphere than the corresponding 7.3 μm channel image, as revealed by the gray shades of the images in Fig. 2.2. Therefore, the WV 6.2 μm imagery is a better tool for analysis of synoptic scale atmospheric circulation.

Interpretation of the WV image gray shades is intended to shed light on the distribution of moisture and temperature in the atmospheric column that contributed to a specific radiation measurement. Usually, such a single value does not provide totally accurate knowledge of the atmospheric moisture that will be discussed in Section 2.2. However, if requirements for accuracy are lowered and additional information is introduced, the knowledge gained from the measurements can be very useful for analysis of the humidity regime in the troposphere as well as the vertical location of moisture layers and the identification of cloud systems. The WV gray shade interpretation usually is associated with the acceptance of some kind of conceptual approach to simplify the problem. Although there is no simple accurate relationship between the image gray shades and the distribution of atmospheric

(A) **(B)**

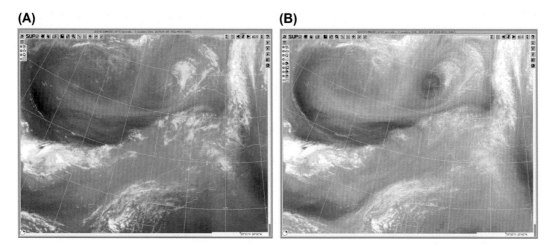

FIGURE 2.2

Water vapor images in (A) 7.3 μm and (B) 6.2 μm.

moisture, the most useful simple concept is the topographic representation, which can be applied under the following conditions or assumptions (Weldon and Holmes, 1991):

(a) Significant moisture is present in the lower troposphere, up to a specific altitude, which is referred to as threshold level.
(b) Above such a threshold level, the air is relatively dry, and if any moisture exists, it is not arranged in layers.
(c) The air temperature decreases upward without significant inversions within the troposphere.
(d) There are no clouds.

Under those conditions, if the top of the moisture layer increases in altitude, the measured brightness temperature decreases, and the image gray shade will become lighter. Using this concept, a WV image may be considered to represent a "moisture terrain," in which the light shades are areas where low-level moisture extends upward to high altitudes, and the dark shades are areas where the high-tropospheric dry air extends downward to the lowest levels. This topographic representation is the most appropriate approach for interpreting WV imagery jointly with potential vorticity (PV) fields for monitoring tropopause foldings, especially those PV fields that reflect the topography of the dynamical tropopause (see Chapters 1 and 3).

The specific threshold, introduced in condition (a), is the level below which the atmosphere becomes relatively opaque to the radiation in the WV absorption band. It is different for the different WV channels and varies primarily with temperature. During the winter seasons or at high latitudes, the threshold for 6.2 μm channel is likely to appear within the lower part of the middle troposphere; during the summer seasons or at low latitudes, it may appear in the upper part of the middle troposphere.

In cases when conditions (a) through (d) are not satisfied, the interpretation of WV image gray shade gives us (qualitative) information about the possible vertical distribution of humidity. Since condition (a) usually is satisfied, the interpretation problem narrows in scope to three primary groups of exceptions to the topographic concept:

- Cold air temperature and inversions
- Earth's surface features and clouds
- Layered moisture conditions that will be presented in Section 2.3

2.1.1.1 Cold Air Temperatures and Inversions

The WV channel radiation reaching the satellite is affected by a complex set of variables, but the basic factors are the amount of WV in the radiation path, the vertical location of the WV, and the temperature of the WV. The first two factors influence the absorption characteristics of the atmosphere. If the water content in the path of radiation and its vertical localization are not changed, the brightness temperature will vary according to the variations of the air temperature. Warming of the entire atmospheric column will result in almost the same increase in the brightness temperature. However, the brightness temperature will not increase identically to the warming of the atmospheric column, since changes of the air temperature also lead to some differences in the absorption characteristics of the atmosphere.

The temperature may dramatically change the origin of the radiation. In the absence of large temperature inversions, the coldness of the air does not violate the topographic concept because relative humidity differences and various vertical distributions of moisture will exist and show on imagery as variations of gray shades. However, because of the low temperature the absolute humidity is low, and the threshold level will be quite low. In such cases, features at lower altitudes will be observed on the imagery.

During winter, large temperature inversions are common in cold-air regimes, especially over continental areas, and these do cause exceptions to the topographic concept. Since the moisture may be present within a relatively deep layer of air with temperature warmer than the surface, the brightness temperature may be warmer than that of the surface. In this situation, patterns of lower tropospheric moisture may be present on a WV image (especially in the channel 7.3 μm) with gray shades similar to adjacent drier regions, or in some cases they may even have darker gray shades than the drier areas.

2.1.1.2 Earth's Surface Features and Clouds

Fig. 2.3 shows a highly variable cloud field in corresponding images over the eastern Atlantic in three Meteosat channels (7.3, 6.2, and 10.8 μm) as well as a vertical cross-section of relative humidity (%) along the pink/green line depicted in the images. Fig. 2.3A illustrates also a case where the Atlas Mountain in North Africa is seen as a dark feature at the position of the green arrow. In specific conditions high-altitude land surface features may appear in the 7.3 μm images: Light gray shades in the 7.3 μm image indicate humid air at low levels; the mountain terrain appears darker (warmer) than the surrounding areas of low-level moist air. In Fig. 2.3B dark gray shades of the 6.2 μm image indicate relatively humid air at upper levels, and the mountain terrain is not visible. This example shows that radiation in the 6.2 μm channel does not originate from Earth's surface features, and this channel is much more useful for observing upper troposphere moisture content than the 7.3 μm channel.

If there are no clouds above a level of about 800 hPa, the WV images can be used qualitatively to identify areas of high and low air humidity (light and dark gray shades respectively). In these conditions, the two MSG WV channels can be used for inspection of specific moisture patterns, which are not seen in the IR channel images, as the mid/upper-level vortex at the black arrow in Fig. 2.3B.

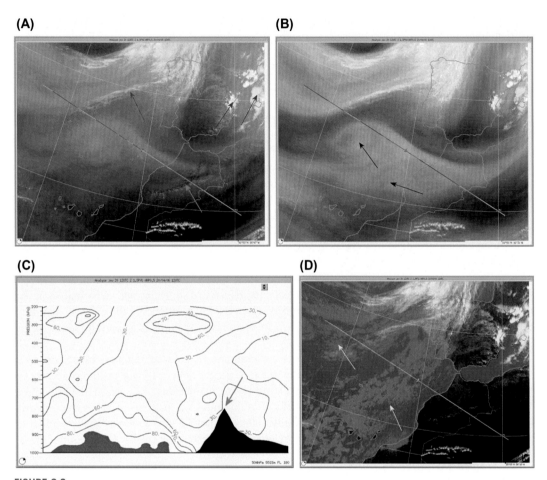

FIGURE 2.3

Meteosat-8 images in (A) 7.3 μm, (B) 6.2 μm, (C) vertical cross-section of relative humidity (%) from ARPEGE NWP model along the pink/green line depicted in the images, and (D) 10.8 μm. The positions of the upper-level vorticity feature (*blue arrow*), cirrus clouds (*magenta arrow*), deep convective clouds (*red arrows*), and low-level clouds (*yellow arrows*) are indicated.

Although clouds are often ignored in discussions of radiation concepts for WV channels, they are frequently significant to operational imagery interpretation and usually obstruct the moisture topography seen in the WV image. Generally, a significant gray shade difference is present between cloud and moisture. However, in a moist environment the change of moisture to cloudiness, although gradual, may be difficult to interpret. This problem lies beyond the scope of this guide, but the best way of detecting such situations is to compare the WV imagery with corresponding window (10—12 μm) IR or visible (VIS) images.

Clouds with tops at levels around or above 400 hPa make a significant contribution to the radiance and are most frequently explicitly observed on the WV imagery as very light or nearly white shades. As depicted in Fig. 2.3, these are the deep-layer cloud systems (marked by red arrows) and high cirrus forms (eg, those at magenta arrow). Such clouds can be distinguished by their nearly white gray shade. Clouds may be implicitly observed on WV imagery, and this is most common with middle clouds located under layers of high-level moisture. Such clouds act as a cold radiation source relative to the surrounding clear air areas. The radiation from the cloud top produces an area of lighter gray shades within the higher-level moisture pattern. Although the high-level moisture layer may be quite uniform, it will appear on the image as an irregular gray pattern consistent with the cloudiness at a lower level. The patterns in the high-level moisture may be mistaken for "moisture parcels" or differences in top height or relative humidity of the high moist layer. Such an example is associated with stratocumulus clouds at the yellow arrows in Fig. 2.3D. In this situation, the low-level convective cells, seen in the IR image, produce different gray shades in the WV images in Fig. 2.3A and B.

Convective cells (open type) often expand and extend upward into the cold mid-tropospheric air within an upper air trough environment, most likely over oceanic or large flat continental areas as shown in Fig. 2.4. Since the associated tropopause is low (at the red arrow in Fig. 2.4B), relatively dry conditions prevail above and the radiation from the top of such clouds often is able to reach the satellite.

On the WV imagery, the area of open convective cells forms a comparatively cold pattern. If the convective cells develop in a relatively deep layer, their cloud tops are high and they produce a cellular appearance. Such an area is seen at the location of the blue arrow, where the cell tops being at a higher altitude appear brighter in Fig. 2.4A and B than the other cellular features at the red arrow seen in the IR image (Fig. 2.4C) and the 7.3 WV channel image (Fig. 2.4A), where the tropopause folding is more pronounced. During later stages of cellular convection, moisture tends to accumulate aloft from dissipated cells and the residual moisture covers the areas among the active cells. The pattern on the image becomes more uniform, with individual cells obscured by the surrounding moisture. Since the cloud tops and residual moisture are in very cold air, the pattern on the image is often of equal gray shades (at location of the blue arrow), or even lighter, than the moisture of the adjacent areas.

2.1.2 SENSITIVITY RANGE OF 6.2 μm AND 7.3 μm CHANNELS

In general, the radiation from WV that reaches the satellite does not arrive from a single surface or level but from some layer of finite depth. WV—in typical concentrations—is semitransparent to the radiation. Therefore, the brightness temperature measured by the satellite is a net temperature of some layer of moisture, not the temperature of any single surface or level. The more densely concentrated the WV is, the more shallow the layer from which the radiation arrives at the satellite. As suggested by Weldon and Holmes (1991), the concept is similar to a person looking into fog: Visibility decreases as the density of the fog increases. We can think of the satellite "seeing" down into the WV in a similar manner, but it senses or "sees" temperature instead of visible light.

Displaying WV radiances in image format for operational applications provides a large-scale view of a set of atmospheric structures depending on the sensitivity of the specific channel to WV and clouds

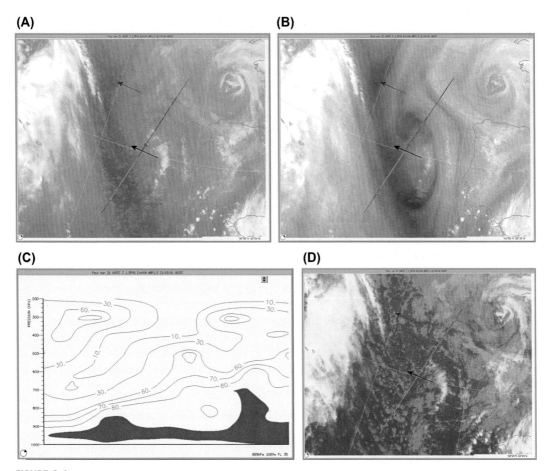

FIGURE 2.4

Meteosat-8 images in (A) 7.3 μm, (B) 6.2 μm, and (C) 10.8 μm. The red and blue arrows indicate convective cells (open type) at different areas of tropopause folding.

at various altitudes. Concerning this, interpreting WV imagery may be considered from two funda-mental points of view:

1. According to the first approach each individual pixel on an image is interpreted as a single value measured by the satellite. The question then becomes determining what information this gray shade provides about the vertical distribution of humidity and temperature in the path of radiation. Section 2.2 addresses this problem, which is further discussed in Appendix A.

2. According to the second approach, many pixels over large areas are considered as patterns and features of gray shades on the image, and their interpretation must relate these patterns and their changes with time to specific atmospheric circulation systems and processes. When using this approach, WV imagery serves operational forecasters with a valuable tool for synoptic-scale analysis. This use is illustrated in Part II.

Applying the first approach calls for knowledge about the response of the WV absorption channels of the satellite radiometer to stratification of humidity in the troposphere. The altitude association of the contribution to the MSG WV channels radiances is considered in Appendix A based on calculations through radiation transfer models and performed for standard temperature profiles of temperature and humidity. This section is focused on the information content of 6.2 µm and of 7.3 µm imagery in the view of their operational applications, based on calculations by the RTTOV8 radiation model (Saunders and Brunel, 2005) with troposphere moisture stratified in 18 individual layers (see Appendix A). From an operational perspective, the sensitivity range is a basic concept (introduced by Weldon and Holmes, 1991) that serves as an indication of the channel's abilities to detect differences of humidity in atmospheric layers at different altitudes (but it is not a measure of those differences). To illustrate the sensitivity range of the three channels, the profile of the differences between the brightness temperatures produced by 10% and 97% relative humidity along the vertical profile as derived by RTTOV8 simulations are drawn in Fig. 2.5. To better understand the specific sensitivity of the MSG WV channels, the 8.7 µm channel is also considered.

The three air mass channels of MSG, presented in Fig. 2.5, show different overall sensitivity as well as a different sensitivity range depending on the altitude of the moisture features. The 6.2 µm channel exhibits large sensitivity at the upper troposphere up to above the 250 hPa isobaric level. The 7.3 µm

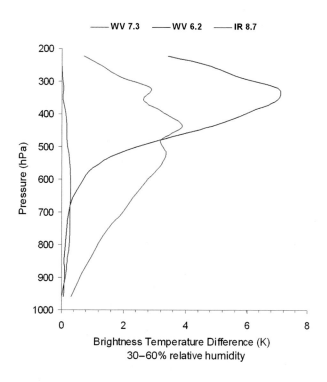

FIGURE 2.5

Sensitivity range of water vapor (WV) 6.2 µm (blue), WV 7.3 µm (magenta), and infrared 8.7 µm (red) channels based on the RTTOV8 model simulations for moderate relative humidity differences at each layer.

channel is sensitive to moisture content in a deep middle troposphere layer centered at about 500 hPa. The larger sensitivity for the 8.7 μm channel is observed in the layer between about 600 and 750 hPa. The radiation in this IR channel is very slightly absorbed by WV and its sensitivity range is very low, producing less than 1°C difference in brightness temperature between dry and moist air at all single layers. Therefore, the overall sensitivity of the 8.7 μm channel is not significant to distinguish moisture features in image gray shades and could be only useful for quantitative applications.

The profiles of the derived differences between the brightness temperatures in WV 6.2 μm and WV 7.3 μm produced by moderate (A) and large (B) relative humidity differences at each layer are drawn in Fig. 2.6.

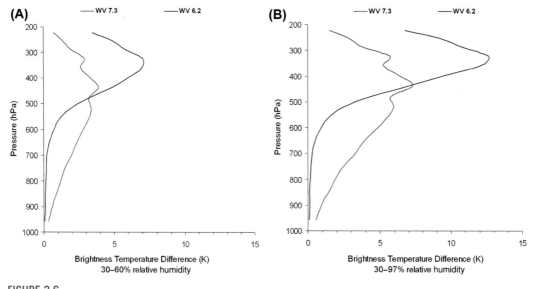

FIGURE 2.6

Sensitivity range of water vapor (WV) 6.2 μm and WV 7.3 μm channels based on the RTTOV8 model simulations for (A) moderate and (B) large relative humidity differences at each layer.

A summary of the sensitivity range for the three channels is presented in Table 2.1. A threshold of 4 K for the difference between brightness temperatures produced by dry (30% relative humidity) and nearly saturated (97% relative humidity) air is applied for definition of sufficiently large sensitivity range.

The 6.2 and 7.3 μm channels of MSG exhibit different overall sensitivity as well as a different sensitivity range, depending on the altitude of the moisture features, as follows:

- The sensitivity range is sufficiently large in the layer 250−500 hPa for the 6.2 μm channel, and for the 7.3 μm channel it is sufficiently large in the layer 350−700 hPa.
- The 6.2 μm channel is very sensitive to distinguish differences in the humidity of middle- and high-level moist layers, and the level of largest sensitivity range for this channel is near 350 hPa.

Table 2.1 The Sensitivity Range for Meteosat Second Generation Water Vapor (WV) Absorption Channels

Meteosat Channel	WV 6.2 μm	WV 7.3 μm	IR 8.7 μm
Overall sensitivity	Very large	Large	Poor
Brightness temperature difference between 30% and 97%	~13 K	~13 K	<1 K
Brightness temperature difference between 30% and 60%	~7 K	~4 K	<1 K
Layer of significant sensitivity range (>4 K difference for 30–97%)	~250–500 hPa	~350–700 hPa	None
Level of the largest sensitivity range	~350 hPa	~450 hPa	Within 850–550 hPa
Lower threshold of sensitivity (>1°C)	~600 hPa	~900 hPa	Not important

- The layer of largest sensitivity of the 7.3 μm band is located in the middle and lower troposphere, and the maximum sensitivity of this channel to detect WV features appears near 450 hPa.
- Differences in water content bellow 600 hPa are not detected by radiation measurements in the 6.2 μm channel.
- Differences in water content down to 900 hPa may be seen by satellite radiation measurements in the 7.3 μm channel.

In effect, depending on the sensitivity range, various layers of moisture and clouds located at high and low levels can be distinguished in the MSG images in Fig. 2.7A and B. The vertical cross-section of relative humidity along the lines AA and BB on these satellite images is shown in Fig. 2.8A and B, respectively. As seen by the IR image (Fig. 2.7C), the cloud field is dominated by a wide area of open-cellular pattern in the rear side of a large frontal cloud band across the right-hand sides of the cross-section lines.

Examples of specific gray shade appearance in the images of the MSG WV channels due to a complex combination of cloud and moisture features are discussed below.

- Between the right-hand sides of the two cross-section lines, the cloud band is formed by low-level clouds (light gray appearance) in Fig. 2.7A and C. There is a stripe of thin cirrus clouds at location of the "Ci." These clouds are usually most distinct in the 7.3 μm image owing to the effect of their complex environment of moisture and lower-level clouds that affects in different ways the radiance in the other channels, as follows:
  - In the 7.3 μm image the low-level clouds appear less bright than in the 10.8 μm image because of absorption by WV in the middle troposphere. This increases the contrast between the cirrus clouds and the adjacent image gray shades on the 7.3 μm image.
  - The upper-level moisture of the frontal band makes it difficult to distinguish the stripe of thin cirrus in the 6.2 μm WV image (Fig. 2.3B).
- On the WV imagery, the area of opened convective cells forms a comparatively cold pattern and can show quite different appearance in the two WV channels. If the convective cells develop below a very dry air above, their cellular appearance is very well distinguished in the 7.3 μm channel (similar to the 10.8 μm window channel), which is the case on the right-hand

(A)   **(B)**

(C)

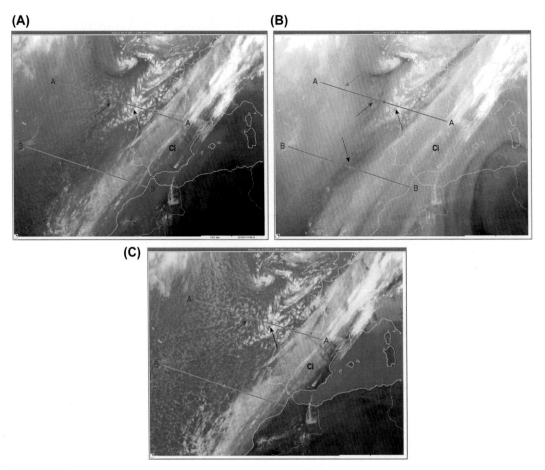

FIGURE 2.7

Meteosat-10 images in (A) water vapor (WV) 7.3 μm, (B) WV 6.2 μm, and (C) infrared 10.8 μm. Positions of convective cloud cells at middle (*red arrow*) and upper (*blue arrow*) troposphere, upper-level dry air (*brown arrow*), upper-air humidity gradient (*green arrow*), as well as the lines of the cross-sections in Fig. 2.8 are indicated.

    side of the green arrows. The cross-sections in Fig. 2.8A and B show that on the left-hand side of the green arrows the mid- to upper-level air is less dry and the pattern on the 7.3 μm image becomes less distinguished, with individual cells obscured by the surrounding moisture (30% relative humidity), which slightly absorbs the 7.3 μm channel radiance.

- At the red arrow in Fig. 2.7, the radiation from the cloud tops of the convective cells produces an area of lighter gray shades within the higher-level moisture pattern on the 6.2 μm channel image (Fig. 2.7B). Although the high-level (300−400 hPa) dry air is quite uniform (less than 10% relative humidity), it will appear on the 6.2 μm image as an irregular gray shade pattern consistent with the cloudiness below the 400 hPa level.

(A) **(B)**

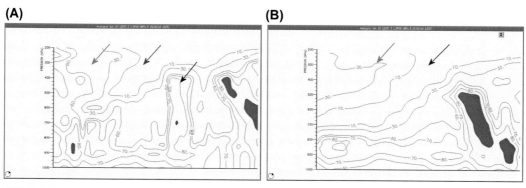

FIGURE 2.8

Vertical cross-section of relative humidity (%) from the ARPEGE NWP model along the lines (A) AA and (B) BB depicted in Fig. 2.7.

- At the blue arrow in Fig. 2.7, the convective cloud tops develop at a higher altitude within the layer of sufficiently large sensitivity range for the 6.2 μm channel (250−500 hPa, as seen in Table 2.1) and dry air above. For that reason, they appear similar in brightness and texture in the images in the 10.8, 6.2, and 7.3 μm channels.

The ability of each one of the two WV channels to reflect complex moisture patterns in a real troposphere depends on the specific moisture stratification as illustrated below.

- The upper-level dry air at the center of the cross-section line BB is seen in the 6.2 μm image (Fig. 2.7B) as a distinct dark band because the relative humidity is less than 10% in the whole layer of sufficiently large sensitivity for this channel (250−500 hPa). However, this very dry layer is not well seen in the 7.3 μm image (Fig. 2.7A), because its layer of large sensitivity (350−700 hPa; see Table 2.1) is sufficiently moist, between 10% and 50% relative humidity (below the position of the brown arrow in Fig. 2.8B).
- At the position of the green arrow in Fig. 2.8B, the two WV channels are able to distinguish the increasing of relative humidity from 10% to 30% in the layer 300−500 hPa (although the difference is more pronounced for the 6.2 μm channel). As seen in Table 2.1, the 7.3 μm channel is sensitive enough to detect moisture at these altitudes as well. That is also the reason why the dry air in the layer 600−700 hPa level below the position of the green arrow in Fig. 2.8B (less than 10% relative humidity) is not visible in Fig. 2.7A, after the 7.3 μm channel radiation from below is absorbed by the moist air above 600 hPa that lightens the image gray shades.

2.1.3 EFFECTS OF LAYERED MOISTURE ON THE RADIANCE

Even when clouds are not a factor, and the air temperature decreases upward with an average lapse rate, moisture is often concentrated in layers (see also Weldon and Holmes, 1991; Santurette and Georgiev, 2005). The complication introduced by layered moisture is of great importance for WV imagery interpretation and will be discussed in detail based on calculations performed by the RTTOV8 radiation model as described in Appendix A.

In Fig. 2.9, areas of moist air (90% relative humidity) are depicted on a vertical scale of pressure and temperature. Layers of high WV content (80–90% relative humidity) at different altitudes are indicated by hatched shading. All other layers of each profile are considered as dry layers of background moisture defined by the minimum value of the mixing ratio accepted by the RTTOV8 code (Saunders and Brunel, 2005) with the fixed temperature sounding so that the relative humidity varies from 0.1% to 0.7%. Eight different cases, or situations, are depicted. The measured brightness temperatures for 45 and 55 degrees latitude (that correspond to satellite zenith angles 51.78 and 62.68 degrees respectively) as well as a qualitative description of image gray shades are shown for each case at the top of the drawings. It is assumed that the relation between brightness temperatures and image gray shades is identical for the two WV channels in order to compare the corresponding radiation effects.

When moisture is present in layers of finite depth above the threshold level, where the radiation is partially transparent for the WV channels, some of the radiation is absorbed and significant amounts of radiation from some warmer origin below pass through the layers. In order to explain this, a set of different vertical arrows is used to depict the transfer of the radiance in Fig. 2.9. As the radiation in WV channels can originate from different tropospheric layers and can be absorbed by WV at different altitudes before reaching the satellite, different types of arrow lines and colors are used as follows:

- Red lines for radiation coming from low-level moisture (from the surface up to 700 hPa).
- Green lines for radiation coming from mid-level moisture (from 700 up to 400 hPa).
- Blue lines for radiation coming from upper-level moisture (from 400 up to 200 hPa).
- Dashed-line arrows, which begin from or below a moist layer and end at the top of the atmosphere, denote radiation coming from below and absorbed partially by moisture in an upper layer, located anywhere in the troposphere.
- Solid-line arrows, which begin from the top of the moist layer and end at the top, denote radiation directly coming to the satellite from the threshold level or from the top of any tropospheric moist layer.

For each case in Fig. 2.9, two kinds of radiation are possible: radiation reaching the satellite coming partly from below (the solid-line arrows) and partly from the moist layer (the dashed-line arrows). Since the atmosphere is never absolutely dry, even in the absence of moist layers above the threshold level the radiation transfer model assumes everywhere presence of background moisture, which absorbs some little part of the radiation in its path to the satellite. Because of this effect, the measured brightness temperature will always be cooler and will not be identical to the air temperature at the top of any moist layer. For the radiation being emitted by high-level clouds near the tropopause, which reaches the satellite without any absorption above the cloud top, the brightness temperature measured by the satellite in WV channels is approximately equal to the cloud top temperature.

The threshold level, below which the atmosphere becomes relatively opaque to the radiation in WV channels, depends on the WV content of the lower troposphere and can be assessed in terms of the portioning effect considered in Appendix A. This is the level at which contribution from below to the radiance becomes zero and varies with the humidity, as follows (as depicted by the portioning effect in Appendix A):

- For the 6.2 μm channel, the threshold level for 30% relative humidity is about 800 hPa and for 97% is near 600 hPa.
- For the 7.3 μm channel, the threshold level is at the ground at 30% relative humidity and only for nearly saturated air (97% relative humidity) is located closely above the surface.

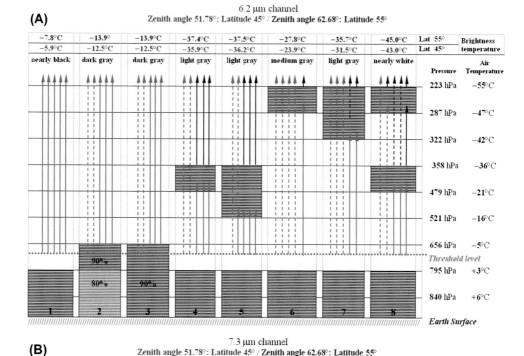

FIGURE 2.9

Layer moisture effects on radiation in (A) 6.2 μm and (B) 7.3 μm channels. Representative air temperatures with labeled pressure surfaces are given to the right. Numbers along the top are brightness temperatures in degrees Celsius for satellite measurements at 55 degrees latitude (first row, blue) and 45 degrees latitude (second row, red). A qualitative description of image gray shades is also shown on the top.

In case (1), the most left one in Fig. 2.9, the air is very dry above 800 hPa, producing a very dark gray shade on the images. Different kinds of radiation effects are present for the two channels as follows:

- For the 6.2 µm channel and 90% relative humidity (Fig. 2.9A), the threshold level is above the top of this moist layer. The resulting brightness temperatures measured in 6.2 µm for middle latitudes are in the range between $-6°C$ and $-8°C$. Taking into account the cooling effect aloft by the absorption of background moisture, it seems that the threshold level is between 650 and 700 hPa. Below the threshold level the moisture has a high mixing ratio in dense air, and most of the 6.2 µm radiation reaching the satellite would originate from the lower middle troposphere (green lines) and then be slightly absorbed by the background moisture through the path of radiation to the satellite.
- The 7.3 µm channel radiation originates from a lower warmer source, but due to the very small transmittance at these altitudes (<0.05; see Appendix A), very little of the radiation comes from below the top of this moist layer.

In cases (2) and (3), the depth of the moisture has been raised to 650 hPa, resulting for the two channels in dark gray image shades that are lighter than that of case (1). For case (3), the relative humidity in the layer 800–650 hPa is 90% as for all other cases. Only a small decrease of relative humidity to 80% is applied from the surface up to 800 hPa for case (2). The resulting brightness temperatures for cases (2) and (3) are very similar, and the following conclusions can be made:

- Since the top of the moist layer is above the threshold level for the 6.2 µm WV channel, it is slightly translucent to the radiation (Fig. 2.9A, green dashed-line arrows).
- However, this lower middle troposphere layer would be more translucent to the radiation in the 7.3 µm WV channel because of the lower altitude of the threshold level. Therefore the brightness temperature measured by the 7.3 µm channel would be warmer than that of the 6.2 µm channel.
- The brightness temperatures would be colder than the air temperature at the top of the moist layer (650 hPa), and this cooling will greatly affect the 6.2 µm channel because of much absorption by the background moisture at the mid- and upper levels.

In case (4), a layer of moisture is present in the upper middle troposphere between 450 and 350 hPa. The image gray shade produced by this case would be significantly lighter than those of the first three cases for the two channels, and their brightness temperature would differ as follows:

- The 6.2 µm brightness temperatures measured by the satellite would be very near to the air temperature of the top of the moist layer. Due to strong absorption of the radiation, a great part of the radiation coming from below will be absorbed by mid-level moisture for this channel (green solid-line arrows in Fig. 2.9A) and then re-emitted at the temperature of the top of the moist layer (blue solid-line arrows). The other, small part of the radiation coming from a warmer lower troposphere will compensate the cooling due to the absorption by the upper-level background moisture.
- For the 7.3 µm the resulting brightness temperatures is higher due to the lower altitude of the threshold level and, accordingly, the warmer origin of the radiation, which will pass through the mid-level moist layer (red dashed-line arrows in Fig. 2.16B) easier than the 6.2 µm radiation.

In case (5), a middle tropospheric moist layer is present between 480 and 520 hPa under the moist layer of case (4). Although the depth of the mid-level moisture is higher than it is in case (4), no significant changes of the gray shades would be observed on the imagery. This is due to the effect that the dominant absorption in cases (4) and (5) comes from the moisture layer above 480 hPa in Fig. 2.9. Also with cases (4) and (5) the following conclusions are noteworthy:

- For cases (4) and (5), the brightness temperature in Fig. 2.9A is much lower than it is in Fig. 2.9B, because for the 6.2 μm channel the transmittance at 400 hPa is twice weaker than this for 7.3 μm channel (see Fig. A3 in Appendix A) and also the 7.3 μm channel radiation originally comes from warmer low-level tropospheric layers.
- For case (5) all resulting brightness temperatures are slightly lower than for case (4), because the radiation is arriving partially from a colder source, located at the base of the higher moist layer.

In cases (6), (7), and (8), significant moisture is present near the top of the troposphere and different moisture profiles below that level. Due to the portioning effect (Appendix A) the resulting brightness temperatures would be warmer than the air temperature of the top of the moist upper-level layer. For these three cases, Fig. 2.9 depicts quite different response of the two WV channels. Because of much more absorption for the 6.2 μm channel radiation in the upper troposphere, the brightness temperatures for this channel significantly differ from case (6) to case (8), and for the 7.3 μm radiation the case-to-case differences are less.

In case (6) with a single upper-level moist layer the brightness temperature would be very far from being representative of the air temperature at the top of the upper tropospheric moist layer. A large amount of the radiation arriving from below passes through this layer, since this moisture has a low mixing ratio and a low density. Only a small portion is absorbed and re-emitted at the low energy level at the very cold air from the higher moist layer. Another much larger portion would originate from the low-level moisture and pass through the upper layer (red dashed-line arrows in Fig. 2.9B). Therefore, the resulting brightness temperature is much higher than the air temperature of the high-level moist layer and would be equal to the air temperature somewhere in the middle troposphere, where the air is dry.

For 7.3 μm imagery the dark gray shade produced by case (6) is very similar to cases (2) and (3). This effect can produce confusion because if such a dark shade is observed on the 7.3 μm imagery, it could result from a high-level layer of moisture or from a moist layer near 650 hPa with dry air above. Therefore, the vertical moisture distribution cannot be estimated from the gray shades of a single-channel image alone. However, the differences between case (3) and case (6) often can be determined by considering the moisture pattern on the imagery of the two WV channels. It is seen that the brightness temperatures of 6.2 and 7.3 μm channel radiation are very similar for case (3) and they are quite different for case (6), and such a comparison can help in solving the interpretation problem.

Case (7) illustrates the difference between the two WV channels regarding the deepness of upper-level moisture. Although the moisture depth is not always a factor, for the upper-level moist layer it significantly contributes to the radiance and decreases the brightness temperature for 6.2 μm as well as for 7.3 μm channels. However, for the 7.3 μm channel the brightness temperature is highly unrepresentative for the air temperature at the top of even such a deep upper-level tropospheric layer. Because of the much less absorption, most of the 7.3 μm radiation reaching the satellite originates within the warmer lower levels, and light gray to white shades are observed on the imagery.

Case (8) illustrates the significant contribution of mid-level moisture to the radiance that is different for the two channels. As in case (6), a large amount of the radiation passes through the high layer, but the brightness temperature is much colder than in case (6), because a great part of the radiation arriving from below in case (8) is mostly originated by a mid-level layer (blue arrows in Fig. 2.9). Only a small portion is absorbed and re-emitted at the lowest energy level at the very cold air. Based on the results presented in Fig. 2.9, the following effects are depicted:

- The resulting brightness temperature would be influenced much by air at 350 hPa (blue arrows in Fig. 2.9A) and will be much cooler than this of the previous two cases. The image gray shade produced by this case would be significantly lighter (nearly white).
- For the 7.3 μm channel medium gray shades are observed on the imagery, because the brightness temperature would be much higher and equals this, resulting from case (5). Therefore, due to the small absorption of 7.3 μm radiation by upper-level moisture, the contribution of the moist layer between 220 and 290 hPa is not significant in case (8).

Finally, Fig. 2.9 provides a general picture of the sensitivity range for the two WV channels to distinguish differences of humidity in the troposphere as follows:

- Since the 6.2 μm WV channel exhibits very high sensitivity (see Fig. 2.6 and Table 2.1), the image gray shades vary from nearly black to nearly white.
- As a whole, no light gray shades are produced by moist layers on the 7.3 μm channel imagery for the profiles presented in Fig. 2.9B, due to the overall lower sensitivity of the 7.3 channel (see Fig. 2.5).

2.1.3.1 Response of WV Channel Radiances to Differences in Humidity Profile

The interpretation of the quantitative results presented in Fig. 2.9 show that different humidity profiles could produce differences in image gray shade and brightness temperature in WV channels.

To illustrate the radiation effects in terms of imagery interpretation, Fig. 2.10A shows the IR 10.8 μm channel image with the brightness temperatures derived from this and from the two WV channels at locations of the release point of three upper-air soundings at stations De-Bilt (at the brown arrow), Brest (at the black arrow), and Ajaccio (at the red arrow). The corresponding vertical profiles of air temperature (T, black curve), wet-bulb potential temperature (θ_w, blue curve), and dew-point (*) are presented in Fig. 2.10B−D. The information content of the MSG WV channels can be illustrated in Fig. 2.11 for three different moisture profiles by considering images and the corresponding upper-air soundings in Fig. 2.10.

As seen in Fig. 2.10B, the upper-air sounding at De-Bilt, Belgium, shows a moisture profile with the following features:

- Dry air in the layer 500−300 hPa
- Saturated air from the Earth's surface up to 550 hPa

In this location (brown arrow in the images), Fig. 2.10A shows IR channel 10.8 μm brightness temperature −15°C, which is representative of the temperature of the cloud top, located at 550 hPa, as seen by the sounding data in Fig. 2.10B. At the same location, the response of the WV channels to this moisture profile is quite different from the response of the IR window channel 10.8 μm:

- Light gray shade in 7.3 μm resulting from a brightness temperature of −28.5°C, which is influenced by the cloud top but is lower than its physical temperature.

(A)

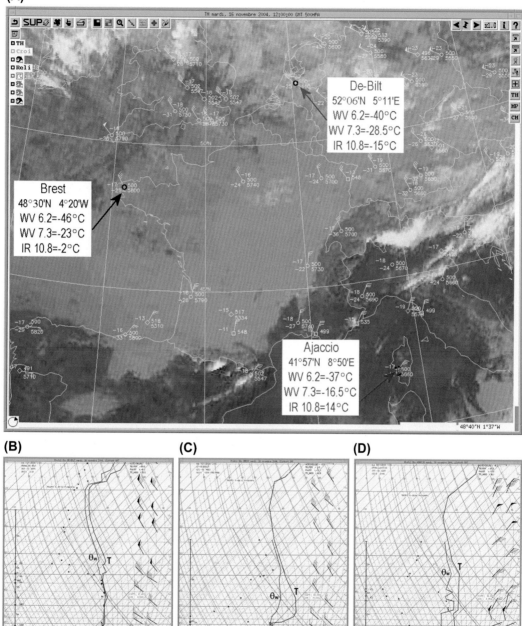

FIGURE 2.10

(A) Meteosat-8 channel image in 10.8 µm with the brightness temperatures from 10.8, 7.3, and 6.2 µm channels at locations of the release points of three upper-air soundings with different vertical profiles as shown in (B) for De-Bilt, (C) for Brest, and (D) for Ajaccio. Air temperature (T, black curve), wet-bulb potential temperature (θ_w, blue curve), and dewpoint (*) are presented in the plots.

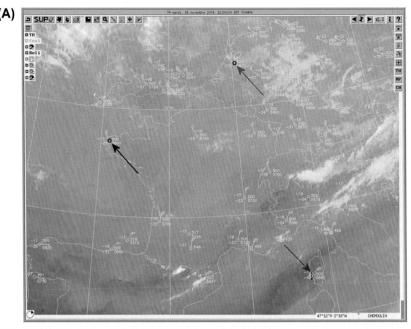

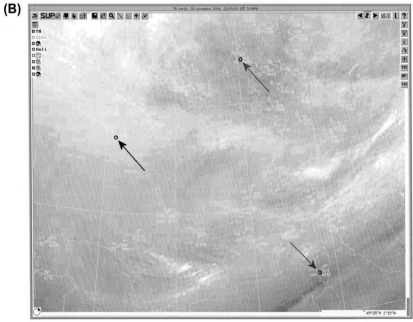

FIGURE 2.11

Response of WV channel radiances to differences in humidity profile: Meteosat-8 image in (A) 7.3 μm and (B) 6.2 μm channel. The arrows indicate the positions of the upper-air soundings from Fig. 2.10.

- Medium gray in 6.2 μm, and brightness temperature −40°C. The radiance originates by the cloud top, and due to higher absorption in 6.2 μm by moisture above, the brightness temperature is lower than the brightness temperature in 7.3 μm.

As seen in Fig. 2.10C, the upper-air sounding at Brest, France, shows a moisture profile with the following features:

- Nearly saturated air at the upper level
- Extremely dry air in 800−500 hPa
- Nearly saturated or saturated air below 850 hPa

Fig. 2.10A shows −2°C brightness temperature from the 10.8 μm channel, which is representative of the temperature of the cloud top, located at 850 hPa, as seen by the sounding profile in Fig. 2.10C. The response of WV channels to this moisture profile (at the positions of the black arrows in the images) exhibits the following characteristics:

- Medium gray shade in 7.3 μm resulting from a brightness temperature of −23°C, which originates by the cloud top but differs from the brightness temperature in the 10.8 μm, so this is not representative for the temperature of the cloud top.
- Light gray in 6.2 μm, and brightness temperature −46°C. The radiance is highly absorbed by nearly saturated air at the upper level, but the brightness temperature is not representative for the physical temperature of this layer.

The upper-air sounding at Ajaccio, France, seen in Fig. 2.10D, shows a moisture profile with the following features:

- A thin layer of humid air at the upper level
- Dry or very dry air from the Earth's surface up to 500 hPa

The brightness temperature derived by IR channel 10.8 μm is +14°C (Fig. 2.10A), which is representative of the Earth's surface temperature in the cloud-free area. The moisture profile in Fig. 2.10D results in the following quite different effects for the two WV channels (Fig. 2.11A and B, at the positions of the magenta arrows):

- Nearly black shade in 7.3 μm resulting from a brightness temperature of −16.5°C. The radiance is absorbed by the small amount of moisture (mostly at 700 hPa), and the brightness temperature is not representative for its temperature because of the portioning effect (see Appendix A).
- Dark gray in 6.2 μm, and brightness temperature −37°C. The radiance is partially absorbed by the upper-level moist layer above 500 hPa, and the brightness temperature is not representative for its temperature due to the portioning effect.

2.2 ABILITY OF 6.2 AND 7.3 μm IMAGES TO REFLECT MOIST/DRY LAYERS, CLOUDS, AND LAND SURFACE FEATURES

In the WV absorption bands 5.35−7.15 μm and 6.85−7.85 μm there are spectral lines of other absorbing gases like CH_4, N_2O, NO_2, and CO_2, but their influence on the radiance signal can be neglected for the used wide spectral intervals (see Fischer et al., 1981). Under this assumption, if the atmosphere contained no moisture in the layer 200–1000 hPa, the brightness temperature in 6.2 and 7.3 μm channels would be equal to the surface blackbody temperature. In such a case of an absolutely dry troposphere, the brightness temperature measured by WV 6.2 μm, WV 7.3 μm, and IR 10.8 μm channels would be identical. If moist air is present at specific levels, it affects in various ways the radiation at these different wavelengths and causes differences in the brightness temperatures derived by these two WV channels depending on the temperature at the low level as follows:

- If the low-level air is cold, a moist layer at a certain altitude causes little difference in brightness temperature measured by the two channels.
- Since the absorption of 7.3 μm channel radiation is weaker than it is for 6.2 μm, the 7.3 μm channel measurements will produce much warmer brightness temperatures.

Clouds are frequently significant to operational imagery interpretation as well and usually obstruct the moisture topography seen in the WV image. This section is aimed to show how different layers of moist/dry air and cloudiness at different vertical locations may produce large differences of brightness temperatures and image gray shades derived by the two MSG WV channels.

2.2.1 UPPER-LEVEL DRY STRUCTURES (200–500 hPa)

The recognition and interpretation of dry structures at the upper level is a critical point in the operational use of WV imagery. The dry intrusions from the upper troposphere are associated with dynamically active areas of mid-tropospheric descending motions and low tropopause heights that will be broadly discussed in this book. In certain conditions, dry structures at the upper level could be reflected in quite different ways in the images of the two WV channels. Fig. 2.12 shows MSG images in WV and IR channels for such cases and a vertical cross-section of relative humidity along a line of quite large moisture differences at upper levels.

Fig. 2.12A and B depicts that the upper-level dry strip indicated by the blue arrow in the vertical cross-section of relative humidity in Fig. 2.12D appears much more distinct in the image of the WV 6.2 μm channel. This effect is due to the quite different moisture stratification in the adjacent area at the position of the red arrow and will be explained below based on the considerations in Section 2.1.3.

- At the blue arrow, the upper-level dry air has penetrated down to the 600 hPa level and it is located above quite moist (10–50% relative humidity) air down to 800 hPa. This tropospheric moisture distribution is very close to case 2 in Fig. 2.9.
- At the red arrow the humidity of the upper-level air is higher (20–30%) and the air below is drier (around 10% relative humidity, Fig. 2.12D) and less cloudy, as seen by the IR image in the 10.8 μm channel in Fig. 2.12C. Such tropospheric moisture distribution is very close to case 7 in Fig. 2.9.

Comparing the responses of the two WV channels depicted in Fig. 2.9 for these two quite different vertical moisture distributions reveals the effects, leading to a quite different appearance of the upper-level dry strip in the corresponding image gray shades. The following quantitative considerations for this region (45 degrees latitude) are applicable:

- In the 6.2 μm channel, the brightness temperature produced by case 2 (at the blue arrow in Fig. 2.12) is 12.5°C and differs significantly from 31.5°C, which is the brightness temperature produced by case 7 (at the red arrow). This difference of 19°C results in quite sharp contrast in the image gray shades that makes the upper-level dry strip distinctly seen in the 6.2 μm image.

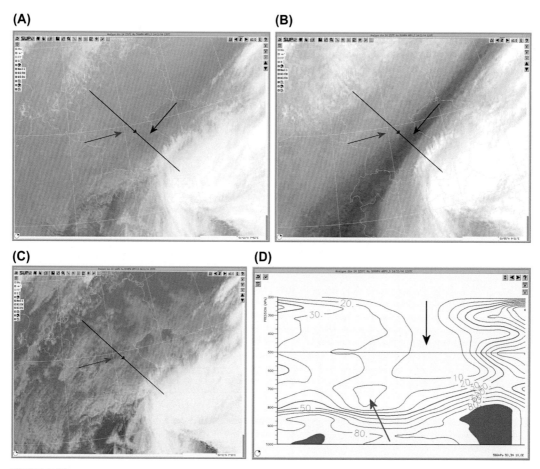

FIGURE 2.12

Upper-level dry structures: Meteosat-8 images in (A) 7.3 μm, (B) 6.2 μm, and (C) 10.8 μm channels and (D) vertical cross-section of relative humidity (%) from ARPEGE NWP model along the black line depicted in the images. The *blue and red arrows* indicate upper-level dry and humid air, respectively.

- In contrast, the radiation in 7.3 μm measured at the satellite shows 9.4°C in case 2 (at the blue arrow) that is very close to the brightness temperature 12.2°C, produced by case 7 (at the red arrow) in this channel. The small difference of less than 3°C in the brightness temperature results in a very slight contrast of the image gray shades on the 7.3 μm image and makes the upper-level dry strip at the blue arrow indistinct in Fig. 2.12A.

This case shows that being not sensitive to the signal arriving from moisture and clouds at low levels, the 6.2 μm image represents in a clear way upper-level dry/moist features. Usually, such large moisture differences at mid–upper levels are produced by significant synoptic-scale perturbations. Therefore, the 6.2 μm channel images are much more useful operational tools for upper-level diagnosis than the 7.3 μm images.

2.2.2 DEEP MOIST LAYERS (200–1000 hPa)

Fig. 2.13D shows a vertical cross-section of relative humidity in a case of a deep tropospheric layer of high moisture content. Such deep layers of moisture produce various gray shades on the images in the two WV channels in cloud-free as well as in cloudy areas. In the upper-level cloud-free area at the position of the blue arrow, there is a fog at the boundary layer seen in Fig. 2.13D between the black arrows. At this area, the IR 10.8 μm image shows a lighter image gray shade (colder brightness temperature than the temperature of the Earth's surface) coming from the top of the foggy layer.

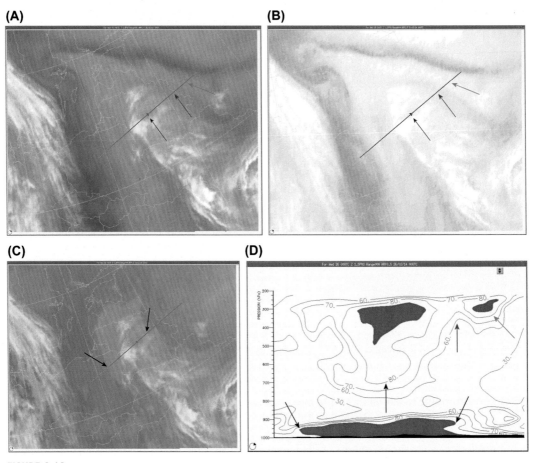

FIGURE 2.13

Deep- and high-level moist layers: Meteosat-10 images in (A) 7.3 μm, (B) 6.2 μm, (C) 10.8 μm, and (D) vertical cross-section of relative humidity (%) from ARPEGE NWP model along the line depicted in the images. Also indicated are the positions of a deep moist layer (*blue arrow*), an upper-level moist layer (*green arrow*), fog in the boundary layer (between the *black arrows*), and a band of decreased deepness and water content of the upper-level moist layer (*red arrow*).

In this case, the following responses of the two WV channels are present:

- The 6.2 μm image (Fig. 2.13B) exhibits the lightest gray shades (lowest brightness temperatures) due to its large sensitivity to the upper-level moisture (see Table 2.1). The 6.2 μm channel is more efficient to distinguish areas of decreased deepness and water content of the upper-level moist layer (at the position of the red arrow) because of its large sensitivity range to distinguish differences in humidity at these altitudes.
- The 7.3 μm image (Fig. 2.13B) reflects this deep moist layer with significantly darker gray shades, because the absorption by upper-level moisture is weaker. Fig. 2.13D shows that the water content of the moist layer is lower at the middle troposphere, where the larger sensitivity of 7.3 μm is located. Therefore, at that location, a significant portion of the radiation originates from the low-level moisture and passes through the upper layer that makes warmer the derived brightness temperatures.
- At the cloudy pixels the IR 10.8 brightness temperature is equal to the physical temperature near the top of the clouds.
- The corresponding gray shades and brightness temperatures in WV 7.3 μm and WV 6.2 μm channels that come from the cloud top are colder than in the IR 10.8 μm channel, due to the absorption of the radiation by moisture above the clouds. The higher the total WV content above the clouds, the lighter the WV channel images become. This effect is much more pronounced for the radiance in the 6.2 μm channel.
- With the increasing of WV content, much of the radiation is absorbed and re-emitted at lower energy levels that produce colder brightness temperatures in the WV channels. In the 7.3 μm radiation, the reduction of brightness temperature is less because of the weaker WV absorption in the upper troposphere.

2.2.3 HIGH-LEVEL MOIST LAYERS (200–400 hPa)

Such a high-level moist layer is present at the position of the green arrows in the vertical cross-section and WV images in Fig. 2.13 and appears in quite different gray shades on the images in the two WV channels, as follows:

- For dry and nearly saturated air at high levels, the 6.2 μm image gray shades (brightness temperatures) differ sufficiently and hence allow distinguishing high-troposphere moisture on the imagery.
- In such cases, the brightness temperatures in the 7.3 μm channel are reduced slightly, and only very large differences in humidity of high-level single layers are likely to be detectable by lighter gray shades in the 7.3 μm image.
- Therefore, in the 7.3 μm channel, the high-level moist layer appears much darker in the images than the high clouds.

In summary, regarding the abilities of 6.2 and 7.3 μm WV channels to detect single moist layers at the upper troposphere (200–400 hPa), the following conclusions can be made:

- High layers of moisture are more detectable by 6.2 μm than by 7.3 μm channel imagery.
- The 6.2 μm radiance is very sensitive to the moisture content in the layer that allows detecting differences in humidity at high levels for operational purposes.

2.2.4 MID-LEVEL MOIST LAYERS (400—650 hPa)

Fig. 2.14D shows a vertical cross-section of relative humidity across a wide band of cloud-free moist air located in the middle troposphere (at the location of the blue arrow). As seen by the IR 10.8 μm image in Fig. 2.14C, in this case of warm air masses the mid-level layer of moisture is cloud-free. Comparing Fig. 2.14A and B, we note the following features of cloud-free layers of moisture at middle altitudes (450—650 hPa) seen in the WV channel's imagery:

- Moisture layers at middle altitudes produce light-gray image shades and cold brightness temperatures in 6.2 and 7.3 μm images. Such layers may be detectable almost in the same way by the two WV channels.

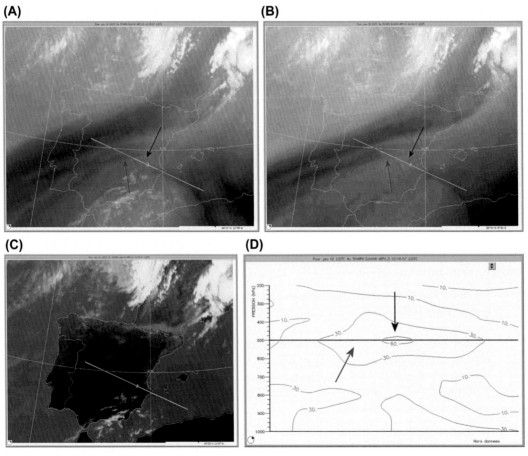

FIGURE 2.14

Mid-level moist layers: Meteosat-8 images in (A) 7.3 μm, (B) 6.2 μm, (C) 10.8 μm, and (D) vertical cross-section of relative humidity (%) from ARPEGE NWP model along the line depicted in the images.

- The 6.2 μm channel images can reflect a deeper mid-level moist layer at slightly colder gray shades (Fig. 2.14B, at the location of the blue arrow), due to the stronger absorption by WV and re-radiation mostly from this layer. At this location a significant portion of the 7.3 μm radiation originates from a warmer source (the low-level air is dry), and it is less absorbed by the mid-level moist air. This effect results in slight darkening of the 7.3 μm image (Fig. 2.14A), but it is not reflected in the 6.2 μm image because differences in water content below 600 hPa are not detected by radiation measurements in the 6.2 μm channel (see Table 2.1).
- Due to its large sensitivity in the layer 350−700 hPa with maximum at 450 hPa (see Table 2.1), the 7.3 μm channel (Fig. 2.14A) may exhibit a better ability to distinguish differences of humidity in the middle troposphere (at the red arrow) by slightly larger variation of the gray shades than the 6.2 μm image (Fig. 2.14B).

2.2.5 LOW-LEVEL MOIST LAYERS (650−800 hPa)

Fig. 2.15D shows a vertical cross-section of relative humidity across a cloudy low-level moist layer (at the position of the blue arrow). The IR 10.8 μm channel image (in Fig. 2.15C) shows that this low-level moist layer is partially cloud-free at the left and right ends of the cross-section line, and the texture of the stratocumulus clouds in the middle is distinctly seen only by this image.

The comparison between images in 6.2 and 7.3 μm WV channels of MSG in Fig. 2.15A and B shows that only one of these WV channels is a tool for detection of moist layers at low levels (650−800 hPa) due to the following considerations:

- As seen by Fig. 2.15A, both the low-level clouds and cloud-free moist layer are seen by the 7.3 μm WV channel. The dry and cloud-free moist areas along the cross-section line appear in a dark gray shade, while the light gray shades of the 7.3 μm channel image are associated with low-level clouds.
- The 7.3 μm image also is able to detect small cloud-free areas of the stratocumulus (around the blue arrow), being more sensitive to detect differences in the WV content at the low troposphere, and may be useful for inspection of moisture commonly found there.
- Low-level moist layers do not affect the 6.2 μm radiation because differences in water content bellow 600 hPa are not detected by radiation measurements in the 6.2 μm channel (see Table 2.1). The darkening gray shade from the left-hand to the right-hand side of the cross-section line on the 6.2 μm image (Fig. 2.15B) is due to the decreasing of humidity above 600 hPa from 45−30% to 30−35% on the right-hand side (at the red arrow, Fig. 2.15D).

2.2.6 MOISTURE/CLOUDINESS IN THE BOUNDARY LAYER (850−950 hPa)

Fig. 2.16D shows a vertical cross-section of relative humidity in a case of high moist content in the boundary layer and less than 10% relative humidity above 700 hPa (at the red arrow). Low-level stratocumulus clouds along the cross-section line are distinguished by the 10.8 μm IR image in Fig. 2.16C.

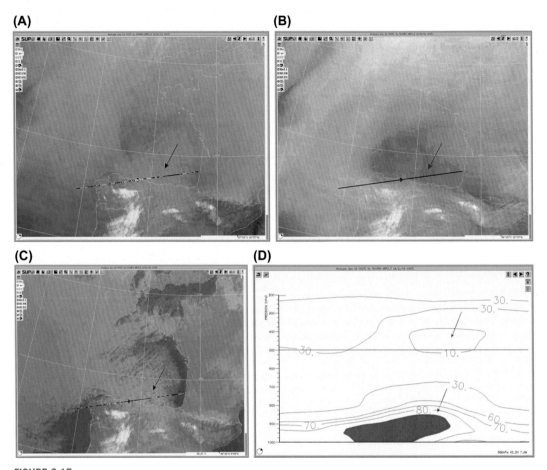

FIGURE 2.15

Low-level moist layers: Meteosat-8 images in (A) 7.3 μm, (B) 6.2 μm, (C) 10.8 μm, and (D) vertical cross-section of relative humidity (%) from ARPEGE NWP model along the line depicted in the images.

Considering Fig. 2.16A and B reveals that the two WV channels of MSG are not tools for detecting moisture in the boundary layer due to the absorption characteristics of the atmosphere in these altitudes as well as the spectral response of these channels. The images can be explained as follows:

- In these altitudes the atmosphere becomes relatively opaque to the radiation due to strong absorption at high WV content in dense air. For that reason, differences in humidity in the boundary layer cannot be distinguished by images in 7.3 and 6.2 μm channels (see Table 2.1).
- The lightening gray shade from the left-hand to the right-hand side of the cross-section line on the WV images (Fig. 2.16A and B) is due to increasing of humidity above 700 hPa from 3−30% (at the red arrow) to 30−40% on the right-hand side (at the blue arrow).

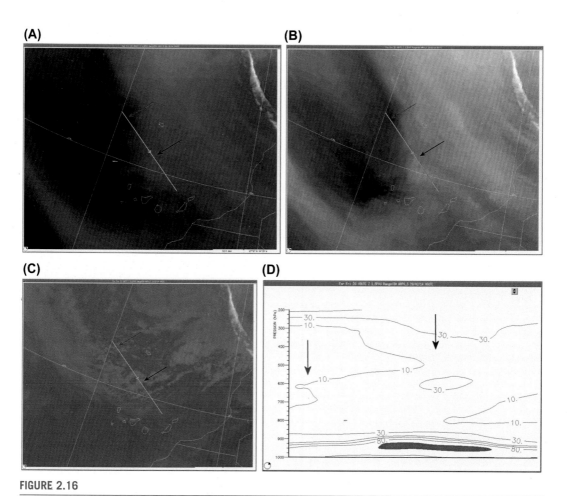

FIGURE 2.16

Moisture/cloudiness in the boundary layer: Meteosat-10 images in (A) 7.3 μm, (B) 6.2 μm, (C) 10.8 μm, and (D) vertical cross-section of relative humidity (%) from ARPEGE NWP model along the line depicted in the images.

2.2.7 EARTH'S SURFACE FEATURES

The ability of MSG channels to distinguish Earth's surface features in a case of very dry atmosphere is illustrated in Fig. 2.17.

The vertical distribution of relative humidity is shown in the cross-section in Fig. 2.17D over the Atlas Mountain in Northwest Africa. The presence of this high-altitude terrain of the Earth's surface is reflected by a quite different appearance in the IR image and the two kinds of satellite WV images as follows:

- On the IR image channel 10.8 μm (Fig. 2.17A) the mountain (perpendicular to the black cross-section line) appears as a colder object in the surrounding warmer land, while the WV image in the 7.3 μm channel (Fig. 2.17C) sees the mountain as a warmer pattern.

(A)   **(B)**

(C)   **(D)**

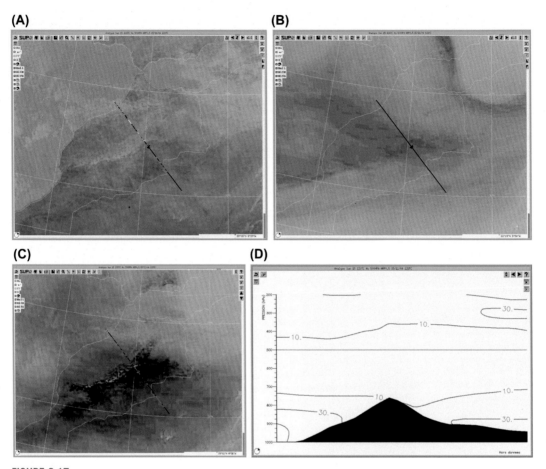

FIGURE 2.17

Abilities of MSG channels to distinguish high-altitude Earth's surface features: Meteosat-8 images in (A) 10.8 μm, (B) 6.2 μm, (C) 7.3 μm, and (D) vertical cross-section of relative humidity (%) from ARPEGE NWP model along the black line depicted in the images.

- Light-gray shades in the 7.3 μm image (Fig. 2.17C) at the left and right sides of the cross-section line indicate humid air at low levels, while the mountain terrain in the middle appears darker (warmer) than the surrounding areas of low-level moist air.
- Dark-gray shades in the 6.2 μm image (Fig. 2.17B) indicate low humidity at mid- and upper levels, but the mountain terrain is not visible.

Therefore, high-altitude mountain terrain is visible in the 7.3 μm channel images in cases of dry mid- and upper troposphere. Even for such a case of very dry troposphere, the radiation in 6.2 μm does not originate from Earth's surface features, and this channel is much more useful for observing upper-troposphere moisture content than 7.3 μm radiance.

2.3 POTENTIAL FOR OPERATIONAL USE OF IMAGES IN 6.2 AND 7.3 μm CHANNELS OF METEOSAT SECOND GENERATION

The radiation characteristics and spectral sensitivity of 6.2 and 7.3 μm WV channels considered in this section provide a basis of methods to use this observational data in air mass analysis, atmospheric circulation pattern recognition, and diagnosis of related processes. Some potential applications of the WV imagery for operational purposes are summarized below:

- Of the two WV channels, the WV absorption is stronger for the 6.2 μm band, and this channel is the most relevant to be displayed and used in image format.
- Since the 6.2 μm radiation is more sensitive to the WV content in the mid- and upper troposphere, the 6.2 μm channel imagery is applied as a tool for upper-level synoptic-scale analysis.
- The 7.3 μm channel is able to detect mid- to low-level moisture and is more sensitive to the moisture content at these altitudes. Therefore single radiation measurements and images in the 7.3 μm channel can be interpreted for studying low-level moisture structures and associated thermodynamic features.
- Data from the two WV channels may be used jointly to serve as additional or complementary information regarding the following operational tasks:
  - To indicate jet streams and wind direction in two tropospheric layers for diagnosis of cyclogenesis, atmospheric fronts, and preconvective situations.
  - To assess atmospheric stability and stability tendency in cloud-free areas to help solve the problem of convection nowcasting.
  - To estimate water content of two deep layers in clear atmosphere to predict convective instability of preconvective environment.

The synoptic-scale interpretation of 6.2 μm (equally 6.3 μm or 6.7 μm) channel images is broadly presented in Weldon and Holmes (1991). WV imagery analysis in the view of PV concept is further considered by Santurette and Georgiev (2005). These operational applications as well as the use of advanced satellite products derived from WV channels data in forecasting convection will be considered in Part II of this book.

PRACTICAL USE OF WATER VAPOR IMAGERY AND THERMODYNAMIC FIELDS

SIGNIFICANT WATER VAPOR IMAGERY FEATURES ASSOCIATED WITH SYNOPTIC THERMODYNAMIC STRUCTURES

CHAPTER OUTLINE

3.1 OPERATIONAL USE OF RADIATION MEASUREMENTS IN WATER VAPOR CHANNELS 6.2 AND 7.3 μm

On a water vapor (WV) image displayed as radiative temperatures in gray shades, the dry areas in a specific tropospheric layer appear warmer (darker) than the areas of higher moisture content. Considering WV as a tracer, the imagery in WV absorption channels offers an observation of the atmospheric flows. Most of the significant structures seen in the imagery are linked to large-scale processes responsible for vertical motion and deformation of the flow.

Of the two Meteosat Second Generation (MSG) WV channels, the absorption of the radiation by WV in the 6.2 μm band is stronger, and this channel is the most useful to be displayed in image format. Since the 6.2 μm radiation is more sensitive to the WV content in the mid- and upper-troposphere, the 6.2 μm channel imagery is applied as a tool for upper-level synoptic-scale analysis. For these reasons, all geostationary satellites perform measurements in any channel with a central-band wavelength between 6.2 and 6.8 μm (see Appendix A).

The 6.2 μm WV image represents the motion field and upper-level dynamics from the middle troposphere up to the tropopause. Areas of low tropopause height, which are associated with descending air and restrict the depth of tropospheric moisture aloft, tend to produce dark-gray shades on the image, whereas areas of ascending air or high tropopause height appear light. The basis for synoptic-scale applications of WV imagery is that moist and dry regions and the boundaries between them often relate to significant upper-level flow features such as troughs, dynamical tropopause anomalies, and jet streams. The boundaries become oriented in the direction of the upper flow of slow-moving weather systems; usually dark regions on the imagery tend to be associated with middle tropospheric troughs and light shades with thermal ridges. Throughout the COMET (2016) lesson water vapor imagery analyses are compared with surface observations to diagnose atmospheric processes and capture forcing for the short-term weather.

The 7.3 μm channel data are available through the MSG Spinning Enhanced Visible InfraRed Imager (SEVIRI) radiometer and the Advanced Himawari Imager (AHI) of the Himawari satellite (see Appendix A). Using this channel we are able to detect moisture at the low to mid-level; the images in 7.3 μm channel often exhibit clear features that can provide relevant information for studying the low-level thermodynamic context. A relevant interpretation of 7.3 μm imagery patterns helps forecasters to better diagnose the atmospheric environment in which weather systems evolve.

Combining data from 6.2 to 7.3 μm WV channels may be used as additional or complementary information regarding the following operational tasks:

- To identify jet streams and to infer wind direction in two tropospheric layers for diagnosis of preconvective situations.
- For air mass analysis as well as for assessing atmospheric stability and stability tendency in cloud-free areas in convection nowcasting.
- To estimate the water content of two deep layers in clear atmosphere to predict preconvective situations.
- For analysis of changes in vertical distribution of humidity to issue early warning and detection of convection.

The WV imagery analysis based on a synoptic-scale interpretation of 6.2 μm (equally 6.3 μm or 6.7 μm) channel images is broadly presented in Weldon and Holmes (1991) and, in the view of the potential vorticity (PV) concept, by Santurette and Georgiev (2005). In this chapter we consider sequences of images superimposed onto various Numerical Weather Prediction (NWP) model fields to gain a dynamical insight into the WV imagery patterns for the purposes of operational forecasting. Possible operational applications will be considered in Chapters 3—5 regarding the use of MSG WV channels in helping to solve the problem of early diagnosis of synoptic development.

3.2 INTERPRETATION OF SYNOPTIC-SCALE IMAGERY FEATURES
3.2.1 MOIST (LIGHT) FEATURES IN 6.2 μm IMAGERY

A number of studies have shown that radiance in WV channel 6.2 μm closely correlates with the humidity field in the layer between 600 and 300 hPa. Since moisture at higher altitudes is supplied from the surface by ascending motions, light features on the imagery correspond to areas of ascent at mid- to upper-levels. Moist features in the WV imagery can be classified into two main groups, according to their appearance in the image gray shades:

- Nearly white to white
- Medium gray to light gray

3.2.1.1 Nearly White to White Features

The nearly white to white features are very cold air masses, produced by large-scale vertical motions, like those marked "R" and "C" in Fig. 3.1A. They are associated with areas of high-altitude clouds within synoptic-scale weather systems. The white-shaded R features of rising air represent the cloud vortex of the low in the northwestern part of Fig. 3.1B. The other feature of this type, located northeast of indication C in Fig. 3.1A, is associated with the cloud system of the eastern trough.

(A)

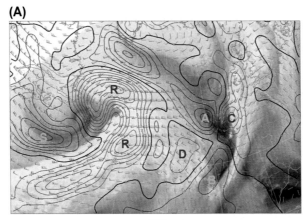

(B)   **(C)**

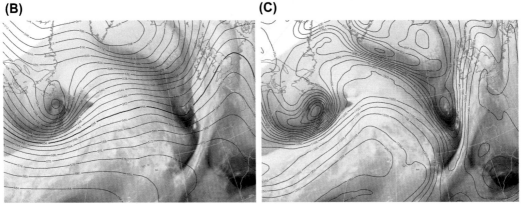

FIGURE 3.1

A water vapor image overlaid by (A) vertical motions (10^{-2} Pa/s, ascending in red, descending in blue) at 400 hPa and wind (*green arrows*) at 300 hPa; (B) 500 hPa heights (dam); (C) geopotential height (dam) of the 1.5 PVU surface.

Such large-scale patterns of nearly white to white image gray shades are associated with areas of high tropopause. As seen in Fig. 3.1C, these features have the following characteristics:

- The western R features are the more pronounced and are associated with significant ascent in the area of very high dynamical tropopause.
- The eastern C feature in the leading diffluent part of a trough is narrow and associated with an area of high gradient but lower tropopause height than the western one.

Such patterns, associated with ascending motion and clouds, also may be identified in the infrared (IR) and visible satellite imagery.

3.2.1.2 Medium-Gray to Light-Gray Features

However, not all the rising air is cloudy, because cloud formation also depends on the relative humidity of the ascending air. Areas lacking cloudy ascent can be identified only in the WV imagery and are interpreted as medium- to light-gray shades. Fig. 3.1 shows such a pattern of accent at location D. The ascending air is associated with a short wave in a slightly diffluent westerly flow at middle troposphere as seen in Fig. 3.1B.

Usually, WV imagery can help in identifying areas of moist ascent during stages of disturbance development in which well-pronounced cloud formation has not yet occurred. As will be discussed in Section 3.5.2, in the very early phase of cyclone development, the process may appear as an area of moist ascent, producing a light image leaf pattern before becoming evident in the cloud field. This is one of the advantages of using WV channel imagery as a forecasting tool compared with either the visible or the window IR channels.

After a light-colored moisture feature has formed, it may be destroyed just as quickly as it formed, or, in the absence of synoptic-scale vertical motions or mixing effects, it may persist. When moisture patterns persist, they may remain as residual features after the circulation and thermal field systems related to their formation are no longer present. Since many moist (light) features on an image are changing with time and others remain as residual features produced by some earlier circulation system, it is often difficult to interpret a single WV image. The task is easier and more accurately accomplished when a sequence of previous images is used, or when images are viewed in animation.

3.2.2 DRY (DARK) FEATURES IN 6.2 μm IMAGERY

Since radiance in WV channel 6.2 μm closely correlates with the upper-level humidity, dry features in the WV imagery, associated with synoptic-scale circulation systems, appear dark gray to nearly black. They are often observed in areas of mid-tropospheric descending motions and low tropopause heights. In Fig. 3.1A, we note the following characteristics:

- The area of large-scale descending motions (S) is associated with a dark zone in the WV image to the rear of the western trough in Fig. 3.1B.
- At locations A and B, areas of mid-troposphere descending motions are also seen. These produce dark-gray shades in the image and are associated with low tropopause heights (see Fig. 3.1C), rearward of the eastern upper-level trough in Fig. 3.1B.

Once an area aloft has become dry, associated with dark shades in the WV image, it tends to remain so until replaced by moist air. This can occur by convergence with adjacent moisture regimes or by moistening via smaller-scale mixing processes such as convection. If no moistening occurs, the residual dry air aloft will move depending on the associated wind fields and can evolve into a variety of shapes. For that reason, not all dry areas are significant from an operational point of view. Also, there are some significant dry areas that cannot be clearly identified in the WV image gray shades since their cold–temperature profiles make them appear light (see Section 2.2). So it is essential to associate the imagery with relevant fields representative of the circulation. An appropriate approach is to classify such dark areas and features according to their association with atmospheric dynamics (eg, upper-level flow, vertical motions, and PV anomalies). There are two main groups of dry features useful to consider:

- Dry (dark) bands/spots
- Dry intrusions

3.2.2.1 Dry (Dark) Bands/Spots

Dry (dark) bands/spots are medium-gray to dark-gray areas shaped like a band or a semicircular spot that are not closed or edged by significant synoptic-scale cloudy areas (nearly white to white-gray shades). Figs. 3.2 and 3.3 below show WV images from the GOES-13 satellite superimposed by NWP dynamical fields over the area of North America. Three kinds of dry bands/spots may be distinguished in the imagery:

1. Latent dry band/spots:
   These features of medium-gray (rarely dark-gray) shades are associated with weak descending motions (or their absence) and a latent tropopause anomaly (a weak PV anomaly) but lack a jet. Such a band is marked "A" in Fig. 3.2, where its association with upper-level dynamics is depicted. These patterns may move in accordance with the upper-level wind field, but they are not associated with further significant disturbance development, as long as they do not interact with a jet.
2. Deformation dry bands:
   These patterns (like those marked "B" in Figs. 3.2 and 3.3B) may be associated with subsidence, but they are not connected with jets and PV anomalies. As seen in Fig. 3.3B, they are associated with deformation zones in the fluid mass shape due to stretching and/or shearing produced by the upper-level flow. In a specific situation, the dry air aloft, associated with such dark features, may be conducive for convective development, by generating instability (see Section 4.3.2).
3. Dynamic dry band/spots:
   These are associated with moderate or strong subsidence, jet streams, and PV anomalies. Such a dynamically significant dry band is marked "C" in Fig. 3.2B, and it is a precursor of subsequent development of a dry intrusion. A dynamic dry spot appears also at location E in Fig. 3.3 associated with significant subsidence and a dynamical tropopause anomaly at the base of a trough.

3.2.2.2 Dry Intrusions

Dry intrusions are specific areas of medium-gray to nearly black shades in the imagery that are bordered by significant synoptic-scale cloudy areas and are associated with any form of cyclonic circulation.

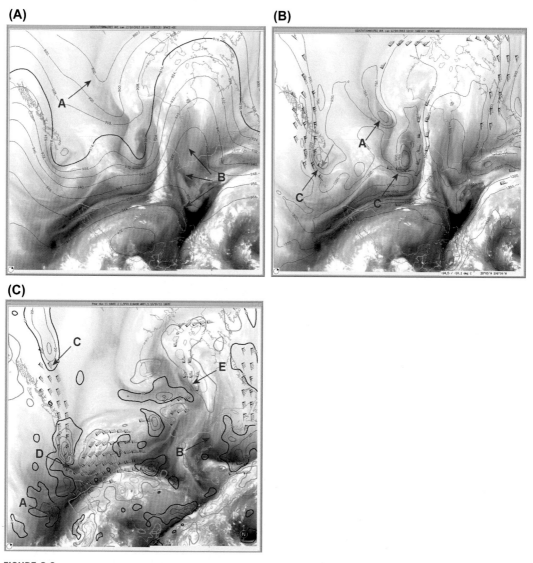

FIGURE 3.2

Water vapor image from GOES-13 over North America overlaid by ARPEGE analysis on October 12, 2013, 1800 UTC of (A) 300 hPa heights and (B) 1.5 PVU surface heights and wind vectors at 300 hPa (only >70 kt) as well as on October 13, 2013, 1800 UTC of (C) vertical velocity (10^{-2} Pa/s, ascent in orange, descent in blue) and wind vectors (only >90 kt) at 300 hPa. Also indicated: "A," a latent dry spot/band, "B," a deformation dry band, "C"/"E," dynamic dry band/spot, and "D," dry intrusion.

Dry intrusions appear generally in the leading zone of dry bands when cyclonic development occurs. They are connected with a jet stream and a PV anomaly as well as with significant synoptic-scale subsiding motions. Such a dry feature is indicated by "D" in Fig. 3.3A and B.

Figs. 3.3A and 3.2C show the dry intrusion at locations D (*pink arrow*) in the leading part of an area of descent. The dynamic dry band from which this dry intrusion originates is shown by the

(A)

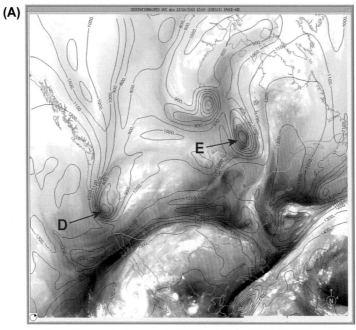

(B)

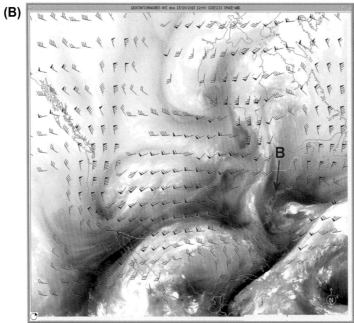

FIGURE 3.3

Water vapor image from GOES-13 over North America overlaid by ARPEGE analysis on October 13, 2013, 1200 UTC of (A) 1.5 PVU surface heights and (B) wind vectors at 300 hPa (only >20 kt). Also indicated: "B" a deformation dry band, "E" dynamic dry band/spot, and "D" dry intrusion.

green arrow in Fig. 3.2B. It is associated with a strong subsidence because a cyclogenesis is in a phase of rapid development.

The identification and monitoring of WV dark and light zones associated with cyclonic disturbances form the basis of validating NWP output (see Santurette and Georgiev, 2005). Analysts concerned with interpreting satellite imagery have revealed two rather symmetrical flows (see Browning, 1997):

- There is a flow of ascending warm moist air originating from both sides of the warm front. This flow is responsible for the cloud head of the cyclone vortex.
- There is a cold dry flow, which comes down from the rear of a developing surface low. This flow forms a tongue of stratospheric/upper-tropospheric descending air referred to as dry intrusion.

The appearance of the dry intrusion in the WV image for cyclone systems over the Eastern Atlantic on December 27, 1999, and North America on October 15, 2013, is shown in Fig. 3.4A and B, respectively. The subsidence of upper-tropospheric air has formed a dry streak on the image. As the descending dry-intrusion air fans out behind the surface cold front (SCF), the ascending moist flow fans out in the middle or upper troposphere behind a front-like boundary corresponding to the convex outer edge of the cloud head. Depending on the magnitude of the ambient stretching deformation, the two fan-shaped flows may be elongated parallel to the cold front like the feature shown in Fig. 3.4B.

As explained in Browning (1997), the moist part of the leading edge of the dry intrusion moves quickly along the cold front from the upper to mid-level; then—after penetrating near the center of the low—the dry intrusion surges perpendicularly to the SCF (in the direction of the mid-level flow) and overruns the low-level moist flow to give an upper cold front (UCF) close to the cyclone center, where horizontal transport plays a major role in producing variability of the upper-troposphere flow. This is associated with a discontinuity between low-level and upper-level structures of the frontal system resulting in a split cold front that will be considered in detail in Section 3.5.5.

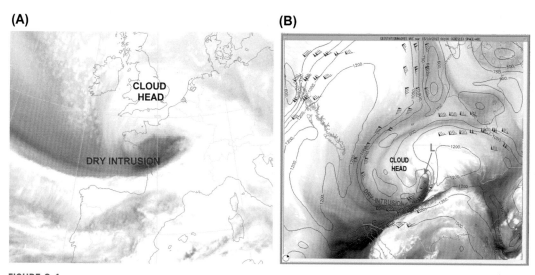

(A)   **(B)**

FIGURE 3.4

Dry intrusion as seen in water vapor images (A) over the Eastern Atlantic on December 27, 1999, 1800 UTC and (B) over North America on October 15, 2013, 0000 UTC: GOES water vapor image, superimposed by 1.5 PVU surface heights and 300 hPa wind (only ≥70 kt). Also marked: "L," the surface low center.

(A) **(B)**

(C) **(D)**

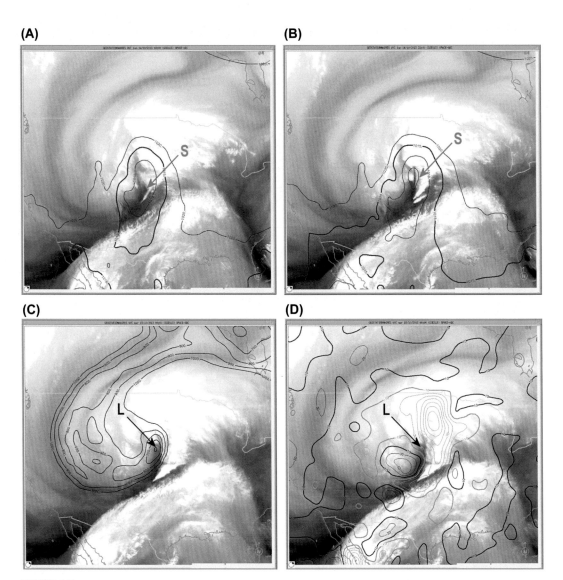

FIGURE 3.5

Dry intrusion over North America as seen in the GOES water vapor images overlaid by ARPEGE analysis fields: for October 14, 2013 of mean sea level pressure (brown, threshold 1020 hPa) (A) at 1800 UTC and (B) at 2100 UTC; for October 15, 2013, 0000 UTC of (C) 1.5 PVU surface heights (dam, blue only ≤1000 dam) and (D) vertical motions at 700 hPa (10^{-2} Pa/s, ascending in orange, descending in blue). Also marked: "L," the surface low center.

Fig. 3.5 shows in more detail the features of the dry intrusion associated with the example of a cyclone over North America on October 15, 2013, seen in the WV image 6.55 μm from the GOES-13 satellite and NWP fields from the ARPEGE model. At the onset of cyclogenesis, the dry intrusion appears as a pronounced dark slot in the imagery owing to the strong subsidence of upper-tropospheric air (in Fig. 3.5A at location S). Just after that (and before the mature stage), a part of the descending dry air overruns low-level moist air that leads to a split cold front development. As a result, the leading part of the dry slot tends to appear lighter, as is obvious to the east of the darker part of the vortex center in Fig. 3.5B at location S.

These illustrations reveal two particularly important properties of the dry intrusion close to the surface center at location L and the association of dark-gray shades on the imagery with the vertical movement of the dry air within the circulation to the rear of a cyclone:

- The dry intrusion is associated with low geopotential of the dynamical tropopause with minimum near the surface low center (in Fig. 3.5C it is 600 dam at location L).
- Close to the cyclone center, some of the dry air that has originally subsided also moves horizontally or even rises. This is obvious at location L in Fig. 3.5D, where the medium-gray shades on the imagery have been produced by ascending middle-level dry air originally associated with the dry intrusion to the rear of the trough and ascending motions below seen in the field of vertical velocity at the 700 hPa level in Fig. 3.5D.

To the rear of the cyclone center, the dry intrusion is associated with upper-level cyclonic PV anomaly (a dynamical tropopause anomaly); such an anomaly promotes cyclogenesis by upper-level forcing (see Section 3.4.2). Prior to the development of the dry slot (ie, before it is flanked on both sides by clouds), the dry intrusion is not evident in the IR or visible (VIS) imagery. However, it is often evident as an expanding darkening zone in the WV channel pictures up to a day before it is detected in IR or VIS images.

WV imagery is a valuable tool to monitor dry intrusions and may be useful in two ways for operational forecasting (Browning, 1997):

1. It helps in identifying and following the upper-level dynamical forcing, which can lead the development. It may also be used to validate numerical model output (considered in Chapter 5).
2. It helps the forecaster to understand what is happening on the mesoscale and to anticipate what may happen over a period of a nowcast and very short-range forecast. This is especially valuable in situations of rapid cyclogenesis and convective activity when local warnings are needed. Cold-frontal rain bands and severe winds often are associated with stratospheric intrusions, which may be studied by using their correspondence to the WV image dark slots.

Fig. 3.6 shows a sequence of Multifunctional Transport Satellite (MTSAT) WV imagery taken every 6 h during the development of a rapid cyclogenesis over Asia (Mongolia and Northeastern China) on October 21 and 22, 2013. It illustrates the evolution of the dry intrusion, during which three important phases may be recognized:

1. Expanding dry zone:
   This dark feature (indicated by "Z" in Fig. 3.6A) appears and expands into a developing white area before the rapid and significant cyclogenesis begins and then a transition from a dry zone to a dry slot (D in Fig. 3.6C) occurs.

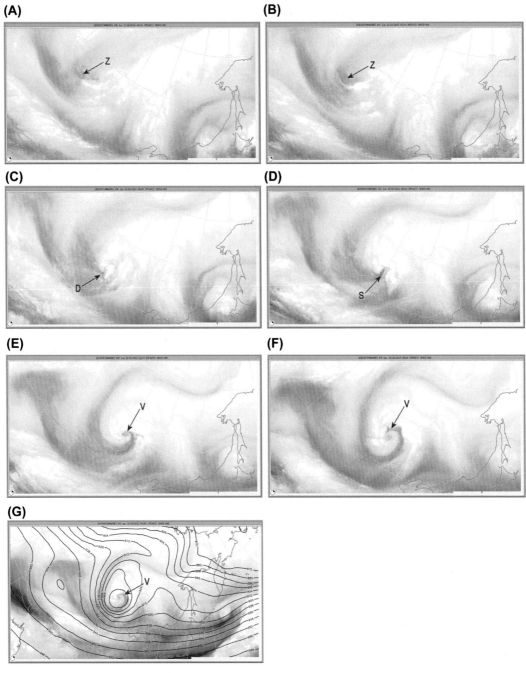

FIGURE 3.6

MTSAT water vapor imagery taken every 6 h during the development of a rapid cyclogenesis over Asia (Mongolia and Northeastern China) on October 21, 2013, at (A) 0000 UTC, (B) 0300 UTC, (C) 0900 UTC, (D) 1500 UTC, and (E) 2100 UTC, and on October 22, 2013, at (F) 0300 UTC and (G) 0000 UTC, superimposed by 300 hPa heights. Also marked: "Z," the expanding dark zone, "D," the transition to a dark slot, "S," the dark slot, and "V," the dark spiral.

2. Dry slot:

   A part of the descending air within the expanding dry zone enters the circulation of the surface depression forming a dry slot southwest of the low (at D in Fig. 3.6C as well as at S in Fig. 3.6D). This pattern exhibits a marked darkening and appears at the beginning of a rapid cyclogenesis associated with upper-level forcing.

3. Dry spiral:

   The spiral (marked "V" in Fig. 3.6E and F) is associated with revolution of the dry intrusion around the upper-level cyclonic circulation. It has been observed that the spiral patterns commonly developed after the surface deepening had nearly finished, and the upper-air cyclone continued to intensify, or maintained its strength (Weldon and Holmes, 1991).

A spiral pattern on the WV imagery indicates the conditions of the upper-air cyclone (as seen in Fig. 3.6G), signifying that the upper low is closed and remains undisturbed for a period sufficient for the dry air to spiral around the center. A spiral pattern that is very well defined and has spiraled rapidly indicates that the associated storm is intense (Weldon and Holmes, 1991). However, the mere fact that a spiral pattern appears on the WV imagery is not necessarily an indication of an intense storm system. Many rapidly deepening storms with low minimum pressure do not develop spiral WV patterns, and spiral patterns are sometimes observed with relatively shallow upper-level lows.

Regarding a dry slot evolution, the most significant stage is the moment when the dry slot appears and eventually when the darkening intensifies, because such evolutions are associated with strong dynamics (with pronounced vertical motion). The dry slot appearance is associated with the onset of significant cyclogenesis, and in Fig. 3.6 it occurs during the period between Fig. 3.6C and D. Fig. 3.7 reveals this element of the dry intrusion evolution associated with the storm development on December 27, 1999. The beginning of cyclogenesis is seen in the image by the formation of a light hook-shaped feature (H) associated with the expanding dry zone Z at 0000 UTC on December 27, 1999. At 1200 UTC, when the spectacular cyclogenesis has already begun, the dry slot S may be distinguished in the WV imagery. Its appearance in the WV image at 1800 UTC was shown in Fig. 3.4A.

(A) **(B)**

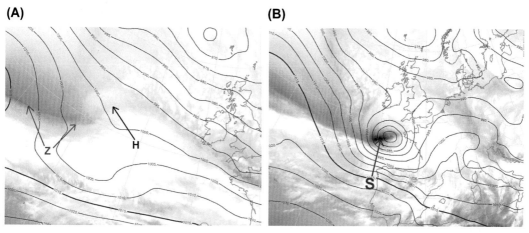

FIGURE 3.7

Water vapor imagery overlaid by corresponding mean sea-level pressure fields. Also indicated: "Z," the expanding dark zone, "S," the dark slot, and "H," the hook-shaped feature. (A) December 27, 1999, at 0000 UTC, ARPEGE analysis; (B) December 27, 1999, at 1200 UTC, subjective analysis, interactively produced and based on the surface observations.

3.2.3 JET STREAM MOISTURE BOUNDARIES SEEN IN 6.2 AND 7.3 μm IMAGERY

The two WV channels 6.2 and 7.3 μm are sensitive to catch moisture boundaries at the zone between the warm/moist and cold/dry side of the jet/wind maximums at two different levels in the troposphere. In most of the situations, channel-type jets are present and the direction of the flow in the jet is approximately parallel to the jet stream moisture boundary. In some situations, by comparing such boundaries in the images in the two WV channels, we can recognize the sign of the wind shear, which is present in the layer between the middle and upper troposphere.

A specific boundary in the mid-level moisture field that appears in the 7.3 μm channel image may be indistinct or even not present in the 6.2 μm image. Fig. 3.8A shows such a moisture boundary indicated by the *blue arrows*. The corresponding 6.2 μm image on Fig. 3.8B superimposed by the 500 hPa surface heights shows how significantly the imagery features in the two MSG WV channels may differ in such a situation of strong troposphere dynamics within the circulation of a mid-level trough. Fig. 3.8C and D shows the zones of maximum winds seen in the wind vectors at 300 hPa (red) and 600 hPa (blue) over the 7.3 μm image as well as in the IR 10.8 μm image. The moisture boundaries at the *blue arrows* (distinctly seen only on 7.3 μm image) and at the *red and brown arrows* (distinctly seen only on 6.2 μm image) are associated with the following quite different dynamical features seen in the images and the cross-section in Fig. 3.8E:

- The upper-level moisture boundary in Fig. 3.8B (at the *red arrow*) that is related to the upper-level jet structure at the *green arrow* in Fig. 3.8E.
- The moisture boundary at the *blue arrow* in the 7.3 μm image on Fig. 3.8A related to the mid-level jet (MLJ) structure at the *blue arrow* in Fig. 3.8E.
- Different orientations of the jets in the upper and middle levels (normal to the moist boundaries at the *red and blue arrows*) that indicate a change in wind direction in the flow between the middle and upper level.
- The moisture boundary at the position of the *brown arrows* in Fig. 3.8B that is associated with the upper-level jet in the forward side of the upper ridge upstream to the trough. The high tilt of this boundary to the west shows a blocking regime, which is seen by the northeasterly winds at the red wind vectors at the upper-left side of Fig. 3.8C.

The dry dark zone in the 6.2 μm image of Fig. 3.8B extended to the southeast of the *brown arrows* is a well-known signature related to an upper-level PV anomaly or a dynamical tropopause anomaly. The cross-section in Fig. 3.8E depicts a folding of the dynamical tropopause (blue thick contour of PV in Fig. 3.8C), located between the *brown and green arrows*.

Fig. 3.8D shows that the image IR channel 10.8 μm is not sensitive to such important features of the troposphere dynamics seen in the vertical cross-section in Fig. 3.8E. This illustrates the value of using in conjunction imagery in the two MSG WV channels as a tool for observing important features of atmospheric dynamics in the upper and middle troposphere.

3.3 MIDDLE- TO UPPER-TROPOSPHERE WIND FIELD FEATURES

The WV images in 6.2 and 7.3 μm channels represent the humidity fields in the upper and middle troposphere, which depend not only on the vertical motions but on the horizontal air movement as well. The moisture ascends from low levels and moves with the wind while the subsidence dries

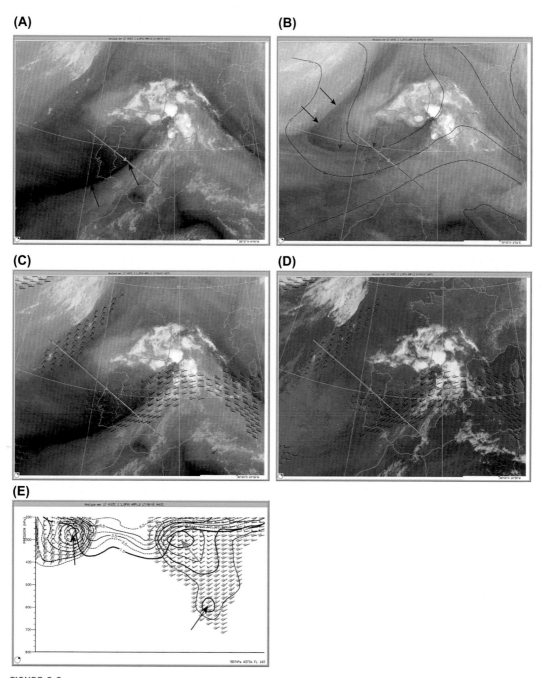

FIGURE 3.8

Jet stream moisture boundaries seen in the water vapor Meteosat-8 images and vertical cross-section for August 17, 2005, at 0000 UTC: (A) in 7.3 μm channel, (B) in 6.2 μm superimposed by the 500 hPa surface heights. Wind vectors at 600 hPa level (blue, only ≥35 kt) and 300 hPa level (red, only ≥50 kt) superimposed by (C) 7.3 μm and (D) 10.8 μm images, respectively. (E) Vertical cross-sections along the axis depicted in (C) and (D) of wind speed (black contours, only ≥35 kt) normal to the cross-section and total wind vectors (red, only ≥35 kt) and potential vorticity (blue contours, only ≥1.5 PVU).

the upper troposphere. Showing the distribution of humidity as a tracer, WV imagery reflects the motion field in deep tropospheric layers.

3.3.1 SPECIFIC UPPER-LEVEL FLOW PATTERNS SEEN IN 6.2 μm IMAGERY

Long waves in the flow are generally associated with the bright bands in a WV image as shown by Fig. 3.9, which represents 6.2 μm WV images covering a large part of the Northern Hemisphere. These long-wave bands often are continuous patterns over thousands of kilometers. Such bands are created and destroyed by divergence of the trajectories and thus show the recent history of the flow. The superimposition of WV image onto the 300 hPa surface geopotential (Fig. 3.9B) and 300 hPa wind (Fig. 3.9C) makes obvious the following circulation features:

- The undulations of the large-scale circulation as well as the large light convex bands at the top of the ridges.
- The large dark concave areas at the bottom of the troughs.

The correlation of the superimposed patterns in Fig. 3.9 shows that overlaying mid- and upper-level dynamical fields onto the WV image provides a powerful way of helping to interpret the imagery. Using this approach, the WV imagery offers a tool to study dynamical processes in terms of the familiar concepts of vorticity and vorticity advection. Figs. 3.10 and 3.11 show the correlation between upper-troposphere dynamical fields and the 6.2 μm WV channel radiance by superimposing the geopotential of the dynamical tropopause and the strong wind at the 300 hPa isobaric surface onto the image over the Eastern Atlantic and the Northwestern Pacific Ocean, respectively.

The 6.2 μm WV imagery is representative of the dynamics some level below the tropopause. At midlatitudes, a pronounced dynamics shows very well-marked signatures on the WV imagery due to the correlation of the radiance field in the 6.2 μm channel with the 1.5 PVU surface heights. In particular, superimposition of the geopotential of the 1.5 PVU surface onto the corresponding WV image in Figs. 3.10A and 3.11A reveals key elements of the relationship:

- Jet streaks and areas of strong geopotential gradient of the tropopause height are generally characterized by strong dark/bright gradients on WV imagery, with the dry air (dark area) on the polar side of the jet (Figs. 3.10B and 3.11B).
- Dynamical tropopause anomalies are associated with well-marked dark areas on the WV images.

However, the exact relationship varies according to the synoptic situation and geographic location:

- Latent tropopause anomalies are not always clearly detectable because of very weak vertical motion.
- In polar regions in winter, the lower atmosphere is so cold (particularly over large continental surfaces) that there is very small contrast in radiance between regions of low tropopause and their surroundings, even if the latter contain clouds. This is the case in Fig. 3.11 where the WV image appears completely white over the cold northeastern part of the Asian continent.
- Consequently, a strong dynamical tropopause anomaly over Eastern Canada, north of China, or over northern areas of Russia in winter may be associated with a much lower radiance and therefore less apparent "dryness" than a weaker anomaly further south and east over the warmer ocean.

(A)

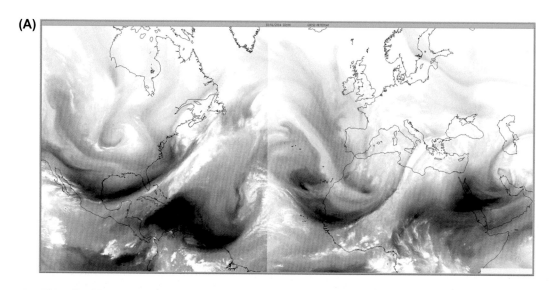

(B)

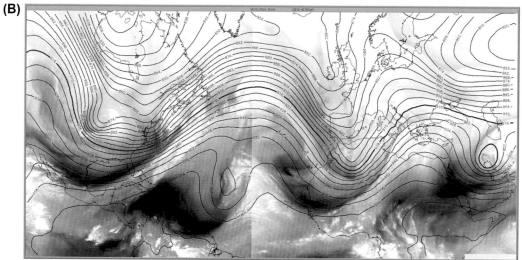

FIGURE 3.9

(A) Water vapor images (composites GOES-METEOSAT) on January 18, 2014, at 1800 UTC, with superimposition of various ARPEGE analysis: (B) 300 hPa geopotential; (C) 300 hPa wind vectors; (D) isotachs of 300 hPa wind (every 25 kt, threshold 75 kt).

(C)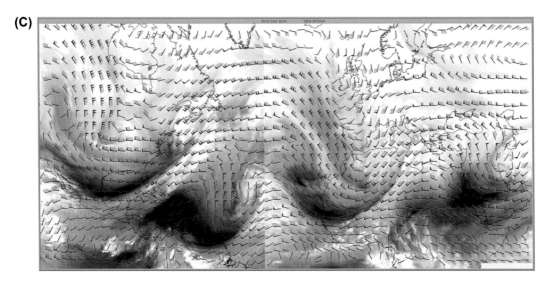

(D)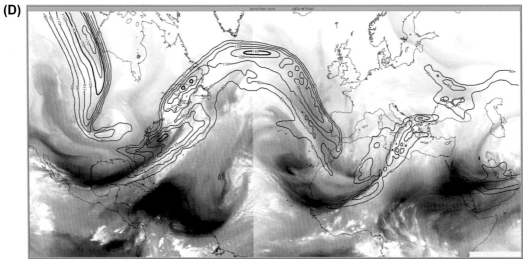

FIGURE 3.9 Cont'd

(A)

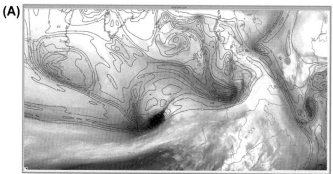

(B)

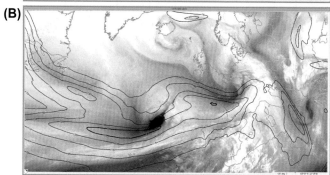

FIGURE 3.10

Water vapor images over the Eastern Atlantic (composites GOES-METEOSAT) on January 23, 2009, at 1200 UTC with superimposition of ARPEGE analysis (A) 1.5 PVU geopotential (every 75 dam, threshold 925 dam); (B) isotachs of 300 hPa wind (every 25 kt, threshold 75 kt).

(A)

(B)

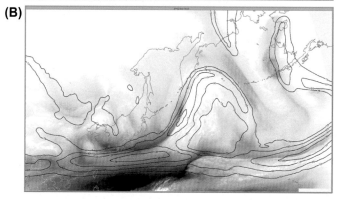

FIGURE 3.11

MTSAT water vapor images over the northwestern Pacific Ocean on January 27, 2014, at 00 UTC with superimposition of ARPEGE analysis (A) 1.5 PVU geopotential (every 75 dam, threshold 925 dam); (B) isotachs of 300 hPa wind (every 25 kt, threshold 75 kt).

In the forecasting process it is efficient to interpret the WV image with a general knowledge of the motion field. We can use the information supplied by the model—either an analysis or sometimes very short-range forecasts when the analysis is not available. Although short-range forecasts are not always perfect, they have enough quality to bring useful information at synoptic scale; then, when combined with the satellite image, the two sources of information complement each other. Maintaining a critical eye when looking at the numerical model fields is crucial, and priority must always be given to the observational data and satellite imagery. We can use the approach to analyze a WV image animation superimposed with the dynamical fields. The motion field can help us to interpret the WV image and to focus on the possible upper-level anomalies. However, the model may have some shortcomings in simulating the upper-level circulation. A special operational part of this book is devoted to this point in Chapter 5.

3.3.2 INTERACTION OF A JET STREAM WITH A DYNAMICAL TROPOPAUSE ANOMALY: JET STREAK STRUCTURE EMERGENCE

Many examples in the real atmosphere (which can be observed through satellite imagery) illustrate the interaction between a jet and a dynamical tropopause anomaly, leading to the emergence of a jet streak. Fig. 3.12 reveals the signatures of the process seen in the WV imagery by using a superposition of wind vectors associated with the jet at 300 hPa and the geopotential of the 1.5 PVU surface. The area of low tropopause, marked "A" in Fig. 3.12A, is isolated from the Atlantic jet stream; at this time the anomaly at location A is not in a dynamical phase: The gradient of geopotential is quasi-regular around it. Progressively, the jet and the low tropopause area approach each other and interact.

This is depicted in Fig. 3.12C–E and is associated with the following dynamical effects:

- The anomaly becomes more dynamic as a consequence of this interaction, and the WV imagery progressively becomes darker.
- The wind in the jet stream increases in the southern part of the anomaly, coinciding with an increase in the gradient of the geopotential of the tropopause.
- As a jet streak forms, a white thin zone develops on the WV image; this zone is closely connected with the axis of the maximum wind.

This example illustrates the designation of "dynamical objects" given to the tropopause anomalies. These anomalies are very essential features due to the following properties:

- They are the main structures leading to formation of a jet streak by interaction with the jet stream that can promote a cyclogenesis (see Sections 1.3.3 and 3.5.3).
- They are quasi-conservative structures that are identifiable as minima (or troughs) of the 1.5 PVU surface height.
- Being quasi-conservative structures, they can be detected before the onset of cyclogenesis.
- They are often well seen as dark zones on the WV imagery and can be tracked over time.

3.3.3 UPPER-LEVEL DIVERGENT FLOW AS A SIGN OF ASCENDING MOTIONS

Upper-level divergent flow is directly related to ascending motion in the middle and upper troposphere. Two types of situations that cause high troposphere divergence can be considered:

1. Perturbations of the upper-level flow produce divergence (and convergence) zones, especially those perturbations associated with a jet (because of the rapid flow) as well as with a PV-Jet interaction (see Sections 1.3.2 and 3.3.2). These divergence areas are associated with

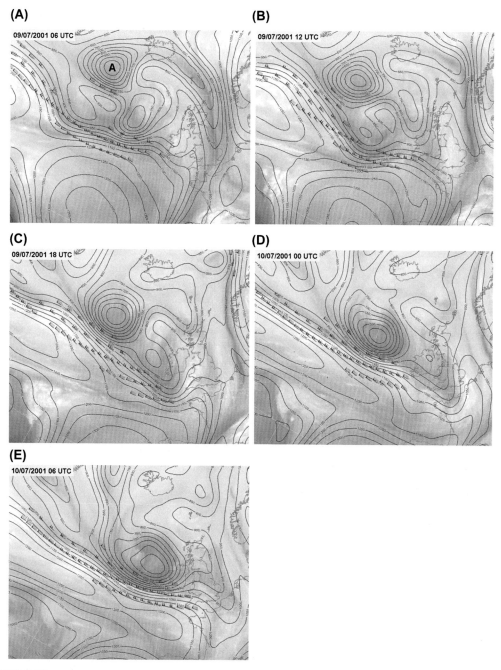

FIGURE 3.12

A sequence of water vapor images from (A) 0600 UTC on July 9 to (E) 0600 UTC on July 10, 2001, at 6 h intervals. Also given is a superimposition of the ARPEGE analysis of wind vectors at 300 hPa, associated with the jet as well as the geopotential of the 1.5 PVU surface (red contours, every 50 dam).

ascending motion in the troposphere below. Such upper-level perturbations with related divergence zone and associated vertical motion can be considered as the main synoptic upper-level forcing element of the atmospheric circulation as they contribute to trigger the development of a majority of bad weather systems at midlatitudes. We note the following examples:

    a. As seen in Chapter 1 (Section 1.3.2), when such an upper-troposphere perturbation (PV-Jet and related divergence zone) encounters a low-level warm anomaly, this leads to cyclogenesis; ascending motion and upper-level divergence intensifies during the cyclone development.

    b. In other situations, such upper-level forcing can occur over potential unstable atmosphere; in this case, mid- and low-level air is favored to rise by the ascending motion related to the upper-level divergent flow that leads or at least helps to release instability and to trigger deep convection.

  2. Upper-troposphere divergent flow can also develop in a uniform upper-level flow pattern when deep moist convection occurs in synoptic situations without upper-level forcing. This mechanism can produce strong divergence near the tropopause. In such cases, the upper-level divergence is a consequence of the deep moist convection. Amplifying convective storms develop their own dynamics and finally increase in size to form a Mesoscale Convective System (MCS). These MCSs can influence and change the synoptic-scale flow field in which they have formed to a greater or lesser extent. Such convective storms can sometimes develop over midlatitude regions under weak (or in the absence of) upper-level forcing. Once MCSs form, the upward motion intensifies in local updrafts through the entire troposphere (Rabin et al., 2004). These updrafts approaching the tropopause, at the base of the very stable stratosphere, are forced to diverge horizontally, resulting in upper-tropospheric divergence at the top of these MCSs, accompanied by the characteristic anvil clouds at the top of thunderstorms.

The divergent synoptic-scale upper-level flow can be detected by following the movement of cloud and moisture features in the animated WV images in the 6.2 μm channel. In this way an experienced observer is able to determine regions where the flow tends to be diffluent. A direct way to diagnose upper-level divergence is using fields of divergence, deduced from satellite-tracked upper-level Atmospheric Motion Vectors (AMVs). Such a source of observational data is the upper-tropospheric divergence (DIV) product (EUMETSAT, 2005) that is calculated hourly at the Meteosat Product Extraction Facilities (MPEF) of the European Organization for the Exploitation of Meteorological Satellites (EUMETSAT) and disseminated operationally. The input to the algorithm consists of all the upper-level AMVs, derived by tracking cloud and humidity features in the 6.2 μm channel imagery in the layer 100–400 hPa, which pass through specific quality control tests. This implies that the MPEF DIV fields may represent the upper-level dynamics of quite a deep atmospheric layer. Some experience on the use of the DIV product given in Schmetz et al. (2005), Georgiev and Santurette (2010), Hofer et al. (2011), and Georgiev (2013) will be presented in the next section as well as in Section 4.2.6.

3.3.3.1 Synoptic-Scale Upper-Level Perturbations

When the divergence of the upper-tropospheric flow is originally related to the upper-air perturbation, it is present before the surface cyclogenesis and can be seen at the origin of the development of the vertical motion. Fig. 3.13A illustrates such a case acting as a synoptic upper-level forcing mechanism in a cyclogenesis situation. It shows a composite image summarizing the initiation stage of an Atlantic cyclogenesis before deepening of the surface low. Note the meeting between the upper-level forcing element and the low-level warm anomaly, as follows:

- The upper-level perturbation (*red arrow*), as the dynamical tropopause anomaly (*red lines*) interacting with the jet (black contours of maximum wind speed).
- The low-level warm anomaly (*color shaded at the blue arrow*) associated with weak vorticity (green contour at the *blue arrow*) and relatively low pressure (blue contours).

At this initial stage, the field of 1.5 PVU surface heights (red contours) shows the presence of strong tropopause folding near a low-level baroclinic zone (the strong θ_w, gradient zone), while the deepening of the surface low has not been started (seen in the Mean Sea-Level Pressure [MSLP] field, blue contours).

Fig. 3.13B shows the corresponding MPEF DIV field. The divergent values are visualized in bluish shades (interval 20×10^{-6} s^{-1}) as well as in a light green color, showing weak divergence in the range [0 to $+20 \times 10^{-6}$ s^{-1}]. The convergence in the range [-20×10^{-6} s^{-1} to 0] is shown in white, and the convergence values lower than -20×10^{-6} s^{-1} are visualized in reddish colors. It can be clearly seen a strong divergence zone (*black arrow*), just downstream the minimum of tropopause height and near the jet-streak exit (black contour noted 110 kt on Fig. 3.13A). This area of divergence is due to the perturbation in the upper-level dynamics and appears prior to the onset of cyclogenesis development.

3.3.3.2 Deep Convection in Midlatitudes

Fig. 3.14A shows a superposition of the cold clouds seen in the IR 10.8 μm channel image (only cloud top brightness temperature (BT) $<-40°$C in cyan, BT $<-50°$C in yellow, and BT $<-60°$C in bright red) on the MPEF DIV field. This visualization scheme was introduced as MSG DIV-IR composite in Georgiev and Santurette (2010) (see also Georgiev, 2013). In Fig. 3.14B and C the DIV field at the time of Fig. 3.14A and its evolution within the next 4 h are shown.

In the view of the interpretation discussed in Section 3.3.3 above, points 1 and 2, an important forecast issue is to distinguish between divergence related to large-scale upper-troposphere dynamics and upper-level divergence in response to a strong convective development. A way to perform such an assessment is to diagnose the upper-troposphere dynamics as a cause in order to evaluate its possible consequence. For that purpose, model analysis of heights of constant PV surface 1.5 PV units (only $\leq$1000 dam) and 300 hPa wind vectors (only $\geq$30 kt) are superimposed on Fig. 3.14A. The use of the PV field allows areas of upper-level divergence produced by the two possible mechanisms in midlatitudes to be distinguished:

1. Divergence produced as a result of the development of MCSs with strong upper troposphere ascent, in the absence of synoptic-scale upper-level jet or PV anomaly advection (at the *blue arrow* in Fig. 3.14A and C).
2. Synoptic-scale divergent upper-level flow, which in this case is related to the ageostrophic wind near the left jet exit (at the *black arrow* in Fig. 3.14A), as described in Carroll (1997a). It is also seen that highly diffluent (divergent) flow is present in the leading part of the PV anomaly, which is marked by low heights of the 1.5 PV unit surface.

According to the experience of the authors, the MPEF DIV product over midlatitudes usually shows divergence between 20 and 80×10^{-6} s^{-1} for the two possible mechanisms depicted in Fig. 3.14A (synoptic-scale divergent flow and divergence produced as a result of deep convection).

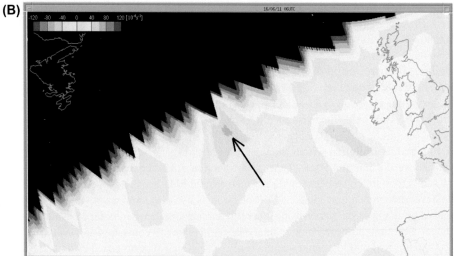

FIGURE 3.13

Upper-level divergence related to upper-level perturbation on June 16, 2011, at 0600 UTC. (A) 850 hPa wet-bulb potential temperature (*color and dashed lines*, °C), MSLP (*blue lines*, every 5 hPa), geopotential height of the 1.5 PVU surface (*red lines*, threshold 1000 dam), wind speed at 300 hPa (*black lines*, threshold 90 kt), absolute vorticity at 850 hPa (*green lines*, every $5 \times 10^{-5}\,\mathrm{s}^{-1}$, threshold $15 \times 10^{-5}\,\mathrm{s}^{-1}$); (B) Meteosat divergence product; the divergence values greater than $+20 \times 10^{-6}\,\mathrm{s}^{-1}$ are visualized in bluish shades (interval $20 \times 10^{-6}\,\mathrm{s}^{-1}$); the light green color represents divergence in the range [0 to $+20 \times 10^{-6}\,\mathrm{s}^{-1}$]. The large black zone in the top left of the figure corresponds to the area that cannot be seen by the Meteosat satellite.

Even in cases of very strong convective development, the divergence values over these regions, derived by the MPEF DIV product, usually do not exceed $+60 \times 10^{-6}\,\mathrm{s}^{-1}$, including cases of intense convection in situations associated with upper-level PV anomalies and jet streams that produce strong dynamical forcing of ascent at the upper troposphere (Santurette and Georgiev, 2005, 2007).

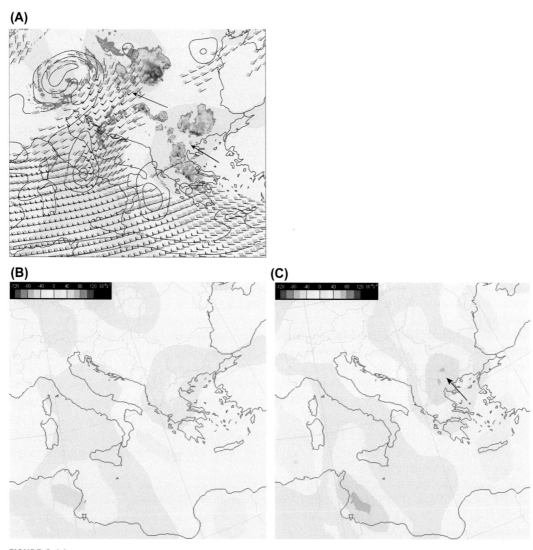

FIGURE 3.14

MPEF DIV product performance. For June 18, 2010: (A) 1245 UTC superimposed by the corresponding coldest infrared 10.8 μm channel radiance (only cloud top BT $<-40°$C) and nearest ARPEGE model analysis of heights of constant potential vorticity surface 1.5 PV units (brown, only $\leq$1000 dam) and 300 hPa wind vectors (only $\geq$30 kt); (B) 1245 UTC; and (C) 1645 UTC. The divergent values greater than $+20 \times 10^{-6}$ s^{-1} are visualized in bluish shades (interval 20×10^{-6} s^{-1}); the light green color represents divergence in the range [0 to $+20 \times 10^{-6}$ s^{-1}]. Also shown is synoptic-scale divergent flow (at the *black arrow*) and divergence produced as a result of deep convection (at the *blue arrows*).

3.3.3.3 Deep Convection in Tropical Areas

In cases of divergence produced as a result of the development of MCSs, the upper-level divergent field derived by satellites offers a tool to diagnose the convective development. Schmetz et al. (2005) reported that the diurnal cycle of the upper tropospheric wind field divergence can be estimated directly from the satellite-derived AMVs at the top of the convective system ascribed to a well-defined layer. Having this information, Schmetz et al. (2005) performed a comparison with corresponding forecast fields from the T511 spectral model of the European Center for Medium-Range Weather Forecasting (ECMWF) with a $1° \times 1°$ grid, and the model level is 150 hPa (Simmons and Hollingsworth, 2002). As a result of their study, Schmetz et al. (2005) argue that the satellite observed divergence offers a useful diagnostic tool to test convective parameterizations in atmospheric models.

Considering large MCSs, which cover areas of approximately $10° \times 10°$ latitude and longitude over Tropical Africa, Schmetz et al. (2005) reported divergence values greater than $450 \times 10^{-6} \, \text{s}^{-1}$ deduced by satellite AMVs from the Meteosat WV channel through the MPEF DIV algorithm. As discussed in Section 3.3.3.2, the upper-level divergence at midlatitudes derived by the MPEF DIV product is much weaker. A possible reason for such lower DIV values could be the smaller scale of the MCS in midlatitude Europe, and the resolution of the current DIV grid may not be able to fully represent the movement of the upper-level cloud/humidity features over midlatitudes (far from the subsatellite point).

The focus of the material in this section is the usefulness of the MPEF DIV product for qualitative applications related to deep moist convection in the Tropics. Concerning this, a smaller Tropical MCS is considered in Fig. 3.15 that covers an area of $10° \times 10°$ latitude and longitude (centered approximately at 13.5°N, 0.5°W). The Meteosat images in the channels IR 10.8 µm and the WV 6.2 µm showing this MCS on June 24, 2015, at 0600 UTC are presented in Fig. 3.15A and B, respectively. The corresponding divergent field is shown in Fig. 3.15C that is operationally available through the MPEF DIV product. This field derived by tracking cloud and moisture features in the 15 min sequence of four WV images (the last of which at Fig. 3.15B) shows maximum divergence values in the range $40-60 \times 10^{-6} \, \text{s}^{-1}$. This result is comparable with the divergence produced by MCS of a similar-scale over the midlatitudes (see, eg, Fig. 3.14C at the *blue arrow*). It is also seen in Fig. 3.15C that the whole area around the MCS (Fig. 3.15A and B) is captured by upper-level divergent flow that confirms the good performance of the MPEF DIV algorithm.

To allocate the pressure level at which the AMVs and the divergence values are derived, the satellite product Cloud Top Pressure (CTP), derived by the NWC-SAF Geo software package of EUMETSAT (NWCSAF, 2015), is used. For the purposes of qualitative applications, the upper-level divergence is relevant to be considered at 200 and 300 hPa levels, which are of significance for synoptic-scale analyses, and these fields derived by NWP models are usually available in the operational forecasting environment. Fig. 3.15D and E are two composite satellite products, which illustrate the altitude allocation of the MPEF DIV values:

- Fig. 3.15D: The WV image from Fig. 3.15B is overlaid only by the cloud tops derived by the NWC-SAF CTP product in the range 225–175 hPa (in brown color), which represents the areas for the divergence derived by the MPEF DIV algorithm by tracking WV imagery features around the 200 hPa level.
- Fig. 3.15E: The divergent field from Fig. 3.15C is overlaid only by the cloud tops derived by the NWC-SAF CTP product in the range 225–325 hPa (in cyan colors), which represents the areas for the divergence derived by the MPEF DIV algorithm by tracking WV imagery features around the 300 hPa level.

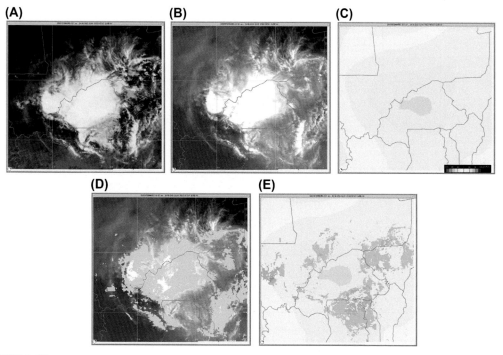

FIGURE 3.15

Satellite view of a tropical Mesoscale Convective System on June 24, 2015, 1200 UTC: Meteosat images in (A) infrared 10.8 μm and (B) water vapor 6.2 μm channels; (C) MPEF DIV product; (D) water vapor 6.2 μm image overlaid by NWC-SAF CTP product in the range 225−175 hPa (brown color); and (E) MPEF DIV product overlaid by NWC-SAF CTP product in the range 225−325 hPa (cyan colors).

The satellite information in Fig. 3.15D and E shows that comparison between divergence fields derived by NWP model simulations and satellite motion vectors can be performed for the flow pattern at two tropospheric layers, as follows:

1. Around the 200 hPa level at the highest divergent values ($40-60 \times 10^{-6}$ s^{-1}) in the central part of the MCS with CTP ranging in the interval 175−225 hPa. Fig. 3.16A and C show compositions of DIV and CTP satellite products overlaid by the corresponding difference fields at 200 hPa derived by the ECMWF operational model (0.5° grid resolution) at 6 h forecast and analysis. In Table 3.1, the digital values at point A and point B on Fig. 3.16C show reasonable agreement between the NWP and satellite data: At location A, a divergence 7.1×10^{-6} s^{-1} simulated by the NWP model corresponds to values in the range $40-60 \times 10^{-5}$ s^{-1} derived by the satellite product; at location B, we see divergence of 2.9×10^{-6} s^{-1} by the NWP model versus $20-40 \times 10^{-5}$ s^{-1} by the satellite data.
2. Around the 300 hPa level, lower divergent values ($20-40 \times 10^{-6}$ s^{-1}) in the outer areas of the tropical MCS with CTP ranging in the interval 225−325 hPa. Fig. 3.16B and D show compositions of DIV and CTP satellite products overlaid by the corresponding ECMWF

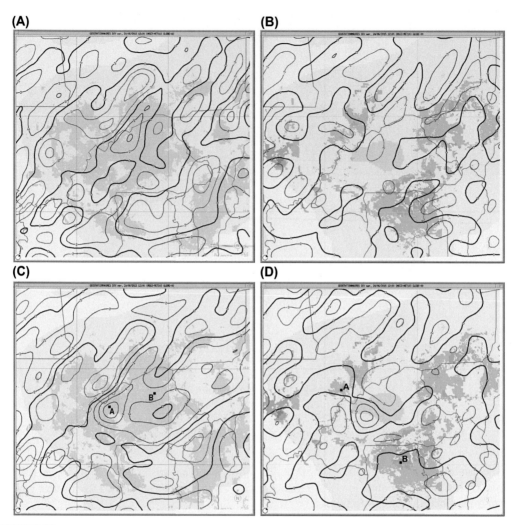

FIGURE 3.16

MPEF DIV product on June 24, 2014, 1200 UTC overlaid by NWC-SAF CTP product in the range 225−175 hPa (brown color) and ECMWF divergence field at 200 hPa (divergence in blue, convergence in red contours) at (A) 6 h forecast and (C) analysis as well as overlaid by NWC-SAF CTP product in the range 225−325 hPa and ECMWF divergence field at 300 hPa at (B) 6 h forecast and (D) analysis.

model fields of difference at 300 hPa at 6 h forecast and analysis. In Table 3.1, the digital values at point A and point B on Fig. 3.16D are shown for comparison. There is a good agreement: at location A, a divergence $2.1 \times 10^{-6} \, s^{-1}$ by the NWP corresponds to values in the range $20-40 \times 10^{-5} \, s^{-1}$ derived by satellite data in the WV channel; at location B the simulated NWP divergence is $0.2 \times 10^{-6} \, s^{-1}$ versus $0-20 \times 10^{-5} \, s^{-1}$ by the satellite data.

Table 3.1 Digital Values of the Compared Parameters at the Points of Comparison A and B in Fig. 3.16.

Point	Coordinates	NWC-SAF CTP	MPEF DIV	ECMWF DIV/Level
A in Fig. 3.16C	13°00′ N 3°12′ W	175–200 hPa	$40-60 \times 10^{-6}$ s^{-1}	7.1×10^{-5} s^{-1} at 200 hPa
B in Fig. 3.16C	13°52′ N 4°07′ W	200–225 hPa	$20-40 \times 10^{-6}$ s^{-1}	2.9×10^{-5} s^{-1} at 200 hPa
A in Fig. 3.16D	13°00′ N 3°12′ W	275–300 hPa	$20-40 \times 10^{-6}$ s^{-1}	1.2×10^{-5} s^{-1} at 300 hPa
B in Fig. 3.16D	10°06′ N 0°55′ W	300–325 hPa	$0-20 \times 10^{-6}$ s^{-1}	1.2×10^{-5} s^{-1} at 300 hPa

Comparing the NWP fields in Fig. 3.16A and C as well as Fig. 3.16B and D also shows that at the forecast run (Fig. 3.16A and B), the NWP model has overestimated the upper-level divergence at 200 and 300 hPa. This confirms that, providing observation data, the EUMETSAT DIV field is useful to estimate the evolution of divergent flow pattern produced by tropical convection in the time between the two runs of the global NWP models.

3.3.4 MID-LEVEL JET SEEN IN 7.3 μm CHANNEL IMAGES

Although radiances in the 7.3 μm WV channel measured by geostationary satellites contain information for mid-level moisture distribution, there is still a lack of practice for using these data in synoptic-scale analyses. This section provides guidance for interpretation of the 7.3 μm channel of MSG satellites as a source of observational data for diagnosing large-scale thermodynamic context.

3.3.4.1 Mid-Level Jet and Associated Synoptic Context

In specific thermodynamic conditions, the atmospheric circulation can exhibit maxima of wind in the middle troposphere. The MLJ is formed as a result of subsynoptic-scale low-level baroclinic processes. Since the 7.3 μm channel radiation is sensible to the moisture content in the middle troposphere, the images in the 7.3 μm channel can be interpreted to study mid- to low-level humidity features associated with wind maxima at mid-level. Georgiev and Santurette (2009) reported more than 16 such cases from 2004 to 2007 over the Mediterranean area, in which an MLJ at about 600/700 hPa is present in the south-westerly–southerly flow seen in the 7.3 μm channel images. In the view of the MSG satellite, from which 7.3 μm channel images are available, the feature has also been present over the North Africa and Eastern Black Sea areas.

Fig. 3.17 shows an example of such an MLJ over the Atlantic and Northwest Africa on May 1, 2010, in the comparison between images in 7.3, 6.2, and 10.8 μm channels. The gray shade appearance of the images around the position of the MLJ feature in the three channels is quite different. The moisture boundary indicated by the *blue arrows* in Fig. 3.17A is closely related to the MLJ feature, with maximum winds seen in Fig. 3.17 in the 600 hPa fields of wind vectors (blue) and wind speed contours (yellow). This gray shade boundary is definitely a moisture structure at mid-level, since it is distinctly seen only in the 7.3 μm image and cannot be recognized on the images in the WV 6.2 μm and IR 10.8 μm channels (Fig. 3.17C and D, respectively).

The MLJ feature is formed along the zone between two large-scale air masses of different origin as seen in the wet-bulb potential temperature θ_w (red contours in Fig. 3.18A) at the 600 hPa isobaric surface. These are:

- The cold, dry air, which descends over the Eastern Atlantic on the rear side of the mid-level trough represented by dark-gray shades in the WV image poleward of the *blue arrows*.
- The warm, moist air, which is present around the coast of Northwest Africa, seen as the light-gray shade area in the WV image equatorward of the *blue arrows*.

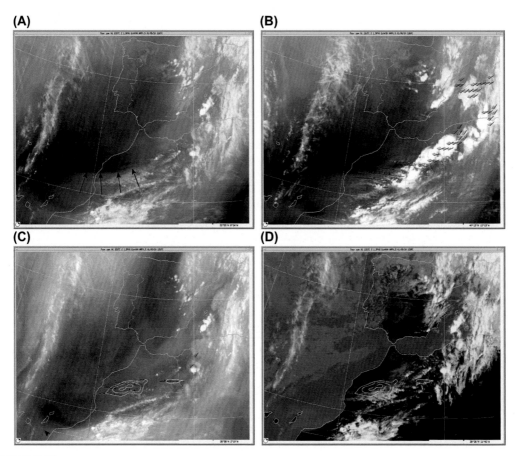

FIGURE 3.17

Mid-level jet stream on May 1, 2010. Meteosat-8 water vapor (WV) 7.3 μm channel image for (A) 1200 UTC and (B) subsequent convective development at 1500 UTC, overlaid by the wind vectors (blue, only ≥35 kt). Meteosat-8 images at 1200 UTC in (C) WV 6.2 μm and (D) infrared 10.8 μm channels overlaid by the wind vectors (blue) and wind speed contours (yellow) at 600 hPa (only ≥44 kt).

In the case of Fig. 3.17, the moisture boundary between the two air masses seen in the 7.3 μm WV image is located in the rear side of a trough in the geopotential of the 500 hPa isobaric surface (Fig. 3.18A). In other cases such a structure can appear on the forward side of a meridional extended upper trough (Georgiev and Santurette, 2009).

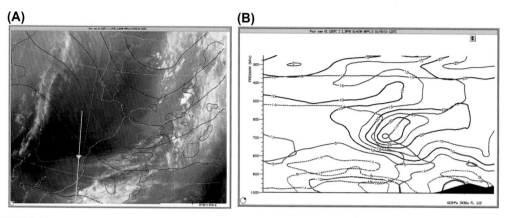

FIGURE 3.18

Mid-level jet stream and related thermodynamic context. (A) Meteosat-8 water vapor 7.3 μm channel image on May 1, 2010, 1200 UTC overlaid by 500 hPa heights (blue) and wet-bulb potential temperature θ_w (red, only $\geq 12°C$) at 600 hPa; (B) vertical cross-sections of wet-bulb potential temperature (red) and wind speed (black, kt) along the *yellow line* in (A).

The origin of the MLJ stream seen in the 7.3 μm WV image is closely associated with the low-level conditions. To illustrate this point, Fig. 3.18B shows vertical cross-sections of wind speed (black contours) and wet-bulb potential temperature (red) along the *yellow line* in Fig. 3.18A. The cross-section axis intersects the MLJ stream at a position, located upstream to the jet streak (see Fig. 3.17C) in order to show the feature over the Atlantic where the thermodynamic conditions are not affected by orography. The MLJ stream is located within a wide area of wind speed maximum down to the lower-middle troposphere. A low-level baroclinic zone is present below the wind maximum seen as a gradient zone in the horizontal distribution of wet-bulb potential temperature (θ_w, red contours).

The appearance of an MLJ feature in the 7.3 μm WV imagery is always associated with the existence of such a low-level baroclinic zone and associated surface θ_w anomaly beneath the wind maximum feature (Georgiev and Santurette, 2009). In order to explain this result, it is useful to consider the thermal wind relation:

$$\frac{g}{\theta_0}\frac{\partial \theta}{\partial x} = f\frac{\partial V_g}{\partial z}$$

[3.1]

where θ is potential temperature with reference value θ_0, g is the acceleration of gravity, f is the Coriolis parameter, and V_g is the y-component of the geostrophic wind. Eq. [3.1] tells that a strong horizontal θ-gradient creates a strong vertical geostrophic wind gradient. Therefore, the origin of the MLJ, seen in the 7.3 μm WV images, is likely a result of strengthening of a baroclinic zone (increasing θ-gradient at the low to mid-level) that is associated with a surface θ-anomaly.

The MLJ moisture boundary seen in 7.3 μm channel imagery is a significant imagery signature that indicates a specific thermodynamic structure. Regarding this structure, the following evolution is under way:

- Enhancement of warm advection leads to increasing of the horizontal θ_w-gradient in the low level that is associated with appearing/strengthening of a low-level baroclinic zone. As a result, an area of wind maximum is created around 600 hPa at the top of the horizontal-temperature gradient zone.
- Low-level moisture convergence exists related to the baroclinic zone at the equatorward side of the 7.3 μm WV boundary.
- Ascending motions are present at the equatorward side of the baroclinic zone that moistens the low- to mid-level air, while a transverse circulation with the MLJ contributes to descent and drying the air at the poleward side. This enables a specific moisture boundary of MLJ to be distinctly seen in the 7.3 μm image.

Jets in the lower-middle troposphere often appear around the Eastern Atlantic and whole Mediterranean area where the thermodynamic conditions associated with such a mid-level wind maximum play important roles in producing deep moist convection downstream (as shown 3 h later in Fig. 3.17B). The Mediterranean Sea, as a reservoir of warm water, is a source of sensitive and latent heat (like the Caribbean Sea). Seasonal effects exist, and the air masses over the East Atlantic and the Mediterranean are frequently close to the instability, especially in late summer and early autumn. The Mediterranean Sea is surrounded by complex orography and mountains (in Southern Europe and Northeast Africa as well as islands, including mountainous islands like Corsica). Therefore, the air masses that affect the Mediterranean countries are strongly influenced by effects of latent heat sources and vertical motion effects due to the orography. This geographical coincidence between the orographic influence and the source of warm and moist air has a crucial role in the mechanisms that lead to heavy convective precipitations over the Mediterranean countries. Over these regions the MLJ is related to a thermodynamic environment favorable for convection, as it is associated with enhancement of advection of low-level warm subtropical air masses. On the other hand, the complex orography surrounding the Western Mediterranean acts as a low-level barrier or brake for the Atlantic fronts entering the Western Mediterranean. As a result, low-level cold air is braked and generally few very pronounced low-level fronts affect Mediterranean countries. At the same time, the associated upper-level perturbations can penetrate more quickly than the low-level ones; this contributes to increasing instability in the warm and moist air flow generally persistent at low levels. So, embedded convection in areas of stratiform precipitation is generally associated with Mediterranean atmospheric perturbations, and heavy rainy systems affecting the Mediterranean countries are usually associated with convective systems.

In a dry air regime as over the North African continent, an MLJ may also be formed as a result of the same thermodynamic mechanism that will be further considered in Section 3.3.4.3. However, the much drier air masses carried by this jet (usually from a southwest direction) do not support an environment favorable for deep convection.

3.3.4.2 Mid-Level Jet and Related Moisture Movement

Fig. 3.19A shows the field of wet-bulb potential temperature at 600 hPa overlaid by the 7.3 μm WV image. The cross-sections in Fig. 3.19B and C help to understand the relation of the feature to the low-level thermodynamic conditions. In Fig. 3.19A the 7.3 μm image shows a moisture band, which is located perpendicular to the cross-section line and is edged by the 16°C contour of θ_w on its polar side. Being a feature of the lower-middle troposphere, the distinct moisture boundary in the satellite image at the *yellow arrows* in Fig. 3.19A corresponds very well to the vertical distribution of the relative humidity in Fig. 3.19B (at the contour of 30% at the position of the *red arrow* with drier air to the northwest). The relative humidity increases to the east above 60% that forms a well-defined mid-level band of moist air advected from the Eastern Atlantic over Spain.

The cross-section of θ_w on Fig. 3.19C shows that the moisture boundary on the 7.3 μm image at the 16°C contour of θ_w on the polar side of the moist feature coincides with a maximum wind speed

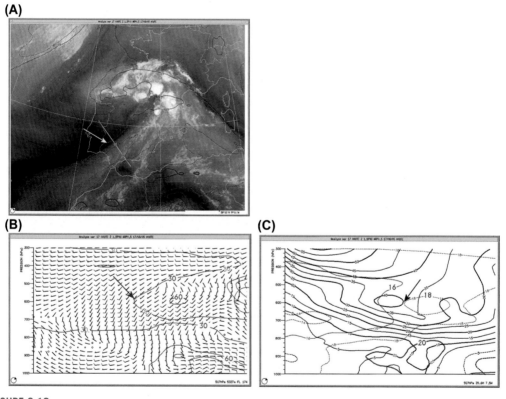

(A)

(B) **(C)**

FIGURE 3.19

Mid-level jet and associated movement of moist air on August 17, 2005, at 0000 UTC seen in: (A) Meteosat-8 image 7.3 μm channel overlaid by wet-bulb potential temperature at 600 hPa. Vertical cross-sections along the axis depicted in (A) of: (B) relative humidity (pink, %) and wind component, tangential to the cross-section plane drawn in (A); (C) wet-bulb potential temperature (red) and wind speed (black, kt).

contour of a jet at about the 600 hPa level. Therefore, we will refer to such features of high wet-bulb potential temperature as mid-level "WV movements" in order to reflect that they represent bands in which warm and moist air masses are moving poleward in the circulation of a mid-level wind maxima or jet streams.

Therefore, Fig. 3.19 shows that movements of moisture in a deep mid-level layer associated with a wind maximum at this level are distinctly seen as specific moisture features in the 7.3 μm image. Similar specific large-scale moist features seen by WV imagery, referred to as upper-level WV plumes, have been recognized for a long time to serve as "fuel" for the mesoscale convective cloud systems (Bader et al., 1995). The term "WV plume" was introduced to indicate large-scale northward movements or surges of moist air masses that appear as specific plume-like shape features in the 6.7 μm of GOES satellites. Thiao et al. (1993) showed that the WV plumes may be coupled with low-level ridges of equivalent potential temperature θ_e (eg, at 700 hPa) and create favorable conditions for severe thunderstorm development over the United States. In terms of the considerations presented here, such an interaction can be classified as coupling between upper-level WV plumes seen by 6.2 μm imagery and low-level baroclinic zones (related to the mid-level WV movements seen by 7.3 μm imagery).

3.3.4.2.1 Moisture Movement Structure

Usually a WV movement associated with an MLJ, seen as a moist band in the 7.3 μm image, can be related to a Warm Conveyor Belt (WCB) according to the classification of Carlson (1980). However, a WCB typically indicates rising flow, while a WV movement can be just associated with horizontal transport of moisture. To illustrate this point, the wind component tangential to the cross-section plane is shown in Fig. 3.19B (represented by black vectors) that depicts the direction of the wind in the vertical plane of the cross-section. It reveals that the moist feature to the southeast of the position of the *red arrow* along the cross-section line is formed in association with ascending motions from the moist surface where the relative humidity is greater than 80%. In other MLJ structures, the humid air of the moisture movement is supplied by horizontal transport of humid air, and the vertical motions are not a significant factor for formation of the MLJ moisture boundary as shown in the case considered below.

Fig. 3.20A shows a WV 7.3 μm channel image overlaid by the wind vectors at 500 hPa equal or greater than 60 kt in speed, and Fig. 3.20B shows the corresponding 6.2 μm image. As in the case in Fig. 3.17, the existence of the MLJ and associated low-level baroclinic zone is definitely confirmed by the significant differences in the moisture field seen by the two WV channels:

- The existence of a specific boundary in the 7.3 μm channel image, which is sensible for mid-level moisture.
- This mid-level moisture feature (and the MLJ, respectively) is not distinctly seen in the 6.2 μm channel image. The 6.2 μm WV radiation is much absorbed by WV at the upper level, and the presence of moist air at the upper troposphere (uniform light image gray shades) obscures the moisture boundary in the region of the MLJ stream.
- Therefore, this is a moisture pattern of the mid-level dynamics that usually corresponds to an MLJ stream.

Cross-sections of wind speed normal to the cross-section plane (black contours) and wet-bulb potential temperature (red contours) are shown in Fig. 3.20C as well as the relative humidity (pink contours) and wind speed normal to the cross-section plane (black vectors) in Fig. 3.20D. Analysis of

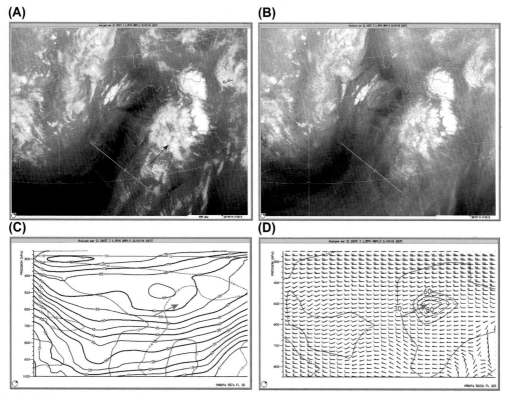

FIGURE 3.20

Mid-level moisture movement. Meteosat-8 water vapor images for March 21, 2006, at 1800 UTC in (A) 7.3 μm overlaid by wind vectors at 500 hPa (red, only ≥65 kt) and (B) 6.2 μm channel. Cross-section along the axis depicted in: (C) of wind speed normal to the cross-section (black contours, kt) and wet-bulb potential temperature (red); (D) of relative humidity (pink, %) and wind vector component, tangential to the cross-section plane (black).

these vertical cross-sections along the zone of maximum wind at 500 hPa shows details of the structures of the MLJ stream and related moisture movement as follows:

- Advection of mid-level moist air is associated with light image gray shades in Fig. 3.20A that correspond to moist mid-level air (80% RH) at the position of the *green arrow* in the cross-section of relative humidity in Fig. 3.20D.
- As seen by the cross-section of relative humidity wind speed normal to the cross-section plane (black vectors in Fig. 3.20D), the mid-level moist air mass at the position of the *green arrow* is advected via strong subsynoptic-scale horizontal transport.
- Over this region of cloud-free moisture features along the cross-section axis, there is no significant vertical transport of moisture from the low level (see Fig. 3.20D). In cases that convective clouds develop downstream (as in Figs. 3.17B and 3.20A, *pink arrow*), the MLJ stream enables permanent moisture supply into the convection environment.

The concept of WV movement is introduced here to indicate that a specific characteristic of such a mid-level moisture boundary is its relation to a horizontal transport of moisture. In Figs. 3.19B and 3.20D the black vectors show the wind vector component, tangential to the cross-section plane, and represent the wind direction in the vertical plane. They illustrate that humidity supply in the mid-level moisture movement may come from two possible mechanisms:

- Both horizontal transport of moist air and ascending motions from the moist surface as with the case in Fig. 3.19B.
- Horizontal transport of moist air as with the case in Fig. 3.20D.

As shown in Figs. 3.19 and 3.20, the 7.3 μm imagery is a tool to observe moisture features and related jet streams at the middle troposphere that can maintain subsynoptic- to large-scale movements of moist air in the convection environment. These thermodynamic features can be observed by WV imagery of MSG, Himawari, and other geostationary satellites, which generate images in a WV channel around 7.3 μm. Similar large-scale moist bands may be seen in the 6.2 μm images and indicate transport of humid air at the upper troposphere (Thiao et al., 1993). The generation and maintenance of conditions favorable for intense convection are present where large-scale WV movements in a deep layer up to the upper middle troposphere are seen simultaneously in the images of the two WV channels. Countries bordering the Northern Mediterranean Sea coast experience situations with subtropical flow (southwesterly to southeasterly) that carry warm and moist low-level air masses toward Europe. Using 6.2 and 7.3 μm WV images can be particularly efficient for these areas since they provide information on the upper-level flow and the low- to mid-level flow, respectively.

3.3.4.3 Mid-Level Jet in Dry Air Mass Over Northeast Africa

The cases of MLJ over Northwest Africa and the Mediterranean area presented in Section 3.3.4.2 are usually related to persistent subsynoptic-scale movements of moisture. In addition, wind maximums at mid-level are frequently seen in the images in the 7.3 μm channel of MSG over the eastern regions of North Africa where the mid-level air is much drier and the low-level air is very warm due to the continental effect of solar heating over Africa.

Fig. 3.21A shows the MSG image in the 7.3 μm channel that is overlaid by vectors of maximum wind ($\geq$25 kt) at 700 hPa, and Fig. 3.21C shows a vertical cross-section of θ_w (red) and the wind speed (black) normal to the cross-section plane depicted by the line in Fig. 3.21A. Looking through the 7.3 μm channel, we are able to indicate the moisture boundary over Egypt at the position of the *yellow arrow* in the image in Fig. 3.21A that is related to the wind maximum at about the 700 hPa level at the position of the *blue arrow* in Fig. 3.21C. Therefore the 7.3 μm channel is sensible to detect wind maxima at lower-middle troposphere in this region as well. The associated thermodynamic context is the same as related to the cases in Sections 3.3.4.1 and 3.3.4.2:

- The jet streak axis is parallel to the moisture boundary, and the zone of maximum winds is located at the side of warm, moist air.
- Also seen in Fig. 3.21C is the close relation of the MLJ to a low-level baroclinic zone along the gradient zone in the horizontal distribution of wet-bulb potential temperature.

However, a dry air mass is present over Northeast Africa (as seen in the cross-section on Fig. 3.21D with less than 50% relative humidity). This supports a much drier moisture regime than in the cases presented in Sections 3.3.4.1 and 3.3.4.2. Accordingly the MLJ system over Sudan, Egypt, and Saudi Arabia in Fig. 3.21 does not consist of sufficient moisture to support a favorable environment for development of deep-moist convection downstream from the mid-level boundary feature in the 7.3 μm image.

This case over continental Africa illustrates the benefits from combined interpretation of the 6.2 and 7.3 μm in order to identify sensitive zones where the images in the two channels exhibit quite different appearances as discussed in Section 3.2.3. In such regions, significant differences between the specific patterns visible through the two WV channels can be interpreted to distinguish different circulation systems and thermodynamic processes in the upper and middle troposphere and their possible interaction. Applying such an interpretation of Fig. 3.21A and B, the following conclusions are made:

- The feature at the *pink arrow* is definitely an upper-level dynamical structure.
- The moisture boundary at the *yellow arrow* is associated with a mid-level feature, which is not visible in the 6.2 μm image and, accordingly, is not related to upper-level dynamics.

In addition, such an interpretation provides a way to diagnose the presence of vertical displacement of significant circulation/thermodynamic systems and related processes. Analysis of Fig. 3.21A and E and where the wind vectors at the 700 hPa level (blue, only ≥25 kt) as well as at the 250 hPa level (red, only ≥70 kt) are superimposed on the 7.3 μm channel image, shows the positions of the dynamically active regions in the middle and upper troposphere:

- The southern part of the moisture boundary (southwest of the position of the *yellow arrow*) in Fig. 3.21A corresponds to a specific thermodynamic system in the lower levels and it is not related to the upper troposphere dynamics. Over this area, the positions of the dynamically active regions in the middle and upper troposphere are far away from each other.
- Over Northeastern Egypt and North Saudi Arabia (northeast of the position of the *yellow arrow*), the part of the mid-level boundary in the 7.3 μm image becomes very close and parallel to the upper-level moisture boundary in the 6.2 μm image. Accordingly, further northwest, as seen in Fig. 3.21E, at the position of the *green arrow* the upper-level jet at 250 hPa (blue) is located above the mid-level wind maximum at 700 hPa (red), associated with the low-level baroclinic zone.
- Considering the orientation of the moisture boundaries seen in the two WV images at the *green arrow*, an experienced observer can recognize rightward wind shift from the middle to upper troposphere (warm advection in the layer, respectively).

3.4 BLOCKING REGIME

Blocking regime involves common large-scale atmospheric patterns and plays a major role in surface weather, mainly due to its persistent nature. Blocking is usually defined as a midlatitude anomalous flow pattern associated with a strong meridional wind component, with a horizontal extension somewhere between synoptic and planetary scales. From a synoptic point of view, a blocking situation is characterized by a stationary positive height anomaly (high value in the upper-level geopotential field, relative to a regional mean), in relatively high latitudes (eg, north of 40° at the Northern

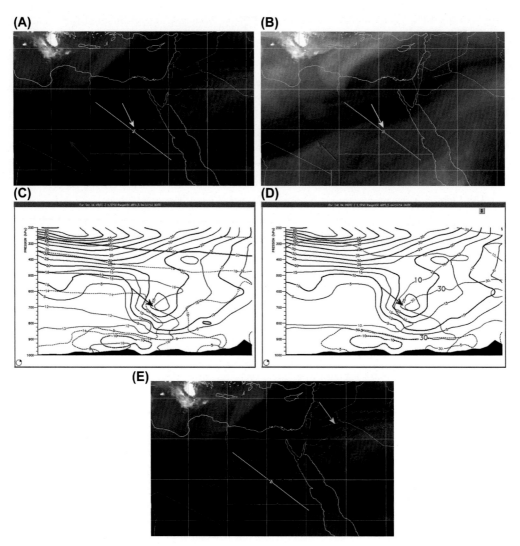

FIGURE 3.21

Mid-level jet in dry air regime over North Africa on October 4, 2014, at 0900 UTC seen in: Meteosat-10 image in 7.3 μm channel overlaid by ARPEGE 3 h forecast of wind vectors at (A) 700 hPa level (blue, only ≥25 kt) and (E) 700 hPa level (blue, only ≥25 kt) as well as at 250 hPa level (red, only ≥70 kt); (B) Meteosat-10 6.2 μm channel image. Vertical cross-sections along the axis depicted in (A) and (B) of: wind speed normal to the cross-section plane (black, kt) and (C) wet-bulb potential temperature (brown, 20°C thick) as well as (D) relative humidity (pink, %).

Hemisphere), in which the normal eastward progression of migrating midlatitude weather systems is blocked. The stationary positive height anomaly is associated with one or two negative height anomalies (low values in the upper-level geopotential field) on its equatorward side. It denotes a breakdown in the prevailing tropospheric westerly flow at midlatitudes that is often associated with a split in the zonal jet and with a more or less persistent ridging at high latitudes.

In general, the blocking regime is characterized by an area of easterly winds within the predominantly westerly large-scale upper-air flow. On the western side of such an area the upstream flow becomes "blocked" and splits into two branches. Blocking regimes occur in the upper-air wind field where warm air in the middle troposphere (associated with some form of upper-air ridge or anticyclone) is present on the poleward side of a cold trough or a cyclone. In terms of PV concept, this corresponds to a reversal of the usual meridional gradient of potential temperature on the dynamical tropopause (1.5 or 2.0 PVU surface) associated with an abnormally "warm" tropopause on the poleward side and "cold" tropopause on the equatorward side. Since such a warm (cold) tropopause is usually associated with anticyclonic (cyclonic) flow respectively over some depth, the reversal of the gradient of potential temperature on the dynamical tropopause is very likely to be associated with easterlies on spatial scales relevant to blocking (see Berrisford et al., 2007). There are different combinations of high-level circulation systems and various ways in which the situation may develop that are associated with different cloud and moisture patterns and their changes with time. Considering these differences, Weldon and Holmes (1991) categorize the blocking regime formation into two basic types (see also Georgiev and Santurette, 2005) and discuss the appearance of two types of related moisture boundaries, "head" and "inside" boundaries. The blocking patterns are illustrated in Fig. 3.22 by superposition of WV images in channel 6.2 μm and the corresponding fields of geopoptential at 300 hPa.

1. Blocking regime formation in which the easterlies result from anticyclogenesis (anticyclonic wave-breaking), as seen at location A in Fig. 3.22A. In such blocking, the easterly wind (downstream from the ridge) is primarily dry in the middle and upper troposphere, as a result of downward vertical motions, and WV image dark zones reflect dynamical patterns on the poleward side of the trough quite well. An "inside" moisture boundary forms on the western side of the easterlies (Weldon and Holmes, 1991). Such a moisture boundary is generated along the deformation zone, which is a result of the newly formed easterly winds (on the equatorward side of the upper-air high-pressure system) in the opposition of the previously dominated upstream westerlies. Usually, the air approaching the deformation zone from the west is moist aloft, which, in combination with the sinking air and drying aloft on the eastern blocking side, forms a well-defined distinct boundary in the WV imagery (indicated as "IB" in Fig. 3.22A).

2. Blocking regime formation in which the easterlies result from cyclogenesis (cyclonic wave-breaking), as seen at location L in Fig. 3.22B. In this kind of blocking the easterly wind regime is primarily moist in the middle and upper troposphere, as a result of upward vertical motions. On the western side of the easterly wind regime, a head moisture boundary forms (indicated as "HB" in Fig. 3.22A) in response to moisture spreading in the upper troposphere from areas of upward vertical motions. The spreading of the upper-level moisture becomes "blocked" on the western side, where it attempts to spread in the opposite

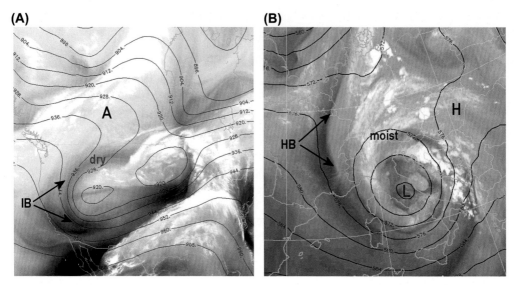

FIGURE 3.22

Two basic types of blocking regime (see Weldon and Holmes, 1991) interpreted from water vapor imagery overlaid by geopotential at 300 hPa. (A) Blocking regime as a result of anticyclogenesis at location "A" (North America, October 15, 2013, at 1800 UTC). (B) Blocking regime as a result of cyclogenesis at location "L" (Central Mediterranean, July 31, 2014, at 1200 UTC). Dry and moist air blocking wind regimes, which appear in darker and lighter gray shades in the imagery, are indicated in (A) and (B), respectively.

direction to the ambient upper-air wind field as seen in Fig. 3.22B. The boundaries formed by this process are well known and are the most common types of boundaries observed on WV imagery (Weldon and Holmes, 1991).

Fig. 3.22 is representative of the mature stage of the two blocking systems, and dynamical features during their developing stages will be further discussed in Sections 3.4.1 and 3.4.2 below.

3.4.1 BLOCKING REGIME FORMATION IN WHICH EASTERLIES RESULT FROM ANTICYCLOGENESIS

Fig. 3.23 shows diagnostic and dynamical fields in a situation of this kind of blocking as result from anticyclogenesis over North America and the corresponding patterns in the WV image. The following main features can be seen in the field of 300 hPa heights (see Figs. 3.22A and 3.23A and B):

- The most southerly part of the trough on the equatorward side of the anticyclone moves westward from the position of the trough originally on the eastern side of the developing ridge.
- During the mature stage, the upper-air high (at location A in Fig. 3.22A) may be closed but not cut off from the upstream westerlies.
- The trough on its equatorward side remains open to the east (east of location L in Figs. 3.23B and 3.22A).

Other dynamical features are revealed by superimposing a WV image onto the fields of 300 hPa absolute vorticity (only the contours of cyclonic vorticity) close to the end of the developing stage in Fig. 3.23C as well as the heights of the 1.5 PVU surface during the mature stage in Fig. 3.23D. The following features are noteworthy:

- A distinct dry zone (dark WV gray shades) is progressively extended on the western side of the large cyclonic vorticity feature, which tends to become blocked with the developing of an easterly wind regime (indicated by "E" in Fig. 3.23C). This process ends with formation of an inside boundary at the mature stage, distinctly seen in Fig. 3.22A.
- The field of 1.5 PVU surface heights (Fig. 3.23D) gives a good representation of the moisture patterns seen in the WV imagery. The zones of high gradient in the imagery gray shades (indicated by the *black arrows*) correspond to the high gradient of the 1.5 PVU surface, following the deformation of the flow in the blocking circulation system. The moist upper-level air in the blocking ridge, which appears very light in the imagery, is associated with the high geopotential of the dynamical tropopause. Conversely, the sinking air and drying aloft on the eastern blocking side (dark image gray shades) are represented by the low geopotential of the 1.5 PVU surface.

3.4.2 BLOCKING REGIME FORMATION IN WHICH EASTERLIES RESULT FROM CYCLOGENESIS

Fig. 3.24A−D shows a Meteosat 6.2 μm WV imagery superimposed onto the corresponding ARPEGE analysis of 300 hPa heights during the development of a blocking regime over the Central Mediterranean. The main features of the blocking regime resulting from the cyclogenesis may be summarized as follows:

- The ridge (marked "H" Fig. 3.24, at the 940 dam contour) "rebuilds" poleward from the position of the ridge originally on the eastern side of the developing trough and remains open to the east.
- During the mature stage (Figs. 3.24D and 3.22B), the upper-level cyclone (marked "L") is closed but is not cut off from the upstream westerlies.

Fig. 3.24E shows the WV image superimposed onto the corresponding ARPEGE analysis of absolute vorticity and wind vectors at the 300 hPa isobaric surface. Interpreting the same 6.2 μm WV image in Fig. 3.24F, an experienced observer can recognize three kinds of moisture boundaries:

- The boundaries between quite different moisture regimes (at the *blue arrows*) and especially their curvature indicate the location where the middle-level flow turns from southwesterly-southerly to southeasterly. The cloud/moist feature to the northwest of this location has been formed as a result of the moist easterly flow of the blocking regime.
- The head boundary at the western edge of the blocking easterly wind regime (at the *black arrow*) has been produced as a result of moisture spreading in the upper troposphere from areas of upward vertical motions that is formed during the cyclogenesis phase of the blocking evolution. Along the "head" boundary, the vorticity field in Fig. 3.24E well represents the area along the shear vorticity pattern associated with downward vertical motions, which produce an image dark zone.

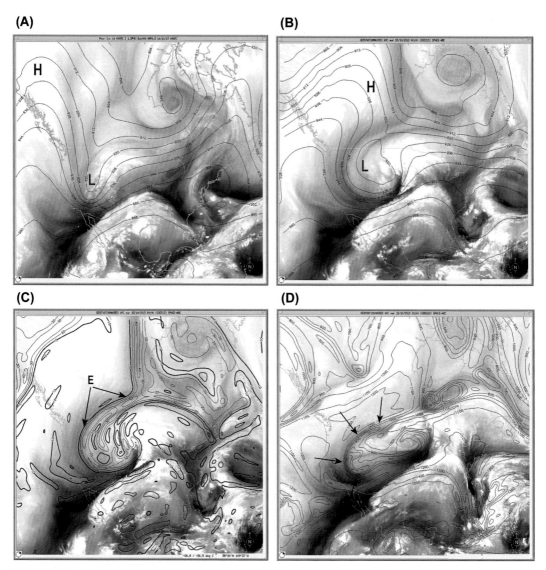

FIGURE 3.23

Development of a blocking regime as a result of anticyclogenesis over North America as seen in diagnostic/dynamical fields from ARPEGE analysis, superimposed onto the corresponding water vapor images: geopotential height (dam) of the 300 hPa isobaric surface for 0000 UTC on (A) October 14 and (B) October 15, 2013. (C) Absolute vorticity at 300 hPa isobaric surface (only $\geq 0 \times 10^{-5}\,s^{-1}$, cyclonic vorticity in blue) for 0000 UTC and (D) 1.5 PVU surface heights for 1800 UTC on October 15, 2013. Also marked: the upper/middle-level ridge, "H," and low, "L."

- The jet-stream moisture boundary at the equatorward side of the low (at the *red arrow* in Fig. 3.24F) is associated with the maximum of curvature vorticity and strong downward vertical motions, which produce darkening of the image gray shade on the polar side of the jet.

3.5 CYCLOGENESIS AND ATMOSPHERIC FRONTS

3.5.1 EXTRATROPICAL AND TROPICAL CYCLONES: ENERGY SOURCE AND MAIN THERMODYNAMIC CHARACTERISTICS

The extratropical and tropical cyclones are associated with quite different atmospheric environments and they are the consequences of different thermodynamics processes. The primary difference between these two kinds of cyclonic systems is the set of energy sources for their initiation and development.

Tropical cyclones draw their energy a from convective instability and from the releasing of latent heat due to condensation of the WV coming from warm sea surface. Extratropical cyclones, on the other hand, get most of their energy from baroclinic instability that prevails in the midlatitude atmosphere; this energy usually gets distributed over larger areas.

Extratropical cyclones require a baroclinic zone to develop. The extratropical lows (especially wintertime) are associated with strong horizontal temperature contrasts (which produce atmospheric fronts) and upper-level perturbation of the wind field. Extratropical cyclones can develop over land or water, even if they are more pronounced over water where release of latent heat due to WV condensation is more important.

Tropical cyclones only develop over warm waters, in the lower latitudes, within a low-level single warm and humid air mass, in a quasi-homogeneous atmosphere regarding temperature field; no fronts are involved. Tropical cyclones occur in a calm upper-level environment, with no preexistent upper-air dynamical features. However, over time, the towering cumulonimbus clouds release enough heat aloft to develop a high-pressure area over the low-level cyclone. This self-developed high-pressure system provides the divergence aloft needed to maintain the surface storm.

Because of these different energy-source and thermodynamic characteristics, tropical cyclones tend to have more compact wind fields, tend to be more symmetric, and have a well-defined inner core of strong winds.

Section 3.5 is focused on the dynamics of extratropical cyclonic systems and application of WV imagery analysis, while related forecasting issues concerning development of cyclonic systems in tropical and subtropical areas will be considered in Section 3.6.

3.5.2 CYCLOGENESIS WITHIN BAROCLINIC TROUGHS: LEAF AND BAROCLINIC LEAF FEATURES IN THE WATER VAPOR IMAGERY

Cyclonic disturbances within midlatitude baroclinic troughs often are associated with clear and typical characteristic "baroclinic leaf" features in satellite imagery, which are very well described in the literature (eg, Weldon and Holmes, 1991; Bader et al., 1995; Santurette and Georgiev, 2005). In this section, considerations will be given to the cloud/moisture pattern associated with the first phase of the baroclinic disturbance development.

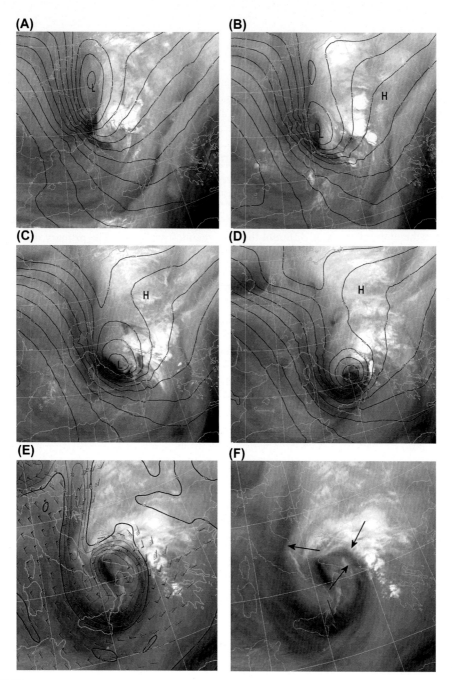

FIGURE 3.24

Development of a blocking regime as a result of cyclogenesis over Western/Central Mediterranean as seen in diagnostic/dynamical fields from ARPEGE analysis, superimposed on the corresponding 6.2 μm water vapor images: geopotential height (dam) of the 300 hPa isobaric surface on September 1, 2014, at (A) 0000 UTC, (B) 0600 UTC, (C) 1200 UTC, and (D) 1800 UTC. Also marked are the upper/middle-level ridge, "H," and low, "L." (E) Absolute vorticity ($\geq 10^{-5}$ s^{-1}) and wind vectors (only ≥ 25 kt at 300 hPa isobaric surface in red) for 0000 UTC on September 2, 2014; (F) the "head" boundary (at the *black arrow*), the jet stream boundary (at the *red arrow*), and the point where the upper-level flow turns southeasterly (at the *blue arrows*), seen in the water vapor image in (E).

Fig. 3.25A shows the 300 hPa heights and wind vectors associated with a trough at the beginning of a baroclinic development over the Northwestern Atlantic. The cyclogenesis is identified in the WV imagery following the evolution of the large-scale moisture/cloud feature at the forward side of high-level troughs located downstream of a PV anomaly. In Fig. 3.25A, the feature of interest appears along and just to the south of the 896 dam contour (thick brown) of the 300 hPa surface heights, and ahead of the dynamical tropopause anomaly at the base of the trough (see Fig. 3.25C). The dynamical deformation of this feature (ie, its evolution over time) provides valuable information on the development of the process. The clearest cases of cyclogenesis are associated with undulation of the large-scale light feature in the leading part of a trough (see Fig. 3.25B), resulting in formation of a typical S-shaped moisture pattern in the WV imagery.

The initial development of such a leaf pattern is indicated by the "L" in Fig. 3.26A, and its S-shaped rear boundary is shown by *arrows* (*red and blue arrows* for the convex and concave parts,

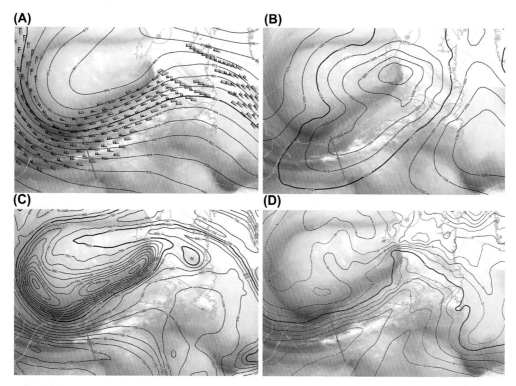

FIGURE 3.25

Water vapor image for February 27, 2002, at 1300 UTC overlaid by ARPEGE analysis fields for 1200 UTC. Also shown is the *green line* of the cross-sections in Fig. 3.26 through the leaf feature of white to nearly white-gray shades on the imagery. (A) Height (brown, the thick 896 dam contour) and wind vectors (blue, threshold at 80 kt) at 300 hPa; (B) mean sea-level pressure; (C) 1.5 PVU surface heights (the thick 800 dam contour); (D) wet-bulb potential temperature at the 925 hPa levels (contours of every 3°C).

respectively). The sharp boundary between the different moisture regimes (perpendicular to the *green line* A—B) at the rear of the developing leaf clearly shows the cyclonic curvature of the feature.

The term "baroclinic leaf" has been introduced to describe precyclogenetic cloud patterns on IR and visible imagery, and such systems are associated with a surface baroclinic zone or a cold front. As shown in Figs. 3.25B and 3.26A, the formation of the leaf signature can be well distinguished in the WV images during very early stages of development (when there is neither a deep surface low nor a cloudy cold front). Studies have shown that cyclogenesis occurs 75% of the time when a leaf is observed (see Bader et al., 1995). For cyclogenesis to occur, a low-level baroclinic zone must be present near the feature in the imagery. Such a zone can be distinguished in Fig. 3.25D by the high gradient area in the wet-bulb potential temperature field.

Since the leaf system appears in the WV imagery as a cold S-shaped pattern associated with a perturbation in the upper-level dynamics, it can be considered as an upper-level structure. Fig. 3.26B shows cross-sections of relative humidity and PV along the *green line* A—B in Fig. 3.26B. The main features of the leaf, seen in the cross-section, are well distinguished in the WV imagery.

- An area of high moisture content appears at upper levels just below the dynamical tropopause (1.5 PVU surface) and just downstream from an area of lower tropopause height (corresponding to approaching the dynamical tropopause anomaly from the base of the trough). In the WV image it is associated with initial cloud formation (at about the middle of the green A—B line).
- Behind this high-moisture area just rearward of the cold leaf feature, high relative humidity air is present at low levels capped by dry air above; together these produce medium-gray shades in the WV image.

Fig. 3.26C is a cross-section of the wind along the A—B axis. Two obvious features are clear:

- Low-level cold advection (a leftward shift of the wind aloft between 1000 and 750 hPa levels) is present to the rear of the concave portion of the leaf boundary (at locations A and R in Fig. 3.26A).
- Beyond the concave part of the leaf boundary (to the east of location F), there is a rightward wind shift in the 900 to 500 hPa levels and corresponding warm advection.

Fig. 3.27 shows the development 12 h later, in its baroclinic leaf phase associated with an enforcement of the low-level baroclinic zone (at the location of the *blue arrow*). Meanwhile, cyclogenesis has already begun, as seen by Fig. 3.27C. On the WV image the baroclinic leaf (indicated by "L" in Fig. 3.27A) appears on the forward side of the upper-air trough as a very cold S-shaped pattern, primarily as a result of radiation originating from cold, high cloud tops. The S-shaped upstream boundary of the baroclinic leaf consists of convex and concave portions separated by an inflection point.

- The convex portion of the baroclinic leaf (*red arrow* in Fig. 3.27A) is likely to be very well defined, but it may be indistinct or become less distinct with time. An elongated dark zone adjacent to the boundary is likely to narrow with time. It may change to a narrow dark band and persist as in the case of Fig. 3.27A and B, or it may disappear entirely. During the cyclogenesis phase, the convex portion moves slowly or reverses its eastward propagation and moves and rebuilds itself to the west.

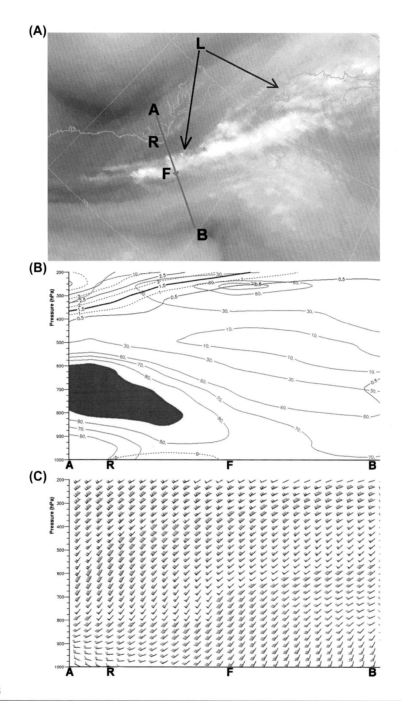

FIGURE 3.26

Initiation of a leaf development over the Northwest Atlantic on February 27, 2002. (A) The leaf feature is indicated by "L" in the water vapor image from 1300 UTC. Also shown: the cross-section line A–B as well as "R" and "F," the areas of different moisture regimes in the rear and the former side of the leaf system, respectively; (B) vertical cross-section of relative humidity (pink) and potential vorticity (blue, the thick 1.5 PVU contour); (C) vertical cross-section of wind vectors.

- The concave portion of the leaf boundary (*blue arrow* in Fig. 3.27A), on its upstream side, usually is associated with a dark spot of synoptic-scale width, corresponding to a dynamical tropopause anomaly. During the evolution of the leaf system, this dark spot is likely to remain as a significant feature on the WV imagery, although it may change shape during the cyclogenesis phase, when the concave portion of leaf boundary usually surges rapidly.

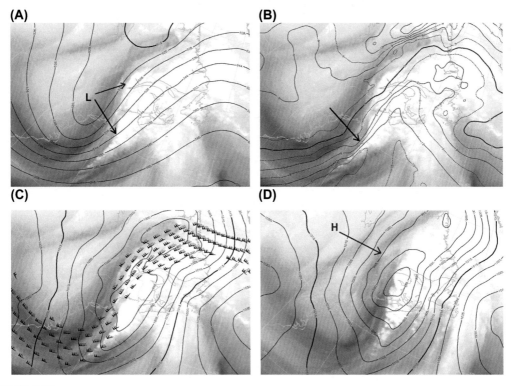

FIGURE 3.27

ARPEGE analysis on February 28, 2002, superimposed on the corresponding water vapor image available (1 h earlier for the 0000 UTC analysis and 1 h later for the 0600 UTC analysis). (A) 500 hPa heights (dam): "L" indicates the baroclinic leaf boundary for 0000 UTC; (B) 925 hPa wet-bulb potential temperature (every 3°C) for 0000 UTC; (C) mean sea-level pressure (brown) and 300 hPa maximum winds (*blue arrows*, threshold at 100 kt) for 0000 UTC; (D) mean sea-level pressure at 0600 UTC: "H" indicates the cloud/moist head.

As shown in Fig. 3.27B and C, the distinct low-level baroclinic zone (along the upstream baroclinic leaf boundary, at the *blue arrow*) is overrun by the upper-level jet. We also see that the jet maximum is located upstream of the baroclinic leaf. The process is associated with cyclogenesis; the surface low is located under the cloud leaf pattern and lies close to its upstream boundary, as depicted in Fig. 3.27C and D. At location H, a moist/cloud hook feature begins to form. This feature is associated with the transition of the baroclinic disturbance from a baroclinic leaf to a "comma" cloud pattern stage (see Bader et al., 1995).

The distinct baroclinic leaf developments are associated with significant dynamical tropopause anomalies seen in the 1.5 PVU surface heights (Fig. 3.28). During early stages of development, when the baroclinic upper-level trough is open, the WV image dark feature associated with the dry intrusion corresponds quite well to the PV maxima as well as to the minima of the dynamical tropopause height.

FIGURE 3.28

ARPEGE analysis on February 28, 2002, at 0000 UTC geopotential height of the 1.5 PVU surface superimposed on the corresponding water vapor image available (valid 1 h earlier).

Figs. 3.27 and 3.28 illustrate a baroclinic leaf associated with an upper-level PV anomaly and surface cyclogenesis. However, the leaf appearance may vary according to the relative importance of the upper trough and jet streak, as Bader et al. (1995) classified for different leaf variations.

3.5.3 CYCLOGENESIS WITH UPPER-LEVEL PRECURSORS

A key element in the understanding and classification of extratropical cyclones is the knowledge of their vertical structure. As shown in Chapter 1, cyclones develop normally along the leading edge of an upper-level trough (associated with a PV anomaly) moving over an intense baroclinic zone connecting the surface-warm anomaly with the upper-level PV anomaly (Hoskins et al., 1985). Early studies emphasized this role of large-amplitude upper-level disturbances and made use of satellite imagery to investigate the structure of cyclones. From the PV perspective, such type of cyclogenesis can be considered as a result of superposition and mutual reinforcement of upper- and lower-level PV anomalies. A cyclonic PV anomaly at the tropopause moves over a region of enhanced baroclinicity at the surface and induces formation of the surface anomaly, and then the anomalies intensify through the mutual interaction. The presence of the diabatically produced low-level PV anomaly intensifies this interaction. When vertically aligned in a mature cyclone, these anomalies can form a PV tower, a structure often found in intense extratropical cyclones.

Therefore, cyclogenesis that occurs as a result of mutual interaction between a low-level warm anomaly (or baroclinic zone) and upper-level perturbation related to a polar jet stream is associated with an upper-level precursor (a clear isolated dynamical tropopause anomaly—or positive PV anomaly—in the initial phase; eg, Bosart and Lin, 1984; Santurette and Georgiev, 2005; Michel and Bouttier, 2006; Iwabe and da Rocha, 2009). This type of a process may produce explosively developing cyclones (often called "bombs"), which are characterized by rapid central pressure reduction and a dramatic increase in intensity. Such characteristics are associated with difficulty of prediction and also with serious threats to human life and property when these cyclones occur off coastal regions. Satellite imagery can be used to identify bomb triggering and for tracking their development:

- Typically the upper-level vorticity maximum and jet max winds involved in a bomb development preexist the surface low by days, and these features can be followed in the WV imagery.
- The appearance of a head/hook moist/cloud pattern in the satellite imagery can be a signal for rapidly deepening surface-low development. The evolution of the head feature differs according to the type of cyclogenesis.

This section is devoted to the understanding of mid- to upper-level circulation patterns seen in WV imagery that are associated with rapid cyclogenesis in polar air streams by using case study examples in the Northern and Southern Hemispheres.

3.5.3.1 Cyclone Development in the Western North Atlantic

The explosive cyclogenesis with upper-level precursor will be illustrated by considering the development of a deep cyclone that affected much of the interior mid-Atlantic, Northeastern United States, and Canadian Maritimes from October 29 to 30, 2011, producing an unusually early season snowstorm. Rapid cyclogenesis occurred as the storm impacted the Eastern United States with a central pressure drop from around 1012 hPa at 0000 UTC October 29 to below 976 hPa by 1200 UTC October 31, 2011.

3.5.3.1.1 Upper-Level Precursors

Fig. 3.29 shows the ARPEGE model analyses fields of MSLP (black contours), low geopotential of the 1.5 PVU surface (brown, every 100 dam, threshold at 1200 dam), and 300 hPa wind vectors (only ≥100 kt) on October 29–30. At 0000 UTC October 29 (Fig. 3.29A), three crucial elements involved in the rapid cyclogenesis are observed over the north of America:

- The westerly jet stream, seen as a specific moisture boundary in the WV image (at the *yellow arrow*).
- The baroclinic zone of a subtropical origin (at the *green arrow*) marked by the strong humidity gradient in the WV image that weakly undulates in an anticyclonic surface pressure field to the north.
- The dynamical tropopause anomaly—or positive PV anomaly—in the initial phase (at the *pink arrow*) to the northwest, evident as a minimum of 1.5 PVU surface heights and a dark band in the WV image. It is advected to the baroclinic zone by a polar jet, associated with high gradient of the dynamical tropopause.

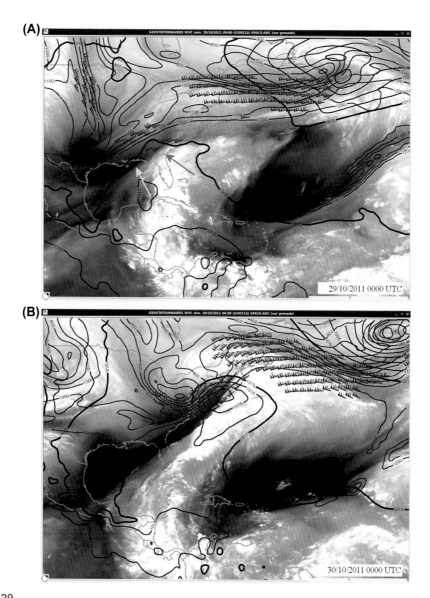

FIGURE 3.29

Rapid cyclogenesis over Western North Atlantic: a sequence of Meteosat water vapor images showing the cyclone deepening at the initial stages of rapid cyclogenesis at (A) 0000 UTC on October 29 and (B) 0000 UTC on October 30, 2011, superimposed on an upper- and lower-level field analysis; mean sea-level pressure in hPa (black contours), geopotential of the 1.5 PVU surface (brown, every 100 dam, threshold at 1200 dam), and 300 hPa wind vectors (threshold 100 kt).

The rapid cyclogenesis resulted from a baroclinic interaction between the dynamical tropopause anomaly (upper-level PV anomaly) and the low-level baroclinic zone. As the two systems interacted, at 0000 UTC on October 30, the surface low associated with the baroclinic zone deepened significantly for 24 h from 1010 to 1000 hPa. This is seen in Fig. 3.29B, where the area of low geopotential of the 1.5 PVU surface (brown contours) marks that this upper-level dynamical structure develops in phase with a low in the surface pressure field and the associated low-level baroclinic wave seen in the cloud field.

3.5.3.1.2 Synoptic Evolution

Fig. 3.30 shows a sequence of WV images overlaid by the same fields as in Fig. 3.29 showing the explosive cyclone evolution from 1200 UTC on October 29 to 1200 UTC on October 31, 2011. The thermodynamic analyses of the process are performed by considering cross-sections through the region of PV anomaly on the western side of the surface low shown in Fig. 3.31. In the cross-sections, the thick 1.5 PVU contour of PV is depicted as an appropriate representation of the tropopause (see Chapter 1). The process can be followed by the WV imagery and the corresponding dynamical fields in Fig. 3.30:

- In Fig. 3.29A, the polar intrusion of strong upper-level vorticity from the pole is seen as a surge moisture boundary at the *magenta arrow.*
- Twelve hours later, in Fig. 3.30A, at 1200 UTC on October 29, the dynamic dry feature is located just upstream of the cross-section axis (*green line*).
- At that time, the associated upper-level PV anomaly is located above the surface low that is a precursor for a mutual interaction between the upper- and low-level features. The PV maximum associated with the upper-level vortex extended from Southern Wisconsin through the Ohio Valley and into Southern Pennsylvania.

The cross-section from the 1200 UTC analysis on October 29 (Fig. 3.31A) indicates a folding of the tropopause (1.5 PVU contour, brown) at around 400 hPa. This is indicative for dry stratospheric air intrusion, associated with a dark zone in the WV image. Additionally, in this region, the isentropic surfaces spread in the vertical. This results in strengthening of the baroclinic zone (ie, increasing of the low-level horizontal gradient of the isentropic surfaces, the green contours) below the folding of the dynamical tropopause (thick brown contour). Such a baroclinic zone is well seen on the satellite images in Figs. 3.29 and 3.30B as a cloud boundary to the east of the cyclone center.

The vertical cross-sections of PV at 1200 UTC on October 30 and 31 (brown contours in Fig. 3.31B and C) depict that low-level PV anomalies (values of 1.5 PVU below 700 hPa level) have been produced in association with the baroclinic zone (low-level horizontal gradient of the isentropes). The role of diabatic processes such as latent heat release (LHR) in extratropical cyclogenesis and related low-level PV anomalies is thoroughly discussed in Brennan and Lackmann (2005). LHR can further enhance the cyclogenesis process in two ways that are readily evident in the PV framework. First, LHR reduces the effective static stability, which enhances the vertical penetration of the circulation associated with the upper- and lower-boundary PV anomalies, increasing the mutual amplification of these waves. Second, a maximum of diabatic heating produces (destroys) PV below (above) the level of maximum heating along the absolute vorticity vector. Owing to the fact that PV maxima are associated with negative geopotential height perturbations, diabatically generated PV maxima in the lower troposphere can thus be linked also to the location and intensity of surface cyclones (see, eg, Stoyanova and Georgiev, 2013). They can be important in the transport of moisture, since these PV maxima are usually located in a region of the atmosphere with high moisture content.

Most of the elements of this case seem very similar to what is well documented in the literature for the development of such a type of US East Coast cyclone (Brennan and Lackmann, 2005). At 1200 UTC on October 29, the baroclinic zone and associated surface low, which carry moist and warm air masses from the coast of the Carolinas (Fig. 3.30A), had propagated far

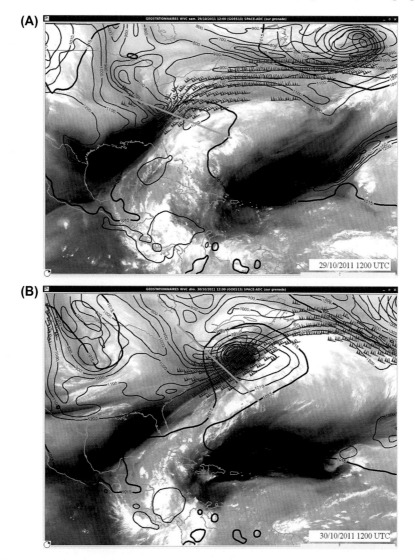

FIGURE 3.30

A sequence of Meteosat water vapor images showing the explosive cyclone evolution, superimposed onto upper- and lower-level field analysis; mean sea-level pressure in hPa (black contours), geopotential of the 1.5 PVU surface (brown, every 100 dam, threshold at 1200 dam), and 300 hPa wind vectors (threshold 100 kt): (A) 1200 UTC on October 29, (B) 1200 UTC on October 30, (C) 0000 UTC, and (D) 1200 UTC on October 31, 2011. Also shown: the cross-section axis (*green line*).

enough north to begin interacting through convection with the upper-level trough and associated PV anomaly. As the PV anomaly captured the surface low, rapid intensification of the storm occurred in the next 36 h. As hypothesized in Brennan and Lackmann (2005), the lower-tropospheric PV maximum initially generated by the LHR in the lower troposphere strengthened the inland moisture transport. The increased horizontal moisture flux and resulting increase in moisture flux convergence favor the inland precipitations over the Northeastern United States and Canadian Maritimes.

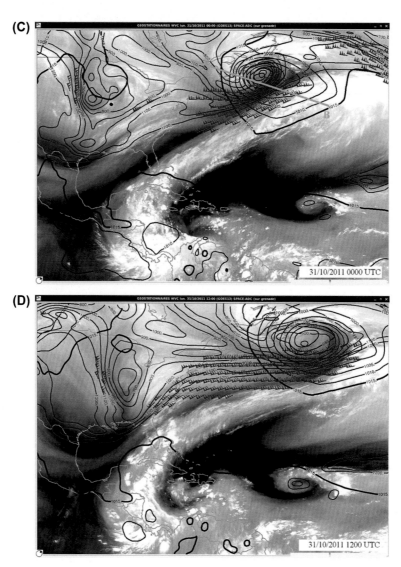

FIGURE 3.30 Cont'd

At 1200 UTC on October 30 the ARPEGE analysis (Fig. 3.30B) showed a progressive interaction of the PV anomaly region with a surface low which had dropped more than 12 hPa over this time to below 983 hPa. The cross-section through the PV anomaly feature in Fig. 3.31B shows strong PV with the 1.5 PVU contour deep into the lower troposphere. At 0000 UTC on October 31, the PV anomaly extended from Ontario through Northern New York (Fig. 3.30C). The central pressure of the surface low had fallen to 980 hPa and continued to interact with the region of enhanced PV. The cross-sections in Fig. 3.31B and C shows the intensification of the upper-level PV anomaly (PV values greater than 1.5 PVU propagates down to below 700 hPa level). This intensifies the surface low deepening, which in turn increases the strengths of the thermal advection and PV generation by the LHR in the lower troposphere that further reinforces the upper-level PV anomaly. The surface low and upper-level jet continued to move northeastward over the next 24 h with a central pressure below 976 hPa on October 31.

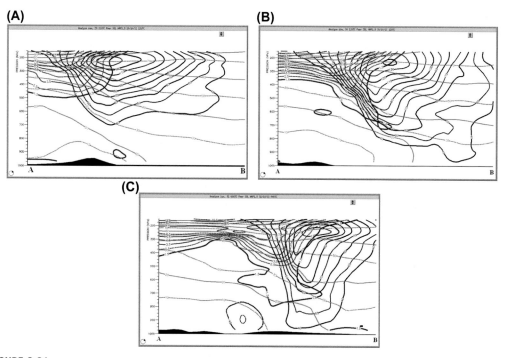

FIGURE 3.31

Vertical cross-section (axis represented by the *green lines* in Fig. 3.30) of potential vorticity (brown contours, PVU), potential temperature (*green lines*, K), and wind normal to the axis (*black lines*, kt) from ARPEGE analysis valid for (A) 1200 UTC October 29, 2011; (B) 1200 UTC October 30, 2011; and (C) 0000 UTC October 31, 2011.

3.5.3.1.3 Water Vapor Imagery Analysis of Explosive Deepening

The sequence of images in Fig. 3.32 demonstrates the utility of the WV imagery in detection of PV anomalies (associated with stratospheric air intrusion) and their interaction with the low-level baroclinic zone that strongly favors the initiation of rapid cyclogenesis:

- A baroclinic system is detected by the WV imagery (Fig. 3.32A, at the *black arrow*) as a broad undulated area of moisture plume (in light-gray shades) and clouds (in white) extending from New England, westward to the Great Lakes, and southward along the Eastern Seaboard.
- The associated WV plume is rooted in the subtropical area (confirmed by the specific boundary in the WV image on its western side, along the *blue arrows* in Fig. 3.32A) and carries a persistent movement of moisture that feeds the process.
- The WV imagery analysis also allows us to detect the PV anomaly as a broad dynamic dry band in dark-gray shades from central Canada south through Wisconsin and Illinois and looping back through the Ohio Valley (Fig. 3.32A, *red arrows*). The location of the PV anomaly in ARPEGE analysis appeared along the northern edge of the dark-gray area in the image.
- At 0000 and 0600 UTC on October 30 (Fig. 3.32B and C), the northern edge of the most intense dark-gray shade feature matched the location of the PV anomaly region from the ARPEGE analysis, and it is closely associated with the leaf development of the low-level baroclinic zone structure.

This case illustrates that the evolution of the gray-shade features in the imagery over time can be tracked to help forecasters evaluate trends of important forcing mechanisms that drive short-wave development for cyclogenesis by determining movement of PV anomalies toward developing downstream baroclinic features.

3.5.3.2 Explosive Cyclogenesis in the Southern West Pacific

Fig. 3.33 shows the genesis and development of a rapidly deepening extratropical low-pressure system, which started over the Coral Sea, to the southeast of Australia at 1200 UTC on July 23, 2008. WV imagery is overlaid by ARPEGE model fields of MSLP (black contour), geopotential of the 1.5 PVU surface (brown), isotachs of the wind on the 300 hPa surface (blue), and 200 hPa divergence (green). The different appearance of the WV imagery features in Fig. 3.33 than from the view over the Northern Hemisphere (Fig. 3.32) is due to the Coriolis force, which acts in opposite directions related to the movement of air masses to the poles in these two parts of the globe. This should be taken into account in synoptic-scale analysis of the upper-level flow patterns over the Southern Hemisphere.

At the time of Fig. 3.33A, a day prior to the initiation of explosive cyclogenesis, three crucial elements leading to this sort of rapid development are observed:

- A low-level baroclinic zone (at the *yellow arrow*) and related surface cyclonic system L. Fig. 3.33B shows that 30 h later the rapid cyclogenesis has started after strengthening this baroclinic zone and associated WV plume (to the east of the *yellow arrows*). This baroclinic structure is extended from the subtropical area and advects moist and warm air in a deep layer (light WV image gray shades).
- A sharp upper-level trough and related jet structure in its leading diffluent part associated with a distinct boundary in the WV image (at the *black dashed arrows*).

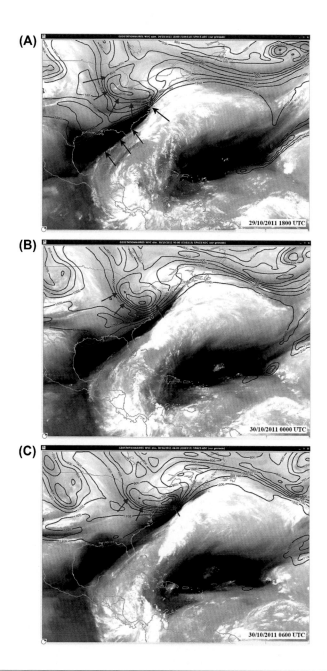

FIGURE 3.32

A sequence of water vapor images for (A) 1800 UTC October 29, 2011; (B) 0000 UTC October 30, 2011; and (C) 0600 UTC October 2011, superimposed onto the ARPEGE analysis of the geopotential of the 1.5 PVU surface (brown contour, every 75 dam, threshold at 1200 dam).

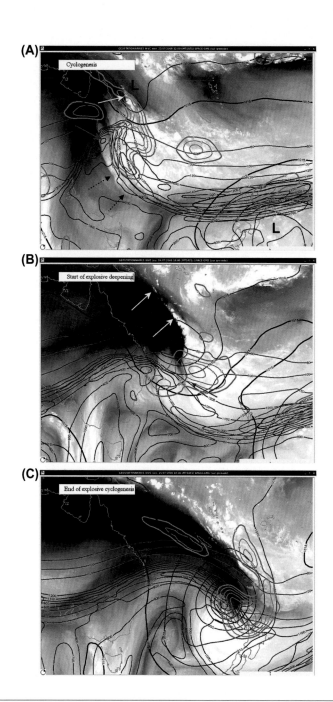

FIGURE 3.33

Mean sea-level pressure (black contour, 5 hPa intervals, only ≤1015 hPa), geopotential of the 1.5 PVU surface (brown, every 75 dam, only ≤975 dam), isotachs of the wind on the 300 hPa surface (blue, every 20 kt), and 200 hPa divergence (green, only ≥2 × 10^{-5} s^{-1}) overlaid by water vapor imagery for (A) July 23, 2008, at 1200 UTC; (B) July 24 at 1800 UTC; and (C) July 25 at 1800 UTC.

- A tropopause folding (at the *pink arrow* in Fig. 3.33A) and related drying of the upper troposphere that appears as a dark dry slot in the WV images at the time of rapid cyclogenesis (Fig. 3.33B, although the NWP model does not match correctly these low heights of the dynamical tropopause with the imagery dark zone).

In Fig. 3.33B, the developing cyclone was located beneath the equatorward entrance region of a jet stream with associated nearby upper-level divergence, which is a sign of enhanced vertical motion in the middle troposphere favoring the deepening of the surface low. This is seen in WV satellite imagery (Fig. 3.33B) as an expansion of the whitening area (around the *red arrow*) due to rising of convective cloud tops developing ahead of the system over the warm waters of the Tasman Sea. The surface cyclone is positioned directly beneath a region of upper-level divergence, which (in the Southern Hemisphere) is located at both the poleward (right) exit region of the upstream jet and the equatorward (left) entrance of the downstream jet (Fig. 3.33C).

As a result of the explosive cyclogenesis, the surface low intensifies during a few hours triggered by a baroclinic interaction with an upper-level coherent feature. The WV images superimposed onto the MSLP (green) and ascending motion (orange) show the main features at the beginning (Figs. 3.34A and 3.35A) of the interaction and during the development phase (Fig. 3.34B−D). The strengthening of the dry slot appearance in the WV image (Fig. 3.35A) is related to the beginning of pronounced tropopause folding seen in the vertical cross-section of PV in Fig. 3.35B that is the precursor of rapid cyclogenesis. The formation and evolution of the "cloud head" (Fig. 3.34D, *blue arrow*) due to ascending motions (orange contours in the cross-section on Fig. 3.35B) over the Tasman Sea are the dynamical features that indicated a subsequent development. The cross-section of relative humidity in Fig. 3.35B confirms the validity of the NWP model to simulate the cloud head (more than 90% relative humidity up to 300 hPa level) as well as the dry slot seen in the WV image (less than 10% relative humidity down to 500 hPa level).

The explosive deepening ended around 1800 UTC on July 25 (Fig. 3.34D). At this time the system was still located beneath a region of upper-level divergence associated with a strong jet stream, seen as a specific WV image boundary, associated with the high gradient of the dynamical tropopause (1.5 PVU surface). The precursor of the deepening end is seen in the WV image as a process of wrapping dry intrusion in the cloud head (Fig. 3.33C). The process finally is implemented in the formation of a hooked cloud head seen in the WV image (to the south of the *blue arrow* in Fig. 3.34D).

3.5.3.3 Water Vapor Imagery Dry Slot as a Precursor of Cyclone Deepening

As discussed in Section 3.2.2, monitoring dry intrusions by WV imagery is a way of identifying upper-level dynamical forcing of cyclone developments. The onset of significant cyclogenesis is associated with the appearance of a pronounced dry slot in the imagery. Figs. 3.36−3.38 present a spectacular example of dry slot evolution associated with a case of pronounced Atlantic cyclogenesis. Two dry slots (*red arrows* in Fig. 3.36A) can be detected around which two cloud heads are wrapped (*red arrows* in Fig. 3.36B). Figs. 3.37 and 3.38 show the relationship between the cloud-head organization, the structure of the MSLP, and the upper-level dynamics seen by dynamical tropopause anomalies (PV anomalies). Fig. 3.37 shows that dry slots are well correlated with minimums of the dynamical tropopause height (*green and yellow arrows* in Fig. 3.37). The MSLP field shows that these dry slots are also associated with two relative minimums of pressure (Fig. 3.38A); and much more, these pressure minimums evolve in close association with the dynamical evolution of these dry slots:

- Darkening of the dry slot in the west associated with the lower tropopause height (*green arrow* in Fig. 3.37A and B) occurs after 0000 UTC; accordingly the pressure minimum associated

with it deepens strongly (from 970 to 945 hPa, just ahead of this dark dry slot; *green arrow* in Fig. 3.38).
- At the same time, the dry slot in the east begins to lessen (*yellow arrow* in Fig. 3.37A and B) and, accordingly, the pressure minimum moving to the north ahead of this dry slot (*yellow arrow* in Fig. 3.38B) deepens slightly (from 960 to 955 hPa for 9 h).

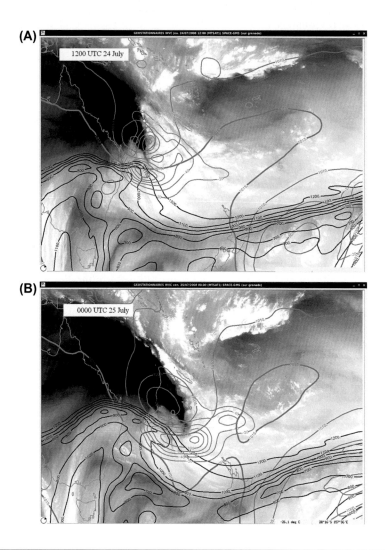

FIGURE 3.34

A sequence of MTSAT water vapor images for (A) 1200 UTC July 24; (B) 0000 UTC on July 25, 2008; (C) 1200 UTC on July 25, 2008 and, (D) 1800 UTC on July 25, 2008, superimposed on ARPEGE analysis fields: mean sea-level pressure (green contours, only $\leq$1020 hPa, every 5 hPa), ascending motion (orange, only $\geq$10 $\times$ 10^{-2} Pa/s) at 400 hPa and heights (dam) of 1.5 PVU surface (brown, only $\leq$1200 dam).

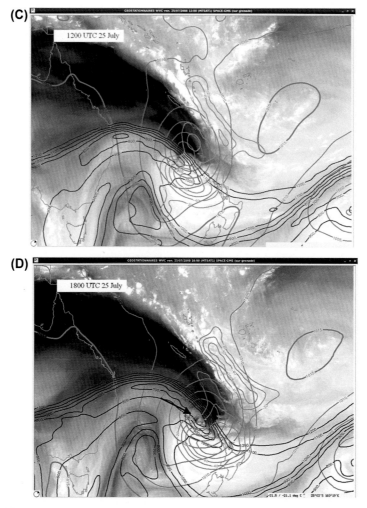

FIGURE 3.34 Cont'd

This case shows that 6.2 μm WV features provide a clear signal for strengthening the dynamics of the surface low; this allows the forecaster to understand the subsynoptic and small-scale structure of the low (the evolution of the two pressure minimums) and to better follow and specify areas of strongest wind in nowcasting. Better still, in such cases over oceans WV imagery is the only high-resolution observation (especially timescale resolution) that can give such a signal regarding the low-pressure dynamical evolution.

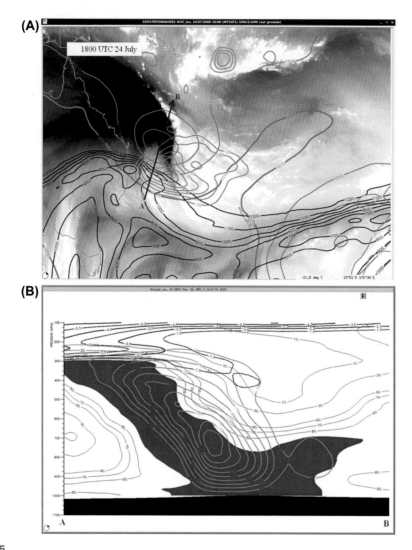

FIGURE 3.35

(A) The same as in Fig. 3.34 for July 24, 2008, 1800 UTC. (B) Cross-section of ascending motion (orange, only the contours $\geq 30 \times 10^{-2}$ Pa/s) relative humidity (pink, values $\geq 90\%$ dashed), and potential vorticity (brown, only ≥ 1.5 PVU) along the line in (A).

3.5.4 USEFULNESS OF WATER VAPOR IMAGERY TO IDENTIFY "STING JET" AND RELATED SURFACE WIND GUSTS

This section illustrates the specific type of extratropical cyclone development over the East Atlantic that have produced damaging winds over Western Europe. As a case study example, the midlatitude cyclonic storm called Klaus is considered, which was formed on January 21, 2009, related to strong upper-level dynamics over the North Atlantic basin.

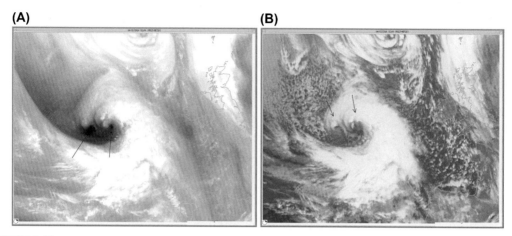

FIGURE 3.36

(A) 6.2 μm water vapor image and (B) infrared image on February 4, 2014, at 0200 UTC.

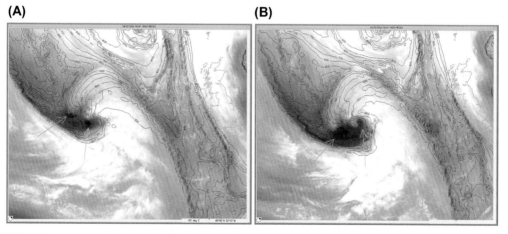

FIGURE 3.37

Height of the dynamical tropopause (red, every 50 dam, threshold at 900 dam) from ECMWF model, superimposed onto the corresponding 6.2 μm water vapor images (A) at 0000 UTC and (B) at 0300 UTC on February 4, 2014.

Fig. 3.39 shows the 6.2 μm WV image from Meteosat in the latest stage of cyclogenesis that reveals two notable signatures:

- The formation of a pronounced area of warm/dry WV channel BTs, which indicated a strongly forced region of rapidly descending middle-tropospheric air.
- The leading zone of the cloud head, which had wrapped around itself in a spiral feature. On the forward edge of this hook (at the *red arrow*), as it turns back eastwards, is the location where a "sting jet" may be present in this type of Atlantic cyclogenesis.

(A)   **(B)**

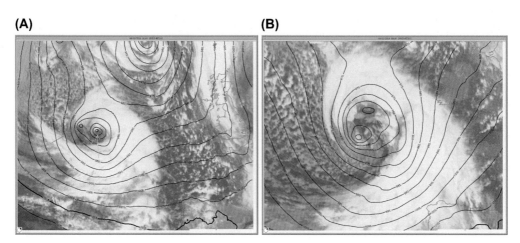

FIGURE 3.38

Infrared imagery overlaid by ECMWF analysis of mean sea-level pressure (blue, every 5 hPa), (A) at 0000 UTC and (B) at 0900 UTC on February 4, 2014 (see text).

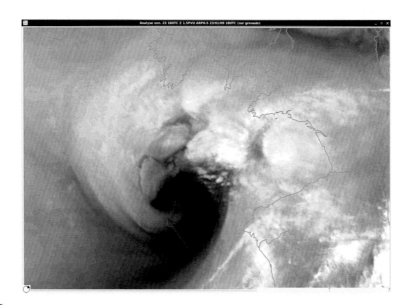

FIGURE 3.39

Meteosat 6.2 μm WV image on January 23, 2009, at 1800 UTC.

A synoptic analysis in another case on October 28, 2013, is shown in Fig. 3.40 where a satellite VIS image is overlaid by available surface observations of 10 m wind speed (wind ≥35 kt in red wind vectors). The corresponding conceptual scheme of a typical sting jet as proposed by Clark et al. (2005) is shown in Fig. 3.41. Sting jets only arise in certain types of storms, called Shapiro-Keyser cyclones, which develop according to the Shapiro and Keyser (1990) model with a specific configuration: frontal fracture, strong bent-back warm front/occlusion, and seclusion. These cyclones have a "T-bone" structure that keeps the warm and cold fronts within the storm from meeting, and a hooked tail of clouds associated with a back-bent front that is just the right shape for triggering sting jets.

The sting jet is a mesoscale descending air stream that can cause strong near-surface winds in the dry slot of the cyclone, a region not usually associated with strong winds (Browning, 2004; Schultz and Sienkiewicz, 2013). A sting jet is a flow of air that originates in the cloud head (see Fig. 3.41) associated with a rapidly deepening area of low pressure, with pressure falls of more than 24 hPa in 24 h.

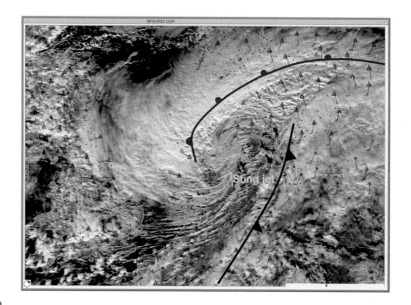

FIGURE 3.40

The sting jet location shown in the Meteosat VIS image for October 28, 2013, 1200 UTC superimposed by available surface observations of 10 m wind speed (wind ≥35 kt in red wind vectors) and the associated surface fronts drawn.

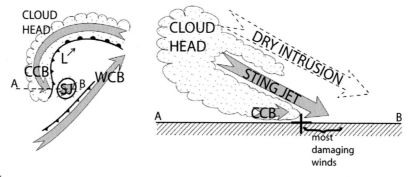

FIGURE 3.41

Left panel: conceptual scheme of the sting jet (SJ). Right panel: the west–east section shows the SJ descending from mid-levels within the cloud head, beneath the descending dry intrusion and above the cold-conveyor-belt (CCB) jet (Clark et al., 2005).

Figs. 3.40 and 3.41 illustrate the characteristics and mechanism of this phenomenon that can be summarized as follows:

- The strongest winds in such Shapiro-Keyser cyclones occur behind the cold front, in the southern quadrant of an extratropical cyclone.
- The sting jet develops on the south side of the hooked tail of the cyclone some 5 km above the ground.
- Air rushes downward from a warm front to a cold front. It can take around 3–4 h to descend toward ground level, and as it does so, it passes through layers of ice crystals that evaporate. The air parcels in the jet acquire negative buoyancy, because as they sink, they remain colder than the environment, and they end up near the surface constituting the cold pool. In this way, the evaporative cooling causes the sting jet to accelerate to high speeds of more than 100 mph (160 km/h or 90 kt).
- The most damaging winds occur in a very small region (perhaps only 50 km across), located close to the "tail" of the cloud head that wraps around the low-pressure center. Hence the term "sting jet" refers to fast-moving air descending from the tip of the cloud head (from the "sting in the tail" of the cyclone) into the dry slot ahead of it.

One of the biggest challenges for forecasters is predicting where and when sting jets will hit the Earth's surface. The point is that the winds are strongest in the cyclone where the back warm front is weakening most intensely. Satellite imagery analysis can tell us specifically the area where we can expect these winds and give forecasters added knowledge about the physical processes that are going on to create this region of strong winds. To illustrate this point, Fig. 3.42 shows IR and WV images overlaid by vectors of strong wind at the 900 hPa level from the 2 h forecast of the French high-resolution numerical model AROME in the case of the sting jet produced by Klaus on January 24, 2009.

(A)   **(B)**

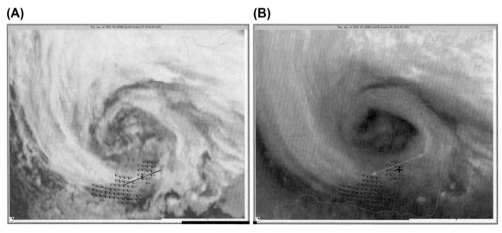

FIGURE 3.42

Wind vectors at 900 hPa level from the 2 h AROME model forecast valid for January 24, 2009, 0200 UTC superimposed on (A) infrared 10.8 μm and (B) water vapor 6.2 μm image. The *black line* in (A) shows the axis of the cross-section in Fig. 3.43. The *blue cross* in (B) indicates the location of the profile shown in Fig. 3.44.

Since the dry air intrusion in the southwest of the low from above is a crucial process in a sting jet formation, WV imagery can help to diagnose the sting jet appearance: Fig. 3.42B shows a darker gray shade area where the 900 hPa winds are stronger (*green double arrow*). This darker zone on the 6.2 μm WV image is the sign of dry cold air above lower-level clouds visible in the IR image in Fig. 3.42A. The cross-section in Fig. 3.43 along the axis of Fig. 3.42A shows the main elements of the thermodynamic environment of the sting jet:

- Low wet-bulb potential temperature, less than 10°C in the middle troposphere (red contours).
- Strong subsidence (blue contours) simulated by the 2 h forecast from AROME mesoscale model.

FIGURE 3.43

Cross-section along the *black line* in Fig. 3.42A of low wet-bulb potential temperature (red contours, °C) and subsidence (blue contours, m/s) simulated by the 2 h forecast from AROME model.

The vertical profile (at the blue cross in Fig. 3.42B) shown in Fig. 3.44 confirms the presence of the typical thermodynamic structure of the sting jet phenomena:

- A very clear layer of dry air above 850 hPa overhanging very moist low-level air.
- Increasing wind speed from the middle troposphere down to the Earth's surface.

Therefore, in nowcasting, to improve the predictability of the wind gusts at the surface, associated with the sting jet, it is required to use WV imagery as a tool to identify changes in upper- and mid-troposphere circulations and moisture content.

- Going along the cloud head through the forward-nose of the cloud hook, the drying of the descending upper-level air becomes more pronounced at the area where the sting jet appears.

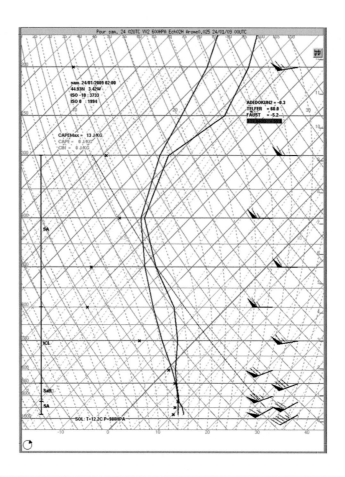

FIGURE 3.44

Vertical profile from 2 h AROME forecast on the location shown by a blue cross in Fig. 3.42B.

- As a result, the WV image gray shades become darker going from the outer part of the cloud head to its inner part, with the darkest image gray shades as a signature for a sting jet formation.
- Such sting jet signatures in the WV images are visible in Fig. 3.39 (from the *red arrow* to the inner part of the moist hook). In Fig. 3.42 this is the zone of the red vectors of the maximum winds as analyzed by the AROME NWP mesoscale model.

3.5.5 SPLIT COLD FRONT SEEN IN WATER VAPOR IMAGERY

Changes in upper- and middle-troposphere circulation patterns can transform atmospheric frontal structures into various forms and alter their activity. Therefore, detecting such changes by interpretation of images in the WV channels is a crucial task in operational forecasting that is especially valuable in situations of split cold front (also called kata-split front) evolutions.

The split cold front is a frontal system whose vertical structure is not only controlled by frontogenesis processes and vertical motion (as in the case of a classical cold front, called an anafront) but is also subjected to an upper-level circulation that produces discontinuity between low-level and upper-level structures. A split cold front is usually an advanced stage of a classical front (see Browning, 1997; COMET, 2016). Fig. 3.45 summarizes the frontal evolutions leading a classical anafront to change into a split front as follows:

- In the structure of an anafront, the main component of the flow is parallel to the front while the weaker perpendicular component is mainly located in the low levels (Fig. 3.45A). Ascending motion of the warm and moist air occurs rearward of the front (Fig. 3.45B).
- At a later stage, a stronger flow practically perpendicular to the front upstream (*wide blue arrow* on the left in Fig. 3.45A) may become predominant (usually along its part near the surface low center).
- Such a strong upper flow approaches the system and then pushes the upper front ahead of its low-level track (Fig. 3.45C and D).
- The postfrontal dry and cold air at upper and middle troposphere is brought above the low-level front, and progressively advances ahead of it (*blue arrow* in Fig. 3.45E and F). By consequence the front is "split," with its low part becoming independent of its upper part. The leading edge of this dry-intrusion aloft is the UCF in Fig. 3.45F where the cloudiness deepens abruptly and often convectively. Ahead of the surface front low-level moist air is capped by dry air above.

In midlatitudes a part of the anafront near the low center frequently turns into a split front sometime after cyclogenesis has been completed. Detection of such evolutions is important for operational forecasting by the following reasons:

- Instability is generated by dry cold air (*blue arrow* in Fig. 3.45E) overhanging the low-level warm moist air (*dashed red line* in Fig. 3.45F).
- Convection and even thunderstorm developments are common in the region between the surface front and the upper front.

Kata-split front development is associated with distinct characteristics in satellite images. Fig. 3.46 schematizes the response of the 6.2 μm WV channel as well as of the 10.8 μm IR channel to the radiation effects, produced by the vertical structure of a split cold front (as presented in Fig. 3.45F) in terms of the considerations presented in Section 2.1.3. The associated field of radiance in WV and IR channels has a "stepped structure," that is, shows three ranges of BT organized like steps, becoming colder and colder from the rear side of the SCF to the forward side of the UCF downstream. The appearance of a distinct kata-split front system in satellite images in WV, IR, and VIS channels is illustrated by a case study example presented in Fig. 3.47.

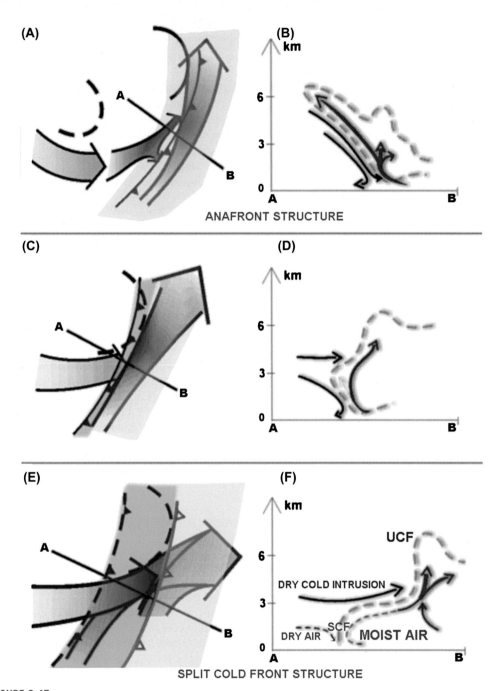

(A)

(B)

ANAFRONT STRUCTURE

(C)

(D)

(E)

(F)

UCF

DRY COLD INTRUSION

DRY AIR

SCF

MOIST AIR

SPLIT COLD FRONT STRUCTURE

FIGURE 3.45

Schematic representation of the evolution of an anafront ([A] and [B]) into a split cold front ([D] and [E]; see text); (A), (C), and (E) are views from above; the high thick clouds are in yellow; the low clouds are in green; the *red arrow* represents ascending warm flow and the *blue arrow* is the dry cold upper-level flow; the *arrows* become thicker with the height (*narrow arrows* at low level). *Thick black line* is the dynamical tropopause anomaly; *thick black dashed line* is a latent tropopause anomaly (not associated with vertical motion). (B), (D), and (F) show cross-sections of the flow relative to the system along the A–B axis; *broken lines* are the limit of the saturated air (clouds); *arrows* represent the flow parallel to the cross-section: The thicker the line, the stronger the vertical motion (moist warm air in red; dry cold air in blue).

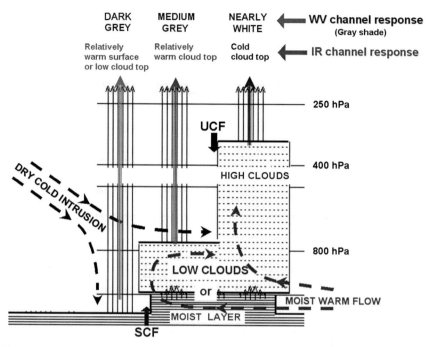

FIGURE 3.46

Schematic representation of a split cold front configuration of cloud/moisture field seen in vertical cross-section as well as the response of 6.2 μm water vapor channel and of 10.8 μm infrared channel (see the text for explanation).

The IR image (Fig. 3.47C) shows the main feature of a split cold front, that is, the presence of two successive cloud bands (indicated by "2" and "3") with very different cloud top temperatures ahead of a warmer region indicated by "1":

- A zone of relatively warm clouds (2 in Fig. 3.47C) corresponding to low-medium clouds.
- A cold cloud band (3 in Fig. 3.47C) corresponding to thick high clouds.

The western boundary of the white band is the upper cold front (*black arrow* in Fig. 3.47), while the boundary between the medium-gray shade band and the darker gray zone reflects the SCF (at the *yellow arrow*).

Fig. 3.48B shows the 7.3 μm WV image with superimposition of wet-bulb potential temperature at 700 hPa. The SCF is clearly associated with the strong temperature gradient (*yellow arrow*); the dry and cold air at its rear side is well marked by low values of wet-bulb potential temperature. The upper-level wind and the wet-bulb potential temperature at 500 hPa superimposed on the 6.2 μm WV image (Fig. 3.48A) show the upper-wind perpendicular to the front that produces dry and cold intrusion aloft, above the SCF, marked by medium imagery gray shades.

Fig. 3.49A presents the wet-bulb potential temperature at 850 (red contours) and 500 hPa (blue contours) from the ARPEGE model superimposed onto the IR image. Around the area indicated by

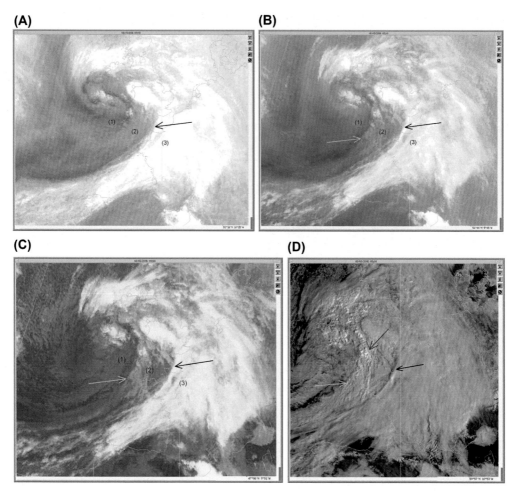

FIGURE 3.47

Split cold front seen in Meteosat images on March 6, 2006, 0900 UTC; (A) 6.2 μm water vapor image; (B) 7.3 μm water vapor image; (C) 10.8 μm infrared image; (D) high-resolution visible image (see text).

"2," which is between SCF and UCF, the θ_w pattern shows clearly the potential instability: area with θ_w less than 11°C at 500 hPa level above a low-level area where θ_w is around 12°C. Fig. 3.49B illustrates the cross-section features, schematized in Fig. 3.45F:

- Dry air (low values of wet-bulb potential temperature) penetrates further forward the system in the mid-level, together with a tendency toward descending motion.
- Ascent (yellow contours in Fig. 3.49B) occurs forward the system.
- The SCF is not pronounced as a θ_w gradient zone (*red arrow* in Fig. 3.49B).
- The UCF is well pronounced as a strong gradient of wet-bulb temperature at mid-level (*blue arrow* in Fig. 3.49B). It is the most active part of the system.
- The wet-bulb temperature gradient zone is tilted forward at lower troposphere.

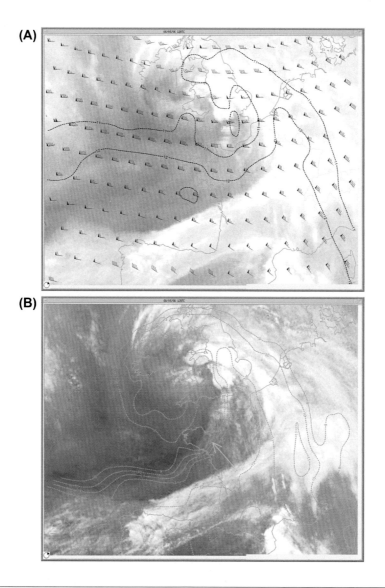

FIGURE 3.48

(A) 6.2 μm water vapor image with superimposition of ARPEGE analysis of 300 hPa wind (red) and wet-bulb potential temperature at 500 hPa (blue, only 11°C and 12°C); (B) 7.3 μm WV image with superimposition of ARPEGE analysis of wet-bulb potential temperature at 700 hPa (green, only 9, 10, 11, and 12°C) on March 8, 2006, 1200 UTC.

The split front structure is an important feature from an operational point of view due to the associated convective environment in a wide range of specific favorable conditions:

- Convective developments are likely to occur near the upper front on the leading edge of the dry cold intrusion in the lower-middle troposphere where forcing on ascending motion exists to release instability.

- The environment is favorable for convection also near the surface front where convergence winds are possible (*red arrows* in Figs. 3.47D and 3.49B).
- Deep convection can develop in the region between the upper cold front and surface front, where potential instability can be large (in the region labeled 2 in Fig. 3.47) as a result of the persistent cold and dry flow aloft overhanging the low-level warm and moist air (see also the schematic in Fig. 3.45F).

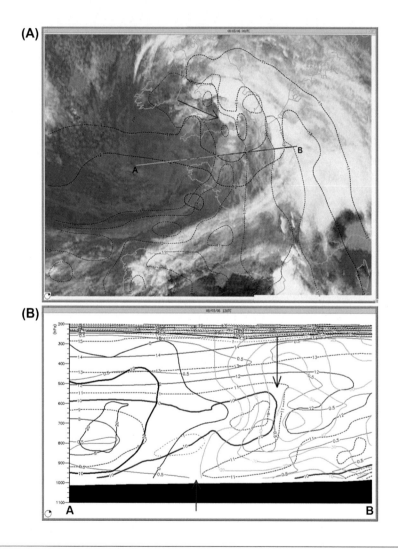

FIGURE 3.49

(A) ARPEGE model analysis of welt-bulb potential temperature, θ_w (only 11 and 12°C) at 500 hPa (blue) and 850 hPa (red) superimposed onto the infrared image at 1200 UTC on March 8, 2006; (B) cross-section along the A–B axis in (A) for vertical velocity (ascent in yellow, descending motion in blue; 10^{-2} Pa/s), potential vorticity (brown, interval 0.5 PVU), and θ_w (red, °C).

In summary, a forecaster can diagnose the kata-split front structure by distinct gray shade configuration in satellite imagery taking the advantage of various associated radiation effects and especially those related to the response of the two WV channels:

- As shown above, the IR image (Fig. 3.47C) reflects the main feature of a split cold front with the presence of two successive cloud bands with sharp differences in their cloud-top temperatures.
- In some cases the position of the upper cold front is seen in the visible channel image along the shadow (dark stripe at the *black arrow* in Fig. 3.47D), which the upper cold front system makes over the zone of relatively low-level clouds.
- The UCF is distinctly seen on WV images (at the *black arrow* in Fig. 3.47A and B).
- The 6.2 μm WV imagery shows very well the UCF but does not reflect clearly the low-level front (Fig. 3.47A) since this channel is representative of the upper-troposphere circulation but does not see low-level structures.
- The 7.3 μm is more sensitive to the presence of dry air in mid- and low levels, so this channel is able to reflect the low-level features, and the SCF (*yellow arrow* in Fig. 3.47B) is clearer on the image in 7.3 μm than on the other channels. The kata−cold front system is identified by three distinctly seen moisture boundaries in the 7.3 μm image, being warmer at the rear side of the SCF and colder at the forward side of the upper cold front downstream.
- The difference between the low-level warm moist air ahead of the surface front and the cold dry air behind is much better seen by 7.3 μm than by visible or IR images.

3.6 INTERACTION OF TROPICAL CYCLONES WITH UPPER-LEVEL DYNAMICAL STRUCTURES

Tropical cyclones (TCs) are massive vertically deep storms developed over the ocean, associated with compact and distinct well-organized convective cloud systems and extremely strong winds exceeding 119 km/h. The same phenomenon is given different names in different parts of the world. In the Atlantic Ocean and Eastern Pacific Ocean they are called hurricanes. In the western Pacific they are called typhoons, and in the southern hemisphere they are called cyclones. No matter where tropical cyclogenesis occurs, improving its prediction remains a goal of the meteorology community, since there are many factors that affect this process. Numerous studies show that a combination of six main conditions is needed for tropical cyclogenesis:

- Sufficiently warm sea surface temperatures (SSTs), at least 26°C.
- Atmospheric instability.
- High humidity in the lower to middle levels of the troposphere.
- Enough strong Coriolis force to develop a low-pressure center.
- A preexisting low-level disturbance.
- Low vertical wind shear (large values of vertical wind shear remove the rising moist air too quickly, preventing the development of the TC). Dynamically, some shear is required in order that advective processes such as differential vorticity advection and the Laplacian of thermal advection can force subsynoptic-scale ascent and help organize deep convection. If a TC has already formed, large vertical wind shear can weaken or destroy it.

Tropical cyclogenesis involves the development of a warm-core cyclone, due to significant convection in a favorable atmospheric environment. Modern methods of meteorological observations involving satellites, radars, etc. allow accurate tracking of the development and paths of TCs. The intensity of convective clusters can be tracked using satellite IR channel BTs. But prediction of TCs' intensity is problematic as they may continually change their intensity evolving and moving into different environments (regarding the upper-air flow pattern vertical wind shear, ocean temperature, and relative humidity). Recent studies on the synoptic-scale processes as indicators of rapid intensification within TC environments are based on statistical approaches. Using logistic regression and support vector machine classification, Grimes and Mercer (2015) identified mid-level vorticity, pressure vertical velocity, 200–850 hPa vertical shear, low-level potential temperature, and specific humidity as the most significant in diagnosing RI, yielding modest skill in identifying rapidly intensifying storms.

The key element that maintains the storm circulation intensity is the maintenance of the deep convection surrounding its core. The warm tropical ocean provides the moisture to drive the convection, increasing the moist static energy of the TC boundary layer. Conversion of this moist static energy into kinetic energy via convection is the mechanism by which a TC intensifies. The TC intensity is regulated by the oceanic surrounding: The SST and the depth of the warm water determine its upper bound on the TC intensity. Whether or not a TC reaches this potential depends on the upper-tropospheric environment (Merrill, 1988). Dry air intrusion or strong vertical wind shear should lead to weakening of the TC. When a TC moves into a region of large-scale upper-level divergence near upper-level jets, its intensity may change because of an enhanced secondary circulation. When an asymmetric upper-level feature, such as a trough, interacts with a TC, it can also affect its intensity through increased vertical wind shear as well as through an enhanced eddy flux convergence of momentum and efficient mass transport. The former process usually weakens the TC while the latter is typically associated with TC intensification. When a strong upper-level PV anomaly interacts with a TC the shear effect may dominate the eddy flux convergence effect, while for a weaker upper-level anomaly the eddy flux convergence may dominate the shear effect, causing TC intensification. Although hurricanes lose strength as they move over land, they still carry vast amounts of moisture onto the land, causing thunderstorms with associated flash floods.

Satellite imagery has been extensively used to identify and observe the TCs (eg, Velden et al., 1998; Olander and Velden, 2009). As shown in Weldon and Holmes (1991), most of the basic concepts for WV imagery interpretation are applicable at tropical latitudes. The primary differences are in the relatively low frequency of occurrence of specific upper-level flow patterns. The image features at the lower latitudes are related to the circulation at higher latitudes, and distinct jet streams are sometimes observed near the equator. However, the upper-air temperature gradients and wind speeds at tropical latitudes are usually less than those commonly observed at middle latitudes. Because of this, changes in the upper-air conditions induced by deep convection are more prevalent and play a more dominant role in the Tropics than in middle latitudes. A summary of WV patterns and features at tropical areas along with useful comments on some specific interpretation issues is presented in Weldon and Holmes (1991). In this section we use WV images to identify some upper-level flow patterns in the environment of TCs that may affect the storm intensity and demonstrate the use of the imagery to observe the upper-air systems applicable to understanding and forecasting the synoptic development in the lower latitudes.

As discussed in Weldon and Holmes (1991), certain common patterns on the WV imagery can be present in the environments of forming TCs. The upper-air circulation systems and their changes are often of significant influence to low-level surface conditions. In certain situations, analyzing specific types of pattern changes on these satellite images might be applied to forecasting or understanding TC developments. This section is aimed to show situations in which the pattern changes in the WV imagery could be used in conjunction with other information to improve the forecasting of TC development.

3.6.1 EFFECTS OF UPPER-LEVEL FLOW PATTERN IN THE SURROUNDING ENVIRONMENT ON THE INTENSITY OF TROPICAL STORMS

In the tropical upper troposphere a mean easterly flow is observed that is associated with hemispheric asymmetry in the mean meridional circulation (Kraucunas and Hartmann, 2005). Moving from the equator to the higher latitudes, the upper-level flow gradually becomes more perturbed and changes direction. In the study of Merrill (1988), it was suggested that the development of a TC is dependent on the upper-level flow of its atmospheric surrounding with radius of about 1000 km in the upper troposphere that is about 9° longitude in the tropical area. Fig. 3.50 shows the proposed schematic of the flow patterns of the surrounding environment of the intensifying and nonintensifying TCs, based on a study of 128 positions of hurricane stages of 28 different TCs over the Atlantic. Diagnosing the upper-level circulation pattern in the area between the generally zonal easterly flow near the equator and the much more perturbed upper-tropospheric motion field in the midlatitudes, the forecaster can anticipate the atmospheric environmental influences on the intensity of the storms considering the significant difference between the two patterns in Fig. 3.50. For an efficient interpretation, the following principles inferred from the conceptual model of Merrill (1988) can be followed:

- For intensifying TCs, an anticyclonic circulation pattern is present to the west, and there is no unidirectional flow near the center.
- For nonintensifying TCs, there is a unidirectional equatorward flow to the west (radius 1000 km) as well as near the center.
- Regarding the outflow pattern for intensifying TCs, the upper-level outflow exhibited open streamlines, allowing the TC to ventilate itself with the surrounding environmental flow, while for nonintensifying TCs, closed streamlines existed.

The most reliable information about the upper-level circulation pattern can be inferred by WV imagery analysis. Even though there are no strong jets in the tropical area, the upper-level flow tends to be parallel to the large-scale moisture boundaries due to the conservative nature of the WV as a tracer of upper-level large-scale atmospheric motion (which, in its great part, is approximately adiabatic and hydrostatic). As shown in Section 3.3, the satellite WV imagery provides clear information on the flow pattern at the tropical and subtropical upper troposphere (generally between 500 and 250 hPa levels). After a TC has formed in association within a favorable low-level ocean and atmospheric environment, the patterns on the WV imagery may be considered as an additional factor for further evolution. Knowledge on the upper-air flow pattern from the WV imagery could be combined with information from other sources to confirm or modify the forecast of TC development.

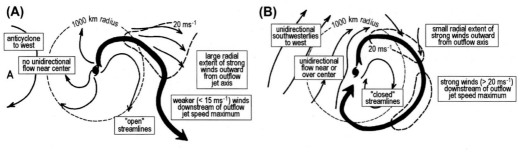

FIGURE 3.50

Schematic view of the upper tropospheric factors affecting tropical cyclone intensity change: (A) Intensifying and (B) Non-intensifying.

From Merrill, R., 1988. Environmental influences on hurricane intensification. J. Atmos. Sci. 45, 1678–1687.

3.6.1.1 Tropical Cyclone Track Satellite Data From National Oceanic and Atmospheric Administration (NOAA) NESDIS

Fig. 3.51 shows the MSLP and Maximum Winds provided by the National Oceanic and Atmospheric Administration (NOAA) NESDIS Center for Satellite Applications and Research for two TC cases considered in this section. The data is available online at: http://rammb.cira.colostate.edu/products/tc_realtime/storm.asp?storm_identifier=WP252013.

The wind analysis is based on information from five satellite data sources to create a mid-level (near 700 hPa) wind analysis and then adjusted to the surface. The MSLP is calculated directly from the azimuthally averaged gradient level tangential winds produced by this multiplatform TC wind analysis. The circular domain for the numerical integration has a 600 km radius. The pressure deficit resulting from the integration is then added to an environmental pressure. The environmental pressure (Penv) is interpolated from analyses of NOAA National Centers for Environmental Prediction (NCEP) in a circle 600 km from the cyclone center.

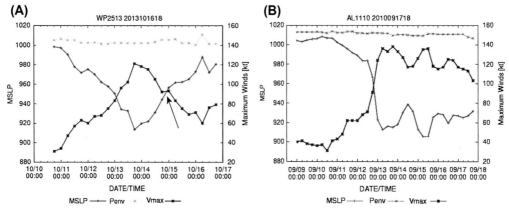

FIGURE 3.51

Track data for evolution of (A) Typhoon Wipha and (B) Hurricane Igor: mean sea-level pressure (red curve), wind maximum (blue curve), surface pressure at the storm environment (gray curve); from NOAA NESDIS Center for Satellite Applications and Research.

The cases shown in Fig. 3.51 are considered here to understand and distinguish general dynamical characteristics of the upper troposphere specific to persistently developing tropical storms and TCs, which encounter fluctuations in intensity.

- Typhoon Wipha over the western Pacific Ocean is considered in Sections 3.6.1−3.6.3, started from October 12, 2015, when the storm reached typhoon status. Moving northwestward and persistently intensifying, Wipha grew into a very large system that ultimately attained its peak intensity late on October 13 (Fig. 3.51A).
- The Major Hurricane Igor over the Atlantic is considered in Section 3.6.1 for the period September 11−15, 2010, when the TC encountered fluctuations in intensity (Fig. 3.51B). This is one of the complex cases that were going through several intensification cycles in the tropical area influenced by the upper-level dynamics at higher latitudes.

3.6.1.2 Water Vapor Imagery Analysis of Typhoon Wipha

The large Typhoon Wipha, which caused extensive damage in Japan and at least 41 fatalities (see Wikipedia Typhoon Wipha, 2013), originated from a tropical depression in the western Pacific Ocean on October 8, 2013. The depression strengthened to a tropical storm on October 11, receiving the name Wipha, and reached typhoon status on October 12. At noon on October 14, Wipha had entered the Philippine area of responsibility, and the TC received the Philippine name Tino.

A sequence of WV imagery, superimposed by the 300 hPa wind vectors during the persistent intensification of Typhoon Wipha is presented in Fig. 3.52. At the first image time, two significant moisture features involved in the TC initiation and development are present in the WV imagery:

- A high-amplitude upper-air ridge, which is extended from the Northwest Pacific Basin southward into the Intertropical Convergence Zone (ITCZ), with its upstream western moisture boundary at the poleward side of the light band "BP" in Fig. 3.52A. As discussed in Weldon and Holmes (1991) the "base ridge" is a more favorable location for tropical cyclogenesis. By that time, two TCs had developed south of location "B," along the ITCZ; Typhoon Wipha is indicated by the *black arrow*.
- A moisture pattern of a low-amplitude open ridge labeled "R" in Fig. 3.52 located to the northwest of Wipha that is going to influence favorably the typhoon development. Typically, on the polar side of such an open ridge system, a well-defined moisture boundary with a convex curved shape is seen on the WV imagery (Weldon and Holmes, 1991). The gray shades on the southern side of the moisture regime darken gradually in the equatorward direction related to the increasing anticyclonic curvature of the streamlines southward, which is indicated by the *blue repeating arrows* in Fig. 3.52A, consistent with the NWP analysis of the wind vectors superimposed on the WV image.

3.6.1.2.1 Intensification

WV imagery analysis of the sequence in Fig. 3.52 provides valuable information about the flow structure on the upper troposphere that is favorable for persistent TC intensification, as follows:

- The open ridge is moving to the east and the base-ridge moist feature BP tends to narrow. Equatorward, at location B on Fig. 3.52B, dominated by the dry regime in the most southern part of the open ridge system, the WV image gray shades tend to become darker.

Strengthening of the anticyclonic circulation on the equatorward side of the open ridge is diagnosed by the distinct convex curved shape of the boundary between the white and light-gray moisture regime in the images south of location R.

- Fig. 3.52C and D shows further moving of the open ridge to the east. The related moisture boundary is amplifying and rotated anticyclonically with time. This leads to strengthening of the anticyclonic circulation to the south of the open ridge on October 13, seen in the WV imagery as a moist anticyclonic gray shade pattern to the west of the storm (around location A in Fig. 3.52C and D). The evolution of the 300 hPa wind confirms the existence of anticyclonic upper-level flow near west of the TC center. This is a signature for intensifying TCs, according to the concept of Merril (1988) shown in the scenario of Fig. 3.50A. The upper-level flow environment to the north of the TC is characterized by an anticyclonic shear that contributes to a ventilation of the outflow. With this kind of circulation pattern during the studied period in this section (October 12−14, 2013), Wipha's strengthening was rather steady as seen in Fig. 3.51A.
- In the end of the period, the open ridge is fast moving to the northeast of Wipha due to amplification of the large-scale long-wave pattern as well as to propagation of the midlatitude upper-level trough deep southward (labeled "T" in Fig 3.52F). As a result, the anticyclonic circulation (previously induced by the upper-air flow in the higher latitudes) to the west of the TC has weakened progressively (Fig. 3.52E and F), which decreases the degree of favorability of the atmospheric conditions for tropical storm development.

3.6.1.3 Water Vapor Imagery Analysis of Hurricane Igor

Igor, the strongest TC of the 2010 Atlantic season, was a very large and intense hurricane that struck Bermuda with Category 1 intensity. Later it hit Newfoundland and was the most damaging hurricane in recent history for that island. The evolution of MSLP (red) and maximum wind speed (blue) in Fig. 3.51B shows significant fluctuations in intensity of TC Igor in the period September 12−16, 2010. According to the report of Pasch and Kimberlain (2011), shortly after the cyclone eventually reached hurricane strength by 0000 UTC on September 12, the central convective cloud pattern became quite symmetric, the upper-level outflow increased markedly, and very rapid intensification took place. The maximum winds increased to an estimated 130 kt by 0000 UTC September 13. Igor slowed its forward speed on September 13, while maintaining a nearly due westward heading across the tropical Atlantic. After exhibiting some hints of concentric eyewall structure, Igor's maximum winds decreased to 115 kt by 0600 UTC on September 14. However, the hurricane soon restrengthened to an estimated peak intensity of 135 kt around 0000 UTC on September 15.

Fig. 3.53 shows satellite WV images overlaid by 300 hPa wind derived by ARPEGE model analysis during the development of Igor from September 13 to 15, 2010. Diagnosis based on the sequence of images can be applied to identify the upper-level flow conditions in the environment, which significantly modify the degree of favorability for a TC development and cause intensification or weakening of the storm.

3.6.1.3.1 Intensification

On September 13, 2010, at 0600 UTC and 1200 UTC (Fig. 3.53A and B), Igor is located to the southeast of the leading part of a mid-level ridge. The WV imagery shows the presence of a dark area within the warm sector of the synoptic-scale wave. This is a sign for anticyclonic flow equatorward

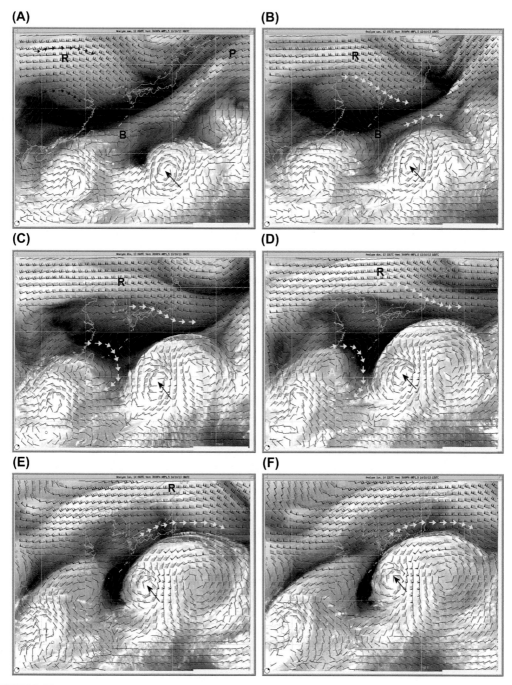

FIGURE 3.52

MTSAT water vapor images of Typhoon Wipha overlaid by 300 hPa wind from ARPEGE analysis:
(A) October 12, 2013, 0600 UTC, (B) October 12, 2013, 1800 UTC, (C) October 13, 2013, 0600 UTC,
(D) October 13, 2013, 1800 UTC, (E) October 14, 2013, 0600 UTC, and (F) October 14, 2013, 1200 UTC.
The *black arrow* indicates the position of Typhoon Wipha; also shown: open ridge, "R," anticyclonic
circulation, "A," and upper-air trough, "T."

confirmed by the NWP analysis of wind superimposed. According to the scenario in Fig. 3.50A of the interpretation scheme of Merrill (1988), this is a signature for atmospheric environment related to an intensifying TC.

3.6.1.4 Weakening

On September 14 at 1200 UTC and 1800 UTC (Fig. 3.53C and D), Igor is located to the southeast of a mid-level trough. The WV image shows midlatitude intense circulation to the northwest, which changes the upper-level surrounding to the west of the TC from an anticyclonic to a cyclonic flow pattern. At location "C" WV imagery analysis shows an upper-air dry cyclone with core convection (Weldon and Holmes, 1991). This is itself a closed low system, and being to the west of the TC (Fig. 3.53C and D) it produces unidirectional flow near the cyclone center. As a result, a transition from the scenario in Fig. 3.50A to the scenario in Fig. 3.50B occurred. Also, the upper-air ridge has rolled over Igor, by that time located just to the north of the TC system, and limits it to deepen due to the insufficient mass evacuation (small radial extent of strong winds outward from outflow jet axis). These changes provide evidence of establishment of an unfavorable upper-level flow environment that explains the decreasing of intensity in Fig. 3.51B. NASA's Aqua satellite passed over Igor on September 14 at 1447 UTC, and its Atmospheric Infrared Sounder (AIRS) instrument got a reading on the SSTs around Igor, which were all warmer than the threshold needed to maintain a TC (NASA, 2010). Therefore, Igor had a good low-level energy source, and the unfavorable upper-air flow pattern can be considered as a significant factor in the system weakening on September 14.

3.6.1.5 Reintensification

The WV images on September 15 at 0000 and 1200 UTC (Fig. 3.53E and F) show that the favorability of the flow pattern is strengthened: Due to intense midlatitude circulation, anticyclonic flow breaking occurs to the northwest of the cut-off low protecting it from the cold advection. This changed the upper-level surrounding of Igor from cyclonic to anticyclonic flow pattern to the west of the TC (Fig. 3.53B and F). This caused a transition from the scenario in Fig. 3.50B to the scenario in Fig. 3.50A that explains the increasing of intensity in Fig. 3.51B. The anticyclonic rotation of the related moisture pattern in Fig. 3.53D−F (labeled "A" to the northwest of the TC) is an indication of further intensification.

The organized upper-level mass evacuation from the storm core (northeast and/or southwest outflow) is an important regulator of TC intensity. Nondeveloping systems often lack this organized outflow or have a small radial extent of strong winds outward from the outflow jet axis (see Fig. 3.50B). The upper-air environment of the intensifying TCs is usually characterized by a flow pattern favorable to increasing upper-level divergence that contributes to a ventilation of the outflow. To illustrate this point, Fig. 3.54 shows the upper-level flow pattern by superimposing WV images and ARPEGE model analysis of isotachs of maximum wind at 250 hPa and divergence at 200 hPa (green contours).

After the storm had reached hurricane strength on September 12, the upper-level flow system (Fig. 3.53A and B) progressively developed a divergent wind field associated with a developing outflow channel to the northeast of the system (Fig. 3.54A and B). The upper-level divergence analyzed by the NWP model on September 13 at 1200 UTC (Fig. 3.54C) shows significant strengthening of the divergent outflow pattern related to the TC system and strong westerly winds to the northeast which appear connected to the core outflow. This is confirmed by the appearance of a distinct outflow spiral band of cirrus clouds on the WV image and related image boundary with the dry dark feature at the poleward side of the jet. This results in a large radial extent of strong winds outward from the outflow jet axis as in the scenario in Fig. 3.50A.

(A)

(B)

(C)

(D)

(E)

(F)

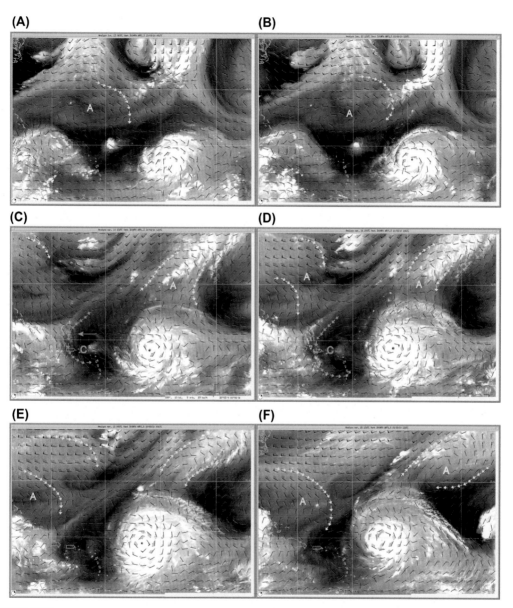

FIGURE 3.53

GOES water vapor images of Hurricane Igor overlaid by 300 hPa wind from ARPEGE analysis: (A) September 13, 2010, 0600 UTC, (B) September 13, 2010, 1200 UTC, (C) September 14, 2010, 1200 UTC, (D) September 14, 2010, 1800 UTC, (E) September 15, 2010, 0000 UTC, and (F) September 15, 2010, 1200 UTC. Also shown are anticyclonic circulation, "A," and upper-air dry cyclone, "C."

Then, on September 14, the upper-level flow pattern had changed (from the anticyclonic one in Fig. 3.53C to the cyclonic one in Fig. 3.53D) in the sensitive area to the west of the TC. At the same time, upper-air convergence dominated close to the northeast of the TC, due to the northerly jet associated with a distinct moisture boundary, which tended to exhibit cyclonic curvature in its equatorward side (Figs. 3.54D and 3.53D). This circulation pattern suppressed the radial outflow winds poleward of the storm. As a result, the area of divergent flow had been progressively disconnected from the TC cloud system and shifted downstream. In the absence of favorable divergent upper-level environmental conditions that can support sufficient outward horizontal mass flux (Fig. 3.53D), the system weakened on September 14.

As shown in Section 3.3.3.3, WV imagery enables a direct view of the divergence of the upper-level flow pattern in the tropical area. This may provide signatures of favorable versus nonfavorable conditions, encountered by the TC regarding the outflow process.

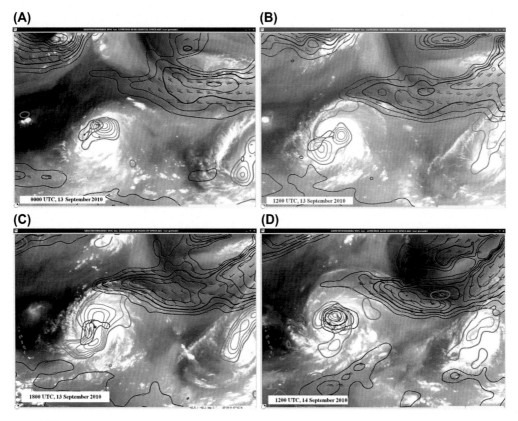

(A) **(B)** **(C)** **(D)**

FIGURE 3.54

GOES water vapor images with superimposition of ARPEGE analysis isotachs of 250 hPa wind (every 5 kt, threshold 25 kt, *blue lines*), 200 hPa divergence (*green lines*), and 250 hPa wind (threshold 30 kt, red): (A) September 13, 2010, at 0000 UTC, (B) September 13, 2010, at 1200 UTC, (C) September 13, 2010, at 1800 UTC, and (D) September 14, 2010, at 1200 UTC.

3.6.2 INTENSIFICATION OF TROPICAL CYCLONES ON THE ANTICYCLONIC SHEAR SIDE OF JET STREAMS

Usually the TCs show a distinct outflow jet toward the pole, where there is weak inertial stability associated with the anticyclonic shear side of a belt of westerlies (see Rappin et al., 2011) as shown in Fig. 3.54B. The outflow pattern is likely a reflection of diabatic warming associated with deep convection well removed from the storm. Divergence in the secondary circulation of the jet entrance region may enhance TC outflow (and the downstream jet through geostrophic adjustment).

Studying the positive effect for TC intensification from an approaching upper-level jet, Rappin et al. (2011) performed a numerical study to examine the asymmetric inertial stability distribution generated by the presence of a jet stream to the north of the TC. They showed that as the system evolves, convective outflow from the TC modifies the jet, resulting in weaker shear and more rapid intensification of the TC—jet couplet. Rappin et al. (2011) argued that an outflow channel is generated as the TC outflow expands into the region of weak inertial stability on the anticyclonic shear side of the jet stream. This minimizes the energy expenditure of forced subsidence by ventilating all outflow in one long narrow path, allowing radiational cooling to lessen the work of subsidence. Furthermore, it is hypothesized that evolving conditions in the outflow layer modulate the TC core structure in such a way that TC outflow can access weak inertial stability in the environment.

Concerning the separation distance between the TC and the jet, if it is too large, the systems may evolve independently of one another. If the separation distance is too small, vertical shear will dominate any influence of the asymmetric environmental inertial stability in the dynamical evolution of the storm. A separation of 900 km was chosen by Rappin et al. (2011) so that a significant interaction with the jet occurred while the vertical shear over the vortex was kept small.

Satellite imagery can be useful for analysis of TCs showing fluctuations in intensity in association with approaching of jets and upper tropospheric troughs and thus providing a tool for reducing the related difficulties for forecasters.

Typhoon Wipha is a typical case of a tropical cyclone, which going through a distinct intensification cycle is moving poleward and reaches subtropical latitudes as a massive well-organized system (Fig. 3.52). The considerations of this section start on October 14 at 18 UTC, when, although the ocean waters surrounding began to cool, Typhoon Wipha (located around 26—27°N latitude) showed a specific fluctuation in intensity: As seen in Fig. 3.51A, at October 15 00 UTC the cyclone system had weakened from 942 to 958 hPa MSLP, while the wind speed had increased (at the *black arrow*). In order to study the related evolution of upper-level flow pattern, Fig. 3.55 shows WV images of MTSAT overlaid by NWP model analyses of wind velocity at 250 hPa isobaric surface. As seen in Fig. 3.55A, this episode was related to the approaching of a subtropical upper-level jet to Wipha from the northwest (seen as a WV moisture boundary at the *red arrow*). Due to the following interaction with the jet, increasing of upper-level divergence and enhancement of the outflow poleward of the storm occurred. Comparing images in Fig. 3.55, the NWP analysis shows increasing outflow jet from 90 to 110 kt (at the *yellow arrow*).

(A) **(B)**

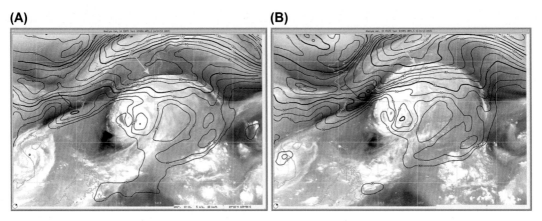

FIGURE 3.55

A jet stream (at the *red arrow*) interacting with Typhoon Wipha. Water vapor images of MTSAT overlaid by ARPEGE analyses of wind velocity at 250 hPa isobaric surface (only ≥25 kt) on (A) October 14, 2013, at 1800 UTC and (B) October 15, 2013, at 0000 UTC.

The sequence of IR images between 23 UTC on October 14 and at 01 UTC on October 15 in Fig. 3.56 shows features that are signatures for a TC intensification as a result of this TC−jet interaction:

- Reforming of the TC eye: Wipha again formed a clear eye, seen in the sequence of satellite images.
- Strengthening of the convection at the polar side of the TC and elongation of the spiral band to the northeast due to increasing the upper-level divergent flow at the jet exit.
- Convection in TC is organized into more distinct "spiral bands."

Although there exist different mechanisms by which the TC eye forms (see, eg, Smith, 1980; Shapiro and Willoughby, 1982), the formation of an eye is almost always an indicator of increasing TC organization and strength. Because of this, forecasters watch developing storms closely for signs of eye formation. An intense TC has an eye visible in satellite imagery and $V_{max} >45$ m s^{-1} as defined by Météo-France Reunion (see Chang-Seng and Jury, 2010). This is confirmed by the data provided by NOAA NESDIS Center for Satellite Applications and Research that show maximum wind stronger than 90 kt as seen in Fig. 3.51A (at the *black arrow*).

The presence of a jet and its locally depressed inertial stability associated with the horizontal anticyclonic shear provides a preferred outflow path. The outflow is ventilated through that path so that further radiation cooling reduces the energy drain of forced subsidence against buoyancy. A cloud-free eye formation in Fig. 3.56B may be due to a combination of dynamically forced centrifuging of mass out of the eye into the eyewall at upper level and to a forced descent caused by the moist convection of the eyewall.

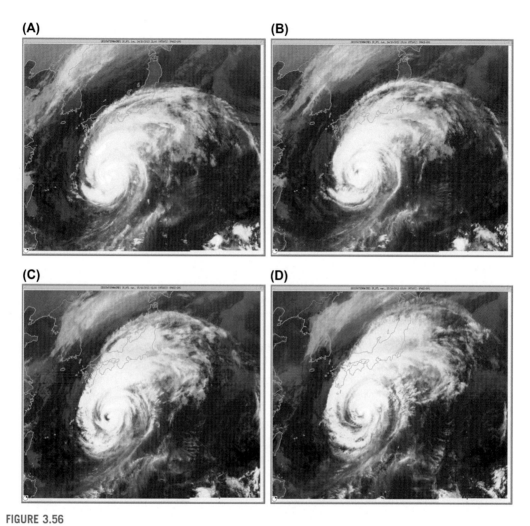

FIGURE 3.56

Intensification of Typhoon Wipha as seen in infrared images of MTSAT on October 14 at (A) 2100 UTC and (B) 2300 UTC and on October 15, 2013, at (C) 0100 UTC and (D) 0300 UTC.

Additional positive factors for the TC intensification can be associated with the passage of the storm across an area of anomalously warm water (Bosart et al., 2000). Fields of SST in the Wipha environment derived by the satellite measurement by WindSat Spaceborne Polarimetric Microwave Radiometer (not presented) are available from the US Naval Research Laboratory (at http://www.nrlmry.navy.mil/tcdat/tc13/WPAC/25W.WIPHA/windsat/sst/). They show that the tropical storm goes through areas of decreasing SSTs in the interval 28−25°C from October 14 2107 UTC to October 15 0823 UTC. Therefore the specific positive fluctuation of the storm intensity on October 15 around 0000 UTC is due to upper-level forcing in the subtropical latitudes.

The WV channel imagery may be useful in the forecasting process to assess elements of the TC—jet couplet and possibility for the TC intensification, as follows:

- The strength of the divergence, associated with the developing outflow channel poleward of the system and with the secondary circulation of the jet-entrance region.
- Strengthening of the outflow jet on the TC polar side as a result of TC—jet interaction leading to elongation of the spiral band to the northeast (in the Northern Hemisphere).
- The position of the approaching jet in regard to the TC.

3.6.3 EFFECTS OF INTERACTION WITH MIDLATITUDE UPPER-LEVEL TROUGHS ON THE INTENSITY OF TROPICAL CYCLONES

Interactions with midlatitude upper-level troughs and associated PV anomalies can strongly affect the intensity and structure of the TCs. Upper-level troughs (or cut-off lows) that interact with TCs usually result from the breaking of a planetary Rossby wave train originating from the midlatitudes that frequently propagates equatorward into the subtropical and tropical latitudes (Leroux et al., 2013). In many cases a TC intensifies in association with the approaching of an upper tropospheric trough and related cyclonic PV anomaly. However an approaching jet or trough can have contradictory effects on the environmental ingredients that are crucial for TC intensification. On the one hand, it may induce significant vertical wind shear, which is usually detrimental to TC intensity. On the other hand, it may increase upper-level divergence and enhance outflow poleward of the storm as well as import cyclonic eddy angular momentum (Molinari et al., 1998), all positive contributions to TC intensification. PV advection associated with a smaller-scale cyclonic PV anomaly can contribute to storm development by forcing subsynoptic-scale ascent over part of the storm (Bosart et al., 2000). Therefore, the configuration and scale of troughs, involved in a TC—PV interaction, relative to the overall upper-level flow and existing low-level disturbances and their evolution are critical to the forecast process.

From a PV perspective, TC-trough interaction can be associated with PV advection from the trough down in the troposphere toward the TC, which is beneficial below the level of the outflow anticyclone ("PV superposition principle"; Molinari et al., 1998). The triggering of convection by upward velocities associated with the PV anomaly, as well as dry air intrusion from the low stratosphere, are other interesting processes that might have a role to play in TC-trough interactions. Leroux et al. (2013) show that the rapid intensification of TC Dora (developed in 2007) in the Southwest Indian Ocean has occured under upper-level trough forcing providing PV coherent structures that interact with the storm. By using numerical simulations, they show that PV is advected toward the TC core with the main tropopause fold extended down to 500 hPa, while the TC upper-level flow turns cyclonically from the continuous import of angular momentum. Considering azimuthal cross-sections, Leroux et al. (2013) show that the PV coherent structure associated with the trough is highly tilted toward the equator. In effect, while the strongest PV values associated with the main stratospheric intrusion do not progress much farther toward the TC core at upper levels, the folding below reaches the storm center. A PV anomaly of smaller size and amplitude detaches from the main trough at mid-levels and is advected toward the storm, feeding its core with cyclonic PV. The stratospheric intrusion of cyclonic PV was associated with a large zone of dry air that gradually encircled the western side of the TC, according to the satellite WV channel imagery.

A critical forecast issue is whether there are any identifiable parameters that can be used to determine whether an upper-tropospheric trough will weaken or strengthen a TC. Although every case will be different, the scale-matching concept mentioned with the results of Hanley et al. (2001) appears to be fundamental to the forecast. According to the assertion of many authors, a good trough should be typified by a fractured PV anomaly that is comparable in scale to the size of the tropical storm and much smaller than the scale of the original trough from which it fractured (see Bosart et al., 2000). Studying all named Atlantic TCs between 1985 and 1996, Hanley et al. (2001) suggest that if a numerical weather prediction model shows the approach of an upper PV maximum of a scale similar to that of a TC, and SST is not subcritical, the TC is likely to intensify as superposition begins to occur. Particular attention needs to be paid to troughs that show evidence of fracturing from the midlatitude westerlies especially when the associated PV coherent structures enter the tropics and interact with storms there. Considering such a case, Leroux et al. (2013) show that the main trough and associated jet stay at a distance greater than 500 km from the TC core at upper levels, so that the interaction benefits from large upper-level divergence with less vertical wind shear at upper levels, where it has been shown to be the most detrimental.

The WV channel imagery as an observational tool for upper-level PV analysis can provide information to help anticipate a possible intensification of the tropical storm due to PV−TC interaction in the view of the considerations of Bosart et al. (2000) and Hanley et al. (2001), as follows:

- The distance between the PV anomaly (upper-level trough) and TC as a measure of the strength of trough interaction.
- The scale of the PV anomaly feature regarding the scale of the TC.
- The strength of the tropopause folding down in the troposphere and associated dry air intrusion from the low stratosphere.

From an operational forecasting perspective, the effect of PV−TC interaction can be considered from two points of view:

- Interaction between a PV anomaly and a TC in the lower latitudes and the effects on the intensity of the tropical storm that has been broadly studied and presented in the literature as considered in this section.
- Interaction between a PV anomaly and a TC in the subtropical latitudes and the effects associated with the extratropical transition (ET) of the tropical storm that will be considered in Section 3.6.4.

3.6.4 ROLE OF THE TROPICAL CYCLONE IN AN EXTRATROPICAL DEVELOPMENT ASSOCIATED WITH AN UPSTREAM UPPER-LEVEL CYCLONIC POTENTIAL VORTICITY ANOMALY

In the higher latitudes, the PV−TC interaction controls the ET, a gradual process in which a storm loses tropical characteristics and becomes more extratropical in nature. In Jones et al. (2003) a brief climatology of ET is given and the challenges associated with forecasting ET are described in terms of the forecast variables (track, intensity, surface winds, precipitation) and their impacts (flooding, bush fires, ocean response).

In the case of Typhoon Wipha, shortly after 06000 UTC on October 15, 2015, the TC rapidly enfeebled as the ocean waters surrounding the typhoon continued to cool (lower than 25°C on October 15 2023 UTC, according to the records of US Naval Research Laboratory). In addition to the decreasing of SST down to subcritical values, the weakening of Wipha and its transitioning to an extratropical system were forced by a subsequent interaction with an approaching upper-level trough and related cyclonic PV anomaly well seen in the WV image as a surge dry zone (marked T in Fig. 3.52F). The evolution of Wipha may be considered as a manifestation of the usual synoptic development when a TC leaves the tropics and encounters interaction with an upper-level trough approaching subtropical latitudes: A jet stream approach may have influenced the TC organization and strength seen in the satellite imagery. The induced strong anticyclonic horizontal shear near a TC core is often followed by significant increases in vertical shear associated with an approaching PV anomaly.

As discussed in Jones et al. (2003), changes in the structure of a system as it evolves from a tropical to an extratropical cyclone during ET necessitate changes in forecast strategies. From the operational forecasting perspective, in cases of upper-level PV anomaly advection in a TC environment associated with a mobile synoptic-scale trough over the subtropical latitudes, the following two issues are important:

- After a transformation period during which the TC loses its convective core, it may develop a frontal structure and progressively undergo a transition to an extratropical cyclone. During this stage, advection of an upper-level PV anomaly can influence and further control the process of ET. This issue will be considered in Section 5.5.1.2, where PV inversion method coupled to the French ARPEGE model is used to demonstrate the role of PV−TC interaction in the case of the ET of Typhoon Wipha.
- The interaction with a TC that a mobile synoptic-scale multitude trough experiences as it moves into the subtropical latitudes can result in the decay or intensification of the related baroclinic cyclone, in conjunction with the ET of the tropical storm (see Agustí-Panareda et al., 2004). This section will focus on the extent to which a TC evolution may be responsible for the explosive extratropical development.

Based on a review of modeling studies and theoretical considerations, Jones et al. (2003) provide a broad idea of the PV structure of a TC as well as knowledge on the nature of the PV anomalies involved in the ET. Fig. 3.57 illustrates the PV perspective of interaction between a TC and an upper-level trough in the first day of the transitioning of the storm Wipha from a tropical into an extratropical system. The WV imagery in Fig. 3.57A and C shows that a characteristic large dark zone of dry air gradually encircled the western side of TC Wipha, which reached extratropical status late on October 16, 2013. The vertical cross-sections of PV in Fig. 3.57B and D show how the upper-level trough and related cyclonic PV anomaly interact with the TC Wipha.

The storm had already began to encounter increasing vertical wind shear and cooler ocean temperatures, conditions not conducive to storm development. At the same time, the intensification of Wipha late on October 14, 2013 (see Section 3.6.2), contributed to its ability to survive the sheared midlatitude environment (related to the approaching PV anomaly) until it had completed the subsequent ET: As the system interacted with a subtropical upper-level jet, this helped to offset the cooler SSTs (as discussed earlier) and maintain higher intensities.

By the time of Fig. 3.57A, the outflow-layer anticyclonic forcing has strengthened the jet and at the same time kept the jet core far enough to the northeast from the storm. Consistent with the numerical simulations of Rappin et al. (2011), this kept the wind shear over the storm core weak enough to be still not detrimental to the storm. Another important consequence of the outflow channel that forms in the presence of the jet is the generation of a downstream anticyclone circulation, as seen in Fig. 3.57A and C by the maximum wind vectors as well as the anticyclonic curvature of the moisture pattern of the WV image in this area (see also Fig. 3.60A).

The vertical cross-section in Fig. 3.57B shows the PV anomalies associated with TC: a positive PV tower (T) and the upper-level negative PV anomaly (N), depicted as an area of PV <0.5 PVU. This

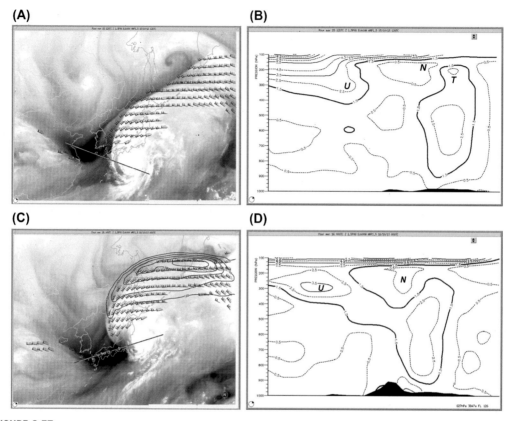

FIGURE 3.57

Water vapor image of MTSAT overlaid by wind vectors at 1.5 PVU surface (only ≥90 kt) for (A) October 15 at 1200 UTC and (C) October 16 at 0000 UTC. (B) and (D) are vertical cross-sections of potential vorticity (PV) across the lines in (A) and (C), respectively. The PV anomalies associated with a tropical cyclone are a positive PV tower "*T*" and an upper-level negative PV anomaly "*N*"; "*U*" indicates the upper-level cyclonic PV anomaly.

negative PV anomaly is associated with the cyclone's outflow with the jet on the eastern flank of the upper-level positive PV anomaly (U), associated with the upper-level trough. The positive PV tower and negative PV anomaly in the upper-level outflow are consequences of the moist convection organized by the cyclone-scale vortex, which characterizes TCs.

Fig. 3.57B and D shows the advection of cyclonic PV, associated with the tropopause folding, which progressively extends down to 500 hPa. While the strongest PV values associated do not progress much farther toward the TC core at upper levels, the folding in the middle troposphere approaches the storm center (Fig. 3.57D). Inspection of the cross-sections confirms that the PV coherent structure associated with the trough is highly tilted toward the storm, feeding its core with cyclonic PV.

Agustí-Panareda et al. (2005) hypothesized that there could be a link between the role of the TC in the extratropical development and the large-scale barotropic shear of the extratropical environment that leads to a different baroclinic life cycle. The concept is shown in Fig. 3.58 and can be considered to illustrate the usefulness of WV imagery as a tool for jet stream and PV analyses that can help in generalizing the large case-to-case variability that exists in ET events. Here the barotropic shear is not relative to the mean zonal jet—as in the idealized experiments of baroclinic life cycles—but to the jet streams that flank planetary waves.

The synoptic situation has been characterized in terms of baroclinic life cycles (labeled as "LC1," "LC2," and "LC3" in Agustí-Panareda et al. [2005] after results and figures of other authors) that depend on the large-scale barotropic shear: zero, positive (ie, cyclonic), or negative (ie, anticyclonic), as follows:

- The life cycle type LC1 is associated with a smaller-scale cyclonic wrapping-up near the surface cyclone and a thinning of the upper-level positive PV equatorward of the cyclone. In this case, as the cyclonic PV anomaly evolves, part of it enters a region of anticyclonic shear equatorward of the jets and part of it remains in the region of cyclonic shear poleward of the polar and subtropical jets at 300 and 200 hPa.
- Type LC2 of baroclinic cyclones is a result of a deep equatorward propagation of the upper-level positive PV anomaly in a large-scale cyclonic wave. LC2 is characterized by a cyclonic wrapping-up of the upper-level PV anomaly and results in broad and deep surface cyclones. In this case, the positive PV anomaly remains on the poleward side of the polar and subtropical jets throughout the life cycle.
- The life cycle type LC3 of baroclinic cyclones is characterized by a dramatic thinning and zonal elongation of the upper-level positive PV anomaly with an insignificant surface cyclonic development along an SCF. In this case, the positive PV anomaly is located equatorward of the strongest (polar) jet to the north.

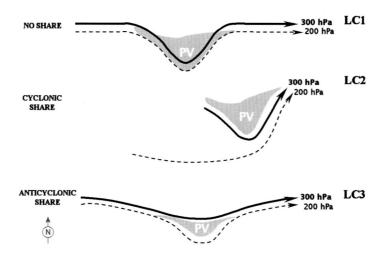

FIGURE 3.58

Schematic showing the planetary wave environment of different baroclinic life cycles at upper levels, considered by Agustí-Panareda et al. (2005), after results and figures of other authors. The relative position of the 300 mb planetary wave and polar jet stream (*solid line with arrow*) and the 200 mb planetary wave and subtropical jet stream (*dashed line with arrow*). The features to the left represent the final shape that the positive potential vorticity (PV) anomaly at upper levels (between 200 and 300 hPa) evolves into at the end of the baroclinic life cycle labeled as "LC1," "LC2," and "LC3" when there is no barotropic shear, cyclonic shear, or anticyclonic shear, respectively. Gray shading depicts PV values greater than 1.5 PVU at upper troposphere.

Agustí-Panareda et al. (2005) proposed a hypothesis of possible scenarios for the intensification stage of a transformed TC that is schematically depicted in Fig. 3.59. In support of such a hypothesis they have shown examples of the different types of development given by a few specific case studies. Some of these studies have been tested by removing the TC from the initial conditions in numerical experiments (based on PV inversion) and others were based on synoptic analyses.

The first factor considered is the interaction with an upstream upper-level positive PV anomaly. According to the concept of Agustí-Panareda et al. (2005), the relative positions of the upper-level positive PV anomaly and the transformed surface cyclone will determine whether the interaction could be significant or not. In the case where there is no interaction with the upper-level positive PV anomaly upstream—because there is no suitably located upper-level positive PV anomaly upstream—the transformed surface cyclone will decay as it is advected downstream subject to the large-scale flow. This would correspond to scenario 1 in Fig. 3.59, and there is no TC contribution.

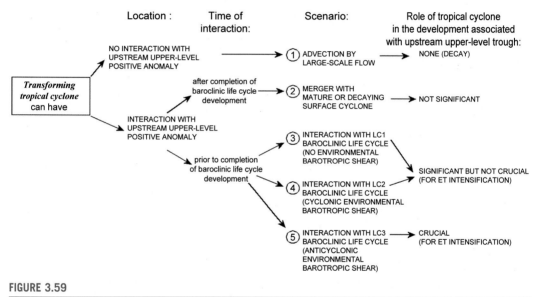

FIGURE 3.59

Schematic showing a hypothesis of possible scenarios for the intensification stage of a transformed tropical cyclone (after Agustí-Panareda et al., 2005). See text for detailed explanation.

When there is an interaction with a preexisting upper-level positive PV anomaly to the west of the cyclone, different outcomes might occur depending on the time of interaction between the transitioning TC and the upper-level cyclonic PV anomaly upstream. If the transformed TC interacts with the upstream trough after its associated upper-level PV anomaly has reached the end of its baroclinic life cycle, then the transitioning cyclone is expected to merge with a broad quasi-barotropic mature low upstream. This corresponds to scenario 2 in Fig. 3.59. The role of the TC is not significant, as the intensity of the resulting cyclone is not significantly affected by the merging of the vortices.

If the upstream upper-level positive PV anomaly is at the beginning of its baroclinic life cycle then there will be an active interaction between the transitioning surface cyclone and the upper-level cyclonic PV anomaly, resulting in intensification (scenarios 3, 4, and 5 in Fig. 3.59). The role of the TC can vary, depending on the environmental barotropic shear that is associated with different baroclinic life cycles, as hypothesized by Agustí-Panareda et al. (2005) and shown in Fig. 3.58:

- LC3 life cycles cannot produce intensification during an ET unless a TC is able to induce a tropopause depression, leading to baroclinic instability. The TC in this case is crucial for the reintensification of the transforming cyclone (see scenario 5 in Fig. 3.58).
- The cases classified as LC2 developments exhibit explosive deepening regardless of the presence of the TC (eg, Agustí-Panareda et al., 2004). Thus, in an environment with cyclonic barotropic shear supporting LC2-type developments, the role of the TC is not thought to be crucial, although it can still be significant to control the process (see scenario 4 in Fig. 3.58). Agustí-Panareda et al. (2004) showed that the outflow of transitioning TC Irene (developed in 1999) played a role in delaying the reintensification period and enhancing the deepening rate of the transitioning cyclone as well as having a strong impact on the extratropical development downstream.
- Given that LC1 life cycles can lead to development of intense cyclones, Agustí-Panareda et al. (2005) speculate that a transforming TC interacting with such a development would also have a secondary role in the ET, that is to say, a significant but not crucial role (see scenario 3 in Fig. 3.58). They also suggest that the ET would have the potential for explosive reintensification and conclude that LC1 life cycles still need to be explored.

The ET scenarios presented in Fig. 3.58 that relate the different baroclinic life cycles to the role of the TC in the development upstream and its potential for intensification are only based on a few case studies (Agustí-Panareda et al., 2005). In addition, there may be other factors that can affect the role of the TC during an ET that have not been considered (eg, the intensity and scale of the TC and upper-level PV anomaly, the advection of moisture by the transitioning TC).

In the section below, it is shown that the hypothesis of Agustí-Panareda et al. (2005) can be a relevant approach in interpretation of WV channel imagery for diagnosis and forecasting ET development. Applying jet stream and PV analyses can help anticipate the ET baroclinic life cycle development and possible ET intensification in the view of the schematic considerations of Agustí-Panareda et al. (2005). Accordingly, the WV imagery, as an observational tool for upper-level diagnosis, can provide duly useful information on the following elements of the synoptic development:

- The evolution of the upper-level trough and the shape of the associated upper-level PV anomaly in order to assess if this cyclonic PV anomaly is in the beginning of its baroclinic life cycle or if it has reached the end of its baroclinic life cycle.
- The relative position of the 300 mb planetary wave (polar jet stream) and the 200 mb planetary wave (subtropical jet stream) that determines the sign of the large-scale barotropic shear.
- The location of the positive PV anomaly at upper levels as well as the barotropic shear that the positive PV anomaly experiences in the end of its baroclinic life cycle.

Fig. 3.60 shows details of the upper-level circulation pattern, surrounding the studied case Wipha at the beginning of its ET as seen by WV imagery, PV, and jet stream analyses. According to the classification of Agustí-Panareda et al. (2005), the extratropical environment of TC Wipha is categorized as class LC2 in the scheme in Fig. 3.58, because of the following considerations:

- Leftward wind shift from 300 to 200 hPa seen in the cross-section of the wind in Fig. 3.60B and therefore a cyclonic large-scale barotropic shear.

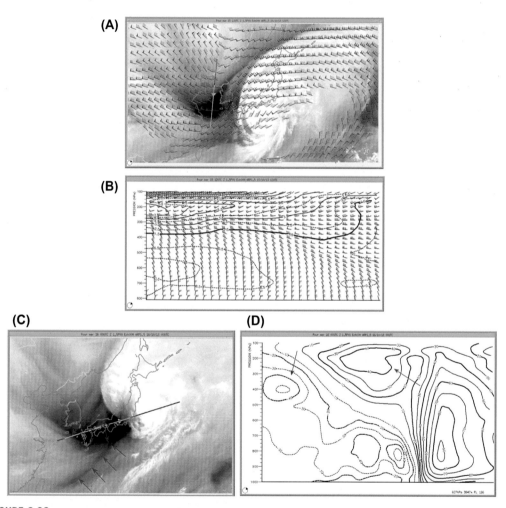

FIGURE 3.60

Water vapor image of MTSAT (A) overlaid by wind vectors at 1.5 PVU surface (only ≥90 kt) on October 15 at 1200 UTC; (B) vertical cross-sections of potential vorticity (blue contours) and wind vectors (*red arrows*) normal to the line in (A); (C) water vapor image on October 16 at 0000 UTC; and (D) vertical cross-section of wind speed normal to the line in (C). In (C) the *red arrows* indicate the moisture boundary, associated with the jet at 300 hPa (*red arrow* in [D]), while the *magenta arrows* indicate the moisture boundary of the jet at 200 hPa (at the *magenta arrow* in [D]).

- This leftward wind shift is confirmed by the location of the moisture boundaries of the polar and subtropical jets seen by WV imagery analysis in Fig. 3.60C along the locations of the *red and magenta arrows* in Fig. 3.60D, respectively.
- The cyclonic anomaly of PV (blue contours in Fig. 3.60B) is located on the poleward side of WV moisture boundaries of the polar and subtropical jets in Fig. 3.60A.

Fig. 3.61 shows the ET of Typhoon Wipha as seen in the IR satellite images from MTSAT. The development is quite similar to that of Hurricane Irene (Agustí-Panareda et al., 2004). They both evolved into LC2-type evolutions, after PV anomaly—TC interaction characterized by large-scale barotropic shear (between 200 and 300 hPa). Accordingly, in the LC2-type extratropical development of Wipha, there is an interaction of the negative PV anomaly (labeled "*N*" in Fig. 3.57D), associated with the cyclone's outflow, with the upper-level jet (at the *magenta arrow* in Fig. 3.60C and D) on the eastern flank of the upper-level positive PV anomaly (labeled "*U*" in Fig. 3.57D).

(A)   **(B)**

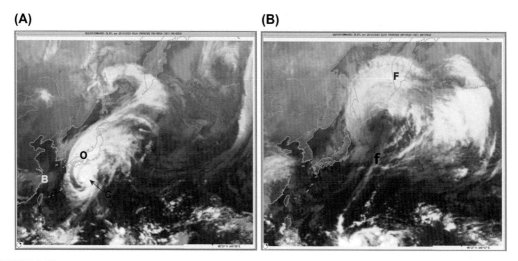

FIGURE 3.61

Extratropical transition of Typhoon Wipha. Infrared satellite images from MTSAT: (A) October 15 at 0900 UTC and (B) October 16 at 1200 UTC. As in Fig. 3.6 of Agustí-Panareda et al. (2004). "B" is the baroclinic zone, "C" is the tropical cyclone's convective core, "O" is the tropical cyclone's outflow, and "F" and "f" are the cloud systems of the warm front and the cold front, respectively.

Similar to Hurricane Irene, Typhoon Wipha was a case of an "explosive" ET, which can experience a TC resulting in a rapid deepening. This is confirmed by the superposition of WV image and NWP analysis of MSLP in Fig. 3.62A showing a distinct image dry slot in the rear side of a very large and deep extratropical cyclone. Fig. 3.62 also shows vertical cross-sections of wind, PV, and wet-bulb potential temperature in the beginning of the extratropical cyclone development. It illustrates that the extratropical system carries its tropical origin and transports warm, moist tropical air to high latitudes. This is evident through specific signatures in terms of wind, wet-bulb potential temperature, and PV:

- The strong southerly jet at the low level (>100 kt, equivalently 50 m s^{-1} down to 850 hPa level seen in Fig. 3.62B) advects very high wet-bulb potential temperature air of tropical origin equatorward to the cold front (indicated as "f" in Fig. 3.61B). This causes a strong horizontal gradient of equivalent potential temperature (see Fig. 3.62D).

- The low-level circulation of the developing extratropical cyclone is associated with the surface thermal anomaly and the low-level cyclonic PV along the front (at the *green arrow* in Fig. 3.62C).

The extratropical cyclogenesis involves vertical interactions between the upper-level cyclonic PV anomaly (Fig. 3.62C at the *black arrow*) associated with the upper-level trough and both the low-level diabatically generated PV anomaly (Fig. 3.62C at the *green arrow*) and the surface thermal anomaly (Fig. 3.62D) along the baroclinic zone (strong horizontal θ_w gradient). The upper-level positive PV anomaly tends then to confine due to enhanced divergent flow (as in Agustí-Panareda et al., 2004). This in turn slows down the cyclonic rotation of the upper-level trough axis and hinders vertical interaction between upper-level and lower-level positive PV anomalies.

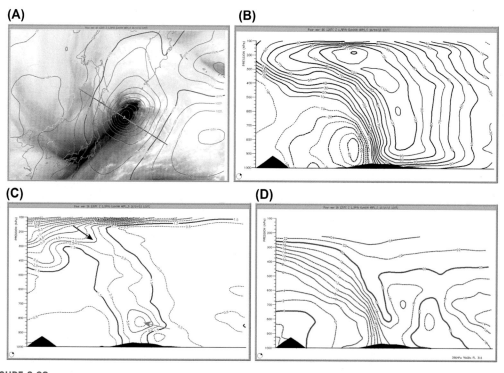

FIGURE 3.62

Extratropical cyclogenesis on October 16 at 1200 UTC after Typhoon Wipha extratropical transition as seen in (A) water vapor image of MTSAT, superimposed by the field of mean sea-level pressure from ARPEGE NWP model. Vertical cross-sections along the line in (A), of (B) wind speed normal to the cross-section line; (C) potential vorticity; and (D) wet-bulb potential temperature.

Agustí-Panareda et al. (2004) studied the extent to which a hurricane LC2-type evolution was responsible for the explosive extratropical development by using Met Office Unified Model to perform forecasts from different initial states with and without the hurricane. The results show that an extratropical cyclogenesis event takes place regardless of the presence of the hurricane in the initial conditions. However, the hurricane makes a significant difference to the track and central mean-sea-level pressure evolution of the resulting extratropical cyclone. When hurricane Irene was present the track of the extratropical cyclone was more zonal and the cyclone deepening rate was twice as fast as when Irene was not present. These effects appear to be particularly associated with a negative potential vorticity anomaly and enhanced divergent flow in the region of the upper-level outflow of the transforming hurricane rather than with the hurricane vortex.

In the case of Wipha, Figs. 3.60D and 3.62B show an enhancement of the upper-level jet. This can be diagnosed by the enhancement of the gray shade distinction of the jet moisture boundary in the WV imagery (Figs. 3.60C and 3.62A). In effect, an enhanced development downstream occurred. As a result, the extratropical cyclone Wipha moving east-northeastward crossed the Pacific Ocean and entered into the Bering Sea, where it began lashing Alaska on October 17 and 18 with strong winds, unseasonably warm temperatures, and heavy rainfall (NASA, 2013).

3.7 SUMMARY

The material considered in this chapter includes different representative features of circulation systems over midlatitude, subtropical, and tropical areas. The case studies presented show the power of the images in the WV channels as observational information for diagnosing synoptic ingredients of typical atmospheric situations. By joint interpretation of WV imagery and dynamical fields, following the principles summarized in this section, operational meteorologists can identify crucial elements responsible for strong development leading to severe weather. The usefulness of such an approach for improving short-range forecasts in an operational environment will be further considered in Chapters 4 and 5.

3.7.1 BASIC PRINCIPLES IN WATER VAPOR IMAGERY INTERPRETATION

Several basic principles regarding the interpretation of WV imagery for synoptic-scale applications may be extracted from the considerations presented in this chapter.

Light and dark areas of gray shades in the 6.2 μm channel image are associated with mid- to upper-tropospheric moist and dry air, respectively.

- Lightening of an imagery feature is a signature of ascent, whereas darkening denotes descending motion.
- The boundaries between synoptic-scale light (moist) and dark (dry) regions are often related to significant upper-level flow patterns.
- The evolution of the boundaries as well as any tendency of interaction between light and dark features in a sequence (or animation) of images indicates the development of important dynamical processes.

3.7.2 LIGHT WATER VAPOR IMAGERY PATTERNS: RELATION TO DYNAMICAL STRUCTURES

Light-gray shade patterns on WV imagery can be associated with various mid- to upper-level dynamical features, including the following:

- Ascending motions (Section 3.2.1).
- High geopotential heights of the dynamical tropopause (Section 3.2.1).
- Planetary waves and undulations of the large-scale circulation with large light convex bands at the top of the ridges (Section 3.3.1).
- The formation of a leaf white/light pattern in the imagery may be a precursor of surface cyclogenesis (Section 3.5.2).

3.7.3 DARK WATER VAPOR IMAGERY PATTERNS: RELATION TO DYNAMICAL STRUCTURES

Dark-gray shade patterns on the 6.2 μm WV imagery may be associated with various mid- to upper-level dynamical features, including the following:

- Descending motions (Section 3.2.2), particularly when darkening occurs.
- Latent tropopause anomalies and low geopotential heights of the dynamical tropopause (Section 3.2.2).
- Dynamical tropopause anomalies that are upper-level precursors of cyclogenesis (Section 3.5.3).
- Dry intrusion regions associated with high PV in parts of them and low θ_w in other parts (Section 3.2.2).
- Monitoring dry intrusions by WV imagery is a way of identifying upper-level dynamical forcing of cyclone developments. The dry slot appearance in the WV imagery is associated with the onset of significant cyclogenesis (Section 3.5.3).

3.7.4 BOUNDARY PATTERNS ON THE WATER VAPOR IMAGERY: RELATION TO DYNAMICAL STRUCTURES

The distinct transition zones from dark (dry) to light (moist) regions on the WV images may be related to various dynamical structures, including the following:

- A pattern of sharp upper-level boundary in 6.2 μm imagery between elongated synoptic-scale light and dark features is related to a jet streak, with the dry air (dark area) on the polar side (Sections 3.2.2, 3.2.3, and 3.3.2).
- Typical boundary patterns ("inside" and "head" boundaries) are associated with blocking regime formation as a result of anticyclogenesis/cyclogenesis (Sections 3.4.1 and 3.4.2).
- A large-scale boundary exists between different moisture regimes on the forward side of upper-level troughs. Its evolution provides useful information regarding the changes of the atmospheric circulation, and any undulations along this boundary may result in wave development associated with the appearance of a baroclinic leaf (Section 3.5.2).

- The baroclinic leaf boundary is associated with maximum winds at the upper troposphere (Section 3.5.2) along its moist (light-gray shade) side.
- The mid-level moisture boundary seen in 7.3 μm imagery is associated with an MLJ and a low-level baroclinic zone (Sections 3.2.3 and 3.3.4).
- Boundaries of large-scale moisture movements in 7.3 μm images are associated with conditions favorable for convection (Section 3.3.4).
- The kata-cold front system is identified by three distinctly seen moisture boundaries in 7.3 μm images, being warmer at the rear side of the SCF and colder at the forward side of the UCF downstream (Sections 3.5.5).

3.7.5 INTERACTION/EVOLUTION OF WATER VAPOR IMAGERY FEATURES: RELATION TO DYNAMICAL PROCESSES

The interaction between dynamical WV imagery features may be related to dynamical processes, including the following:

- A jet streak forms by interaction between the jet stream (large-scale dark/light boundary, usually not well pronounced) and a dynamical tropopause anomaly (WV image dark zone; Section 3.3.2). The appearance of the jet streak is associated with sharpening of the dark/light boundary.
- An upper-level PV anomaly acts as a driving force of cyclogenesis, which subsequently occurs when a polar jet stream and a low-level warm anomaly interact: The dynamical tropopause anomaly (WV image dark spot) approaches a baroclinic zone (an undulating wide light band), and the two WV features merge, contributing to a cloud head formation (Section 3.5.3).
- Upper-level divergent flow (detected by the movement of cloud and moisture features in the animated 6.2 μm WV images) is a sign of ascending motions in the middle and upper troposphere (Section 3.3.3).
- Significant differences between the specific patterns visible through the WV channels 6.2 and 7.3 μm can be interpreted to distinguish different circulation systems and thermodynamic processes in the upper and middle troposphere and their possible interaction (Sections 3.2.3 and 3.3.4).

3.7.6 UPPER-TROPOSPHERIC FLOW PATTERNS AFFECTING TROPICAL CYCLONE DEVELOPMENT

Interactions with upper-tropospheric flow can strongly affect the intensity and structure of the TCs, as follows:

- Whether or not a TC reaches its upper bound of intensity (determined by the SST and the depth of the warm water) depends on the upper-tropospheric environment. In cases of little change in the oceanic influences, diagnosing the potential for the storm intensification or weakening can be performed through WV imagery analysis of the upper-air flow (Section 3.6.1).
  - For intensifying TCs, an anticyclonic circulation pattern is present to the west, and there is no unidirectional flow near the center.

- Regarding the outflow pattern for intensifying TCs, the upper-level outflow exhibits open streamlines, allowing the TC to ventilate itself with the surrounding environmental flow, while for nonintensifying TCs, closed streamlines exist.
- Interaction of an upper-level jet stream boundary in the 6.2 μm channel imagery with a TC leading to storm intensification by increasing upper-level divergence and enhancement of the outflow poleward of the tropical system. The WV channel imagery may be useful in the forecasting process to assess elements of the TC—jet couplet and the possibility for storm intensification (Section 3.6.2):
    - The strength of the divergence, associated with the developing outflow channel poleward of the system and with the secondary circulation of the jet-entrance region.
    - Strengthening of the outflow jet on the TC polar side as a result of TC—jet interaction leading to elongation of the spiral band to the northeast (in the Northern Hemisphere).
    - The position of the approaching jet in regard to the TC.
- In the tropical latitudes, an approach of upper-level cyclonic PV maximum (seen as WV image dark feature) of a scale similar to that of a TC leads to storm intensification as superposition begins to occur and SST is not subcritical (Section 3.6.3).
- Interaction of a TC with dynamical tropopause anomaly (cyclonic WV image dark feature) in the subtropical area is followed by rapid storm decay as strong vertical shear disrupts the TC core and transition to an extratropical system occurs (Section 3.6.4).

3.7.7 SUPERPOSITION OF WATER VAPOR IMAGERY AND DYNAMICAL FIELDS: A TOOL FOR SYNOPTIC-SCALE ANALYSIS

A sequence of WV images superimposed with upper-level dynamical fields derived by NWP models potentially provides dynamical insight into the imagery interpretation and allows observed features of gray shades to be associated with significant synoptic-scale dynamical structures:

- It provides knowledge of the motion field that may help in interpreting WV imagery and in focusing on the possible upper-level PV anomalies.
- It highlights important elements of interaction between significant dynamical features that may be precursors for subsequent developments (Sections 3.3–3.5).
- Knowledge on the upper-air flow pattern from the WV imagery may be combined with information from other sources to confirm or modify the forecast of TC development (Section 3.6).
- Superimposing appropriate NWP model fields on WV images gives warnings of any possible shortcomings of the numerical model in simulating the upper-level circulation (to be considered in Chapter 5).

When using this approach for interpreting WV imagery, the following principles are important:

- Look at an animation of WV images to see changes in the dynamical gray-shade features.
- Superimpose various fields of the forecasting environment onto the image to gain insight into the nature of the PV—WV image relationship, which depends on the synoptic situation (Section 3.3.1, to be further considered in Chapter 5).
- Keep a critical mind when considering the model fields; priority must always be given to the observational data and satellite imagery.

DIAGNOSIS OF THERMODYNAMIC ENVIRONMENT OF DEEP CONVECTION

CHAPTER OUTLINE

Weather Analysis and Forecasting. http://dx.doi.org/10.1016/B978-0-12-800194-3.00004-2

4.1 INTRODUCTION

Convective systems are among the most dangerous atmospheric phenomena, frequently causing serious damages due to flash floods, strong gusts, and hail resulting sometimes in natural disasters. Deep moist convection is a difficult topic as convective systems involve different scales processes; many factors play a role in convective development (mesoscale and synoptic-scale forcing at lower and upper troposphere, orography, urban effects, soil types, and moisture). This leads to a complexity of its forecasting. Chapter 4 concentrates on the aspects of convective development as dynamical forcing and temperature/humidity structures associated with mid- and upper-level patterns that can be identified in water vapor (WV) imagery and potential vorticity (PV) fields. Different authors have stressed specific meso- and large-scale features related to the environment of severe convection that are seen on WV imagery in 6.3/6.7 μm channels (eg, Ellrod, 1990; Thiao et al., 1993; Browning, 1997; Roberts, 2000; Georgiev, 2003; Krennert and Zwatz-Meise, 2003; Rabin et al., 2004; Georgiev and Kozinarova, 2009). Images and advanced products based on data in the 6.2 μm (equally 6.3 and 6.7 μm) channel are efficiently used for upper-level diagnosis (Weldon and Holmes, 1991; Santurette and Georgiev, 2005; Rabin et al., 2004; Schmetz et al., 2005; Georgiev, 2013). Although radiances in the 7.3 μm WV channel measured by geostationary satellites contain information for low- to mid-level moisture distribution, which is a crucial element of the convective environment, there is still a lack of reports regarding the way to use these data in synoptic-scale analyses. As shown in Section 3.3.4, 7.3 μm images are useful to distinguish typical moisture features associated with specific low- to mid-level thermodynamic conditions favorable for convection (Georgiev and Santurette, 2009).

Numerical forecasts have made important progress recently, and mesoscale models do well in predicting strong convection. However, they are far from being perfect, and their performance on a daily use in operational forecasting environment still has to be evaluated by forecasters. The main tool to monitor the numerical production is the satellite imagery. This section is focused on the basic aspects of the atmospheric environment favorable for strong convection and the relevant approach for their interpretation from a dynamical point of view using WV images. It is shown that combining 6.2 and 7.3 μm channels' information in a specific way together with relevant numerical model fields can help to detect areas favorable to imminent deep convective developments and to anticipate the evolution of existing systems. The synoptic development is interpreted in terms of typical circulation patterns that may be clearly identified by imagery in WV channels: These are dynamical tropopause anomalies, jet streams, blocking regime, upper-level divergence, advection of moist air in a deep layer, and vertical displacement of air masses.

4.2 ATMOSPHERIC ENVIRONMENT FAVORABLE FOR DEEP CONVECTION
4.2.1 THE CONVECTIVE INGREDIENTS

Deep, moist convection can develop in many different synoptic environments that have in common the potential for vertical motion in a deep layer up to and above the tropopause. The atmosphere is not spontaneously unstable (especially at midlatitudes), but potentially (or conditionally) unstable, which is not sufficient to produce deep convection. In addition to a potential instability, it calls for ascent of the air to produce a convection outbreak. Numerous authors have shown specific meso- and large-scale features related to the environment of severe convection, but in the domain of operational forecasting the most efficient approach to interpret a synoptic development is the "method of ingredients" as described by Doswell et al. (1996). From the perspective of the ingredients-based approach, the necessary and sufficient "basic" atmospheric conditions for deep moist convection are moisture, potential instability, and ascent of air parcels to their level of free convection (LFC) by some lifting mechanism (Doswell and Bosart, 2001). These convective ingredients are normally present during the warm season when high moisture content is available and buoyant instability promotes strong upward vertical motions. The establishment of a preconvective environment depends largely on the synoptic circulation and especially on the upper-level circulation, which determines the synoptic-scale flow at lower levels and can cause forcing on vertical motion. Studies dealing with intense convection show the importance of synoptic context at midlatitudes and how much the upper-level circulation governs the low-level flow and the possibilities for mesoscale features to set up. As discussed in Doswell (1987), the lift required to raise a parcel to its LFC generally must be supplied by some process operating on subsynoptic scales, because the rising motions associated with synoptic-scale processes usually are too slow to lift a potentially buoyant parcel to its LFC in the required time. However, Browning et al. (2007) presented examples that illustrate forcing from low levels as well as upper-level influences on initiation of convection in midlatitudes that will be further considered in the section below.

When deep, moist convection is already under way, it is obvious that the ingredients are already in place (Doswell et al., 1996). In situations where convection is not happening at forecast time, the forecasters must determine whether or not those ingredients will be in place at some time in the future. This involves assessing the possibility that the missing ingredient will become available, while the other ingredients will remain in place. Existing convection's future evolution should be considered in the same light: The existing convection will continue as long as the ingredients remain present and will cease when one or more of the ingredients is no longer favorable.

Therefore, it is important for the operational forecasters to recognize the characteristic elements of an atmospheric situation, to detect the potential of the situation, and to be aware of which kind of systems could develop. In general terms, as ingredients of an atmospheric phenomenon one can consider all the elements inducing the known physical processes that produce and make the phenomenon evolve. In operational terms, the critical point is to define corresponding characteristic marks, which represent these elements and processes on the relevant parameters and fields available in the operational forecasting environment. Taking into account the considerations of Doswell et al. (1996) and Browning et al. (2007) as main ingredients for strong convective development at mid-latitudes the following processes/characteristics can be defined:

1. Potential instability
2. A layer of convective inhibition, which assists in increasing convective available potential energy (CAPE)
3. Low-level warm and moist air advection: The presence of a low- to mid-level jet (MLJ) is an aggravating factor
4. Low-level convergence flow
5. Upper-level divergence (in some specific contexts of pronounced instability, a neutral atmosphere—without divergence but not convergence either—can be sufficient)
6. A cold and dry mid-upper troposphere advection helps to increase instability (and is favorable to maintain the convective system)
7. Vertical wind shear

Potential instability (that combines moisture and instability in one measure) occurs when the wet-bulb potential temperature (θ_w) decreases with height, that is, when:

$$\partial\theta\omega/\partial z < 0$$

This usually occurs when dry air overlays moist air, but this is not the only condition for convection to develop. Typically, deep convection takes place where sufficient ascent occurs in the areas, which are potentially unstable, and identifying precisely where the convective development can possibly break through is an important point for prediction.

A characteristic feature of the atmosphere in situations leading to the outbreak of deep convection is the stable-layer phenomenon referred to in Browning et al. (2007) as a lid. A low-level layer of warm, dry air can trap air of high θ_w, usually in the boundary layer, beneath potentially colder air in the middle and upper troposphere. Such low-level stable layers give rise to so-called convective inhibition (CIN), and they act as lids that tend to inhibit the onset of deep convection. Such a lid (Browning et al., 2007) assists in the buildup of latent instability by allowing warm, moist air to be bottled up at low levels, thereby increasing CAPE. However, to realize the potential for deep convection, it is necessary for the low-level air eventually to be able to penetrate the lid. The site of the initial outbreak is influenced partly by any spatial variability in the temperature and/or humidity that allows the anticipation of convection initiation at the zones of moisture convergence. Experience has shown, however, that the moisture convergence is not always the mechanism for the initial outbreak for deep convection. Sometimes the variability is due to the effects of variable terrain height or differing land (or sea) surface characteristics, or differential heating along discontinuity lines in fields of moisture/cloudiness that are important for nonsynoptically forced environments. Lifting can also result from ascent induced via upper-level forcing beneath an advancing dynamical tropopause anomaly (or PV anomaly). Examples that illustrate forcing from low levels as well as upper-level influences on the process of destabilization and initiation of the first outbreak for deep convection are discussed in detail by Browning et al. (2007).

For maintenance of durable strong convective systems, persistent warm and moist air supply is needed to feed the convective cells. We can think of the convective system as a machine for which the warm moist air is the fuel. In the frame of this concept, in addition to the fuel an engine is needed, first to trigger convective motion and then to maintain and favor strong vertical movements through the whole depth of the troposphere. This engine is brought by the dynamics of the circulation pattern, that is to say, by specific configurations of the wind field involving jets, wind shear, and convergence/divergence zones. The low-level convergence of the flow is the main lifting mechanism that allows potentially unstable air parcels to reach their level of free convection, and large-scale upper-level divergent flow produces forcing of ascending motion aloft. The low-level jet is also an element enhancing warm moist advection and acting to strengthen and maintain durable convergence. Therefore, deep convection, especially the strong convective development, is not only determined by instability and moisture; the three-dimensional wind field configuration is a crucial factor as well.

An essential forecast issue is to analyze if a situation carries these ingredients of strong convective development. However, the main ingredients for deep convection can occur in various patterns of the atmospheric environment to produce strong convective systems. For instance, some supercell or tornado cases are not associated with very pronounced instability but with strong helicity (low-level vertical wind shear) and low-level convergence. Some other severe convective cases occur with strong instability and lifting mechanisms with a very weak low-level wind shear. The development of strong and sustainable convective systems requires specific circulation conditions for moist air supply and for the establishment of an environment favorable to the amplification of convective vertical motions.

Accordingly, deep intense convection can develop in many different synoptic situations, which are able to produce the convective ingredients and to combine them efficiently. Therefore, a description of the typical synoptic situations for severe convection might prove to be an elusive goal because the ingredients for development of a convective event can be brought together in an untypical pattern. Hence, more than recognizing a typical synoptic situation favorable to thunderstorms, it is crucial in operational forecasting to be aware of the convective potential of the situation, to keep in mind the ingredients method, and to recognize how convective ingredients can fit together with a favorable circulation pattern. WV imagery reflects the mid- to upper-level moisture flow and offers a relevant tool to diagnose the conditions regarding the upper-troposphere dynamics. Typical thermodynamic structures, which can be monitored by the moisture channel imagery in the context of convection forecasting will be considered in the following sections.

4.2.2 A DYNAMICAL TROPOPAUSE ANOMALY (UPPER-LEVEL CYCLONIC POTENTIAL VORTICITY MAXIMUM) FAVORS DEEP CONVECTION

Strong PV aloft has a great effect on the lower troposphere as seen in Section 1.2.4: It produces a decrease of static stability and an increase of vorticity below the PV anomaly. As shown in Griffith et al. (1998), the tropopause fold acts to increase the generation of potential instability in its vicinity. Then dry/dark zones seen in 6.2 µm WV images, associated with cyclonic PV anomaly aloft, are regions of potential instability. Fig. 4.1 summarizes how a dynamical tropopause anomaly (an upper-troposphere PV maximum) moving in westerly or northwesterly flow contributes to produce a favorable environment for deep convection.

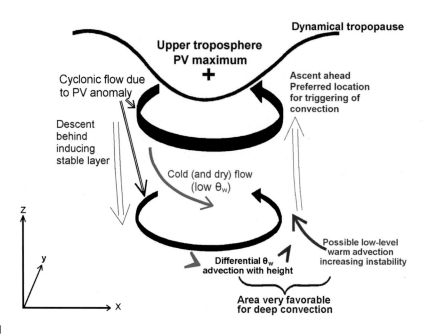

FIGURE 4.1

A schematic representation (in the North Hemisphere) of various processes involved in the establishment of an area very favorable for deep convection due to the arrival of an upper-level potential vorticity maximum (see text).

It is well known that a way of generating potential instability is by the process of differential advection of θ_w, that is, advection low-θ_w air aloft that overlays an area of high-θ_w advection in low levels. This occurs especially when a cyclonic circulation affects a large thickness of the troposphere, as in the case of the presence of a dynamical tropopause anomaly. Fig. 4.1 schematizes the main processes accompanying such an upper-level perturbation related to a development of deep moist convection. First, a dynamical tropopause anomaly is associated with low static stability in the troposphere below it (Section 1.2.4). So, in case of latent tropopause anomaly (ie, no pronounced descending motion in the rear part of the anomaly; see Section 1.3.1), and if favorable conditions at the surface (moist and hot soil) are present, a development of deep convection is possible everywhere below the anomaly. In case of dynamical tropopause anomaly, the following processes dominate to establish a preferable area for deep convection on the forward part of the anomaly:

- The cyclonic circulation induced by the arrival of the upper-level PV anomaly creates/ increases advection of low θ_w in the northwesterly flow (light blue arrow in Fig. 4.1) and high θ_w advection in the southerly flow ahead of the anomaly (red arrow in Fig. 4.1).
- The vertical wind shear ahead of the anomaly can act across this differential advection established in the horizontal plane to create a differential advection in the vertical that increases instability.

- Moreover, dynamical forcing can occur by interaction between the upper-level PV anomaly and a jet streak; this perturbation produces divergence aloft ahead of the PV anomaly (see Section 1.3.2) and then ascending motion (open red arrow in Fig. 4.1) which helps to trigger and maintain convective ascent, and sometimes acts to increase instability in case of dry air aloft (see Section 4.2.3). On the contrary, the pronounced descending motions in the rear part of the anomaly inhibit convective development (see Section 4.4.1). Obviously this dynamical effect can be more or less pronounced according to the jet strength and the PV anomaly magnitude.

A relevant approach based on a joint interpretation of WV imagery and thermodynamic fields to diagnose the most part of these processes involved in deep convection development (dynamical forcing, differential advection of wet-bulb potential temperature with height, destabilization, convective inhibition and moisture flows), is considered in Section 4.4.

4.2.3 DRY AIR ALOFT INCREASES INSTABILITY AND FAVORS CONVECTIVE DEVELOPMENT

Dark zones in WV imagery mark dry air that can be more or less cold at the middle and upper levels. It is important to be aware that the presence of dry air in the middle and/or upper troposphere (even if it is not associated with PV anomaly and not especially cold) is favorable for convective development. Air masses in which moist low-level air is capped by dry layer aloft have been noted to be particularly conducive to the rapid growth of deep convective cells. A convective situation rapidly develops under these conditions if a lifting of the total air column occurs (caused by upper-level forcing or orography, for instance). Because of this lifting, the dry air aloft cools dry-adiabatically and then becomes very cold, while moist air in the low level cools moist-adiabatically, that is, cools less than the upper air. In the lower moist layer, the release of condensation heat helps mitigate the cooling caused by the rising air, whereas the dry-adiabatic lifting of the dry air aloft further reduces the temperature. On the contrary, in cases of very moist air in middle and upper levels (lighter gray-shade zones on the WV imagery), the layers aloft are cooling moist-adiabatically and do not cool more than low-level air.

Fig. 4.2A shows a real profile presenting moist air in the low level topped by dry air aloft. Let us consider a 100 hPa adiabatic lifting of this total air mass; then the new temperature profile becomes the red one in Fig. 4.2B: no change in the low-level temperature where the air mass is very moist (red line) but a clear change of the temperature profile in the layers above 620 hPa (dashed red line) where the air mass is cooling because of the dry air aloft that produces an increase in instability. If we consider a rise of the surface temperature, reaching, for example, about 14°C thanks to the diurnal effect, the surface air becomes instable and can lift adiabatically (green curve in Fig. 4.2B) up to about 650 hPa in the case of no change aloft in the initial profile (black arrow in Fig. 4.2B); however, if this air mass is forced adiabatically to rise (by upper-level forcing or mesoscale lifting by orography), the surface parcels of air will meet colder air in the mid- and upper-level environment (red dashed line) and then will be always in instability—up to 400 hPa in this case (red arrow in Fig. 4.2B). Such dry air aloft increases the potential instability, which is the case in a split cold front (see Section 3.5.5), and can be identified by WV imagery in 6.2 and 7.3 µm channels.

(A)   **(B)**

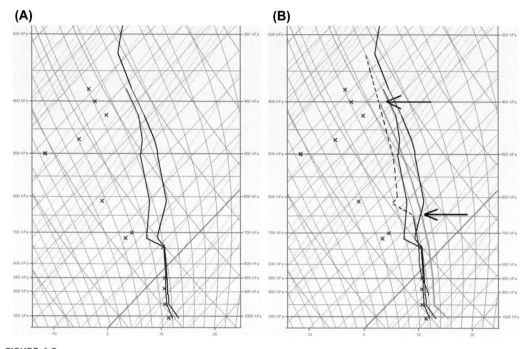

FIGURE 4.2

Real vertical profile in the mid-morning over France (skew *T-log p* diagram) with moisture in low levels and dry air aloft: (A) initial profile (temperature in black, dew point as pink crosses, wet-bulb potential temperature in blue); (B) the same with superimposition of a new temperature *curve* (red) resulting from a 100 hPa adiabatic lifting, and the track of an air parcel at 14°C adiabatically lifted from the surface (light green *curve* in (B)). See text for explanation.

Finally, the less moist the air aloft is, the larger the difference in cooling of the two different layers (the dry one and the moist one) and the larger the increase in the instability. It should also be noted that the presence of mid-level dry layers might suppress convective activity in the case of very dry air able to produce evaporation of the developing cloud. However, when convective clouds develop through such mid-level dry layers, this dry air can favor the formation of intense convective downdrafts that might lead to strong surface wind gusts. Moreover, these gusts can contribute to new cell developments by lifting the surrounding warm air they encounter.

4.2.4 DIVERGENT/CONVERGENT UPPER-LEVEL FLOW AS A POSITIVE/NEGATIVE FACTOR FOR DEEP CONVECTION

Over midlatitudes, divergent flow at the tropopause is a sign of rising motion below. As discussed in Section 3.3.3, this vertical motion can be triggered by ascent due to deep convection in areas without upper-level forcing. Alternatively, the ascent related to diffluent synoptic-scale flow near the jet axis aloft can accelerate the lower-level upward motion and thus favor an intense convective development.

In such situations, the divergent flow in the upper troposphere can be a preconvective positive dynamical factor in the formation of deep convective storms over Europe (Georgiev and Santurette, 2010; Georgiev, 2013).

A divergence-based procedure for diagnosing convective development, assuming a two-layer model of the atmosphere, is discussed by Carroll (1997a). To illustrate this approach, areas of vertical velocity deduced from consideration of upper-level divergence of the ageostrophic wind associated with a westerly jet due to advective (ie, downwind) changes in velocity are shown. The following simplified assumptions are made:

- The increasing pressure gradient force experienced by an air parcel entering the jet dominates over the Coriolis force, which, being proportional to the wind velocity, lags behind until the velocity increases to restore the geostrophic balance.
- At the jet exit, the pressure gradient force decreases while the Coriolis force takes time to decrease to the geostrophic balance.
- The result is a geostrophic departure down the pressure gradient at the jet entrance and up the pressure gradient at the jet exit, and an accompanying divergence pattern.

These considerations concern upper-level ascending motions at the right entrance as well as at the left exit of the jet (in the Northern Hemisphere). Carroll (1997a) pointed out that thinking in terms of vorticity advection and thermal advection is directly equivalent to considering divergence due to advective and isallobaric components of the ageostrophic wind. He argued that such an approach is similar in basis to the quasi-geostrophic omega equation and it is a more suitable approach for subjective application. Consistent with Carroll (1997a), several authors have confirmed that the synoptic-scale divergent (diffluent) upper-level flow is a positive factor (related to the upper-level dynamics) for convective development in a deep layer (Rabin et al., 2004; Santurette and Georgiev, 2005; Ghosh et al., 2008; Hofer et al., 2011). Regarding the operational application of the relationship between upper-level divergence and convective storm occurrence, Rabin et al. (2004) concluded, "Divergence patterns are sometimes useful in diagnosing areas of vertical air motion prior to possible convective development, especially when synoptic forcing and surface fronts are lacking. In addition, the satellite winds capture the divergence that develops as a consequence of storm updrafts and can be useful in diagnosing the extent and intensity of storm clusters." As shown in Section 3.3.3, these two possible applications are based on the two mechanisms, which control the establishment of this relationship associated with absence or presence of upper forcing for vertical motions.

The divergent and convergent synoptic-scale upper-level flow can be operationally detected by following the movement of cloud and moisture features in the animated WV images in the 6.2 μm channel. Another way to diagnose upper-level divergence/convergence and thus to find regions where the vertical air motion and the convective development might be enhanced/suppressed aloft is using fields of divergence, deduced from satellite tracked upper-level Atmospheric Motion Vectors (AMVs). For the coverage of Meteosat Second Generation (MSG) satellites, such a satellite product is the upper-troposphere divergence (DIV) product, derived by WV channel 6.2 μm, that is operationally available at hourly basis (EUMETSAT, 2005).

As favored areas for the initiation of deep moist convection, the transition zones from dark (dry) to light (moist) regions on Meteosat WV images have been identified (Krennert and Zwatz-Meise, 2003; Ghosh et al., 2008). To distinguish the upper-level moisture boundaries associated with favorable convective environments, the corresponding dry stripes seen in the WV imagery are considered with respect to the strength of the upper-level dynamics that is essential in helping to forecast the development of deep convection (see, eg, Georgiev and Kozinarova, 2009). In this context, the Meteosat Production Extraction Facilities (MPEF) DIV product is useful to recognize which of the WV-image moisture boundaries, being related to a convergent upper-troposphere flow, are not favorable for deep convective development. Such moisture boundaries are shown at the blue arrows on the 6.2 μm WV image in Fig. 4.3A.

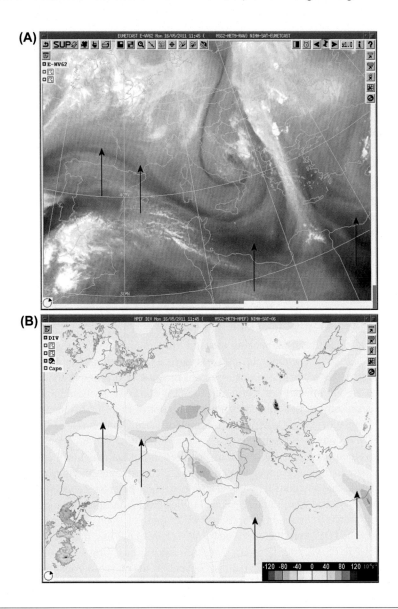

FIGURE 4.3

(A) Water vapor image channel 6.2 μm and (B) MPEF DIV product superimposed by infrared 10.8 μm channel image (only cloud-top BT $< -40°C$) for May 16, 2011, 1145 UTC. Also shown on (B): the color palette for satellite DIV product.

Fig. 4.3B shows a composition of the satellite DIV product overlaid by the infrared (IR) 10.8 µm channel image (only cloud-top BT $<-40°C$ in cyan, BT $<-50°C$ in yellow, and BT $<-60°C$ in bright red) introduced as MSG DIV-IR composite in Section 3.3.3. The divergent values are visualized in bluish shades (interval $20 \times 10^{-6}\,s^{-1}$) as well as in a light green color, showing weak divergence in the range $[0 \text{ to } +20 \times 10^{-6}\,s^{-1}]$. The convergence in the range $[-20 \times 10^{-6}\,s^{-1} \text{ to } 0]$ is not colored, and the convergence values lower than $-20 \times 10^{-6}\,s^{-1}$ are visualized in reddish colors.

Comparing Fig. 4.3A and the corresponding MSG DIV-IR composite in Fig. 4.3B shows the relation between the divergent/convergent upper-troposphere flow and the appearance of deep moist convection.

- Convective cells developed up to the $-60°C$ (IR channel radiance in red) level in areas of upper-level divergence, as seen by the MPEF DIV product.
- There are no deep convective cells with cloud-top temperature lower than $-40°C$ (IR channel radiance in green) at the areas of upper-level convergence.

This is consistent with the work of Rabin et al. (2004), who illustrated examples of the upper-level wind fields deduced from WV imagery for summertime thunderstorm events. They used the atmospheric motion wind vectors to track upper-air features such as jet maxima, as well as divergent regions, where vertical air motion and convective development might be enhanced. They demonstrated how the winds derived from the 6.7 µm WV channel of the Geostationary Operational Environmental Satellite (GOES) can be used to capture features on a more detailed scale than can be obtained from conventional meteorological observations. While distinct upper troughs can be easily identified from the weather balloon network over land, weak disturbances, which can be a factor in thunderstorm formation, are more difficult to detect. Using two case studies over Colorado and Texas, Rabin et al. (2004) further illustrate the existence of rapid convective evolution in areas of upper-level divergence diagnosed from the satellite wind analyses. In each case, low-level forcing mechanisms were lacking (strong convergence was absent at the surface) and the forecast of thunderstorm development was especially difficult. The storms formed near a "dry line" along the western edge of a region of sustained divergence. The divergence, as derived by the satellite WV channel data, was present 4–5 h prior to thunderstorm development. It was expected that the upward air motion associated with this divergence would contribute to weakening of the capping inversion and deepening of the moist boundary layer in a localized region near the dry line. In contrast to the previous day's activity, thunderstorms failed to develop over West Texas on June 12, 2003, despite the presence of similar thermodynamic conditions and low-level forcing mechanisms. In this case, the mid- and upper-level flow was even more zonal (straight west–east) than on the previous day, making identification and tracking of waves in the southern jet particularly difficult. Rabin et al. (2004) explained this failure of convective development by using satellite-derived divergence fields, which indicated a region of sustained upper-level convergence over west Texas in the afternoon and evening of June 12, 2003. It was suggested that the downward air motion associated with the convergence was at least partly responsible for the absence of convective development across the region.

Most of the studies on deep moist convection have identified instability characteristics of the environment (such as those measured in terms of CAPE and instability indexes, moisture distribution via precipitable water content) as well as dynamical forcing parameters (related to low-level jet, upper-level jet, surface convergence lines, vertical motion, and low- to mid-level vertical wind shear). Useful signals for the occurrence of severe weather seem to be provided by a chosen combination of these parameters. For the purposes of nowcasting, convection instability parameters derived by using satellite data (Köenig and de Coning, 2009; Conte et al., 2011) are also useful.

Therefore, in midlatitudes the synoptic-scale vertical motion in the upper troposphere associated with divergent flow is an additional positive factor that can contribute to a deep moist convective development (Ghosh et al., 2008; Hofer et al., 2011). The MPEF DIV product can provide useful information for nowcasting since in many situations divergent areas are present in the images over midlatitudes a few hours before the initiation of deep convective cells, especially in the areas of strong upper-level dynamics near the jet streams as well as in the leading part of an upper tropospheric PV anomaly (see Rabin et al., 2004; Santurette and Georgiev, 2005, 2007). Also, convergent upper-level flow deduced by satellite data is a signal for convection inhibition at the middle and upper-troposphere (see Section 4.4.1).

The concept presented in this section is that before starting a deep convective development it is valuable to diagnose also the upper-level divergence in the areas of instability that can help the forecaster to anticipate where the convective updraft is going to be accelerated or inhibited at the upper level. The possibility for upper-level diagnosis in this context by using the MPEF DIV product is illustrated in Fig. 4.4. Fig. 4.4A shows the MPEF Global Instability Indices (GII) product for the Lifted Index on May 16, 2011, at 1145 UTC derived from an advanced nowcasting tool using satellite data (Köenig and de Coning, 2009). The corresponding MSG DIV-IR composite superimposed onto the nearest analysis of CAPE (black contours) derived from the ARPEGE Numerical Weather Prediction (NWP) model for 1200 UTC is shown in Fig. 4.4B. The satellite and NWP instability products in Fig. 4.4A indicate an area of instability west of the Black Sea (at the black arrow) and another area of instability south of the Black Sea (at the blue arrow). Combining the MPEF DIV product with information for atmospheric instability is a way to determine where the instability and upper-level divergent flow are coupled. Using CAPE from the ARPEGE model in this context, care must be taken that this CAPE parameter is derived only at the areas where the NWP model convective scheme releases instability. CAPE is considered one of the most important parameters related to severe weather forecasting since it includes the effects of the properties of the surface layer and the upper air. Hence, CAPE is a measure of how much the lifted parcel would be positively buoyant in the deep layer that is essential for the occurrence of deep moist convection. However, CAPE provides only a general guide to the strength of convection and has its limitations (Prasad, 2006). Other elements, in particular those related to low-level forcing mechanisms, can be critical to produce convection initiation at a specific location.

Regarding the convective evolution, the results presented in Georgiev (2013) show that 24% of the studied deep convective cells initiated (appeared) in areas of upper-level convergence, while only 8% of these deep convective cells developed in areas of convergence as derived by MPEF DIV product. The fields in Fig. 4.4 are useful to help recognize which one of the areas of tropospheric instability diagnosed by NWP models or satellite data is related to upper-level convergent flow and thus is not favorable for strong deep convection. The fact that the convergent upper-level flow is a negative factor for occurrence of deep convection can be seen in the evolution of the MSG DIV-IR composite in Fig. 4.4 at the position of the blue arrow. The sequence in Fig. 4.4B–D shows a strengthening of the upper-level convergence (observed by the satellite MPEF DIV product, reddish colors at the blue arrows) over the area of instability (predicted by the NWP model and seen in the satellite GII Lifted Index). This convergence can be taken as observational evidence for descending motions in the upper troposphere and hence indicates that there are less favorable conditions to release the predicted convective available potential energy within a deep layer reaching the middle and upper troposphere. Fig. 4.4C shows that in the southern area (location of the blue arrow) very few convective clouds have reached heights of brightness temperature $-40°C$. As discussed in

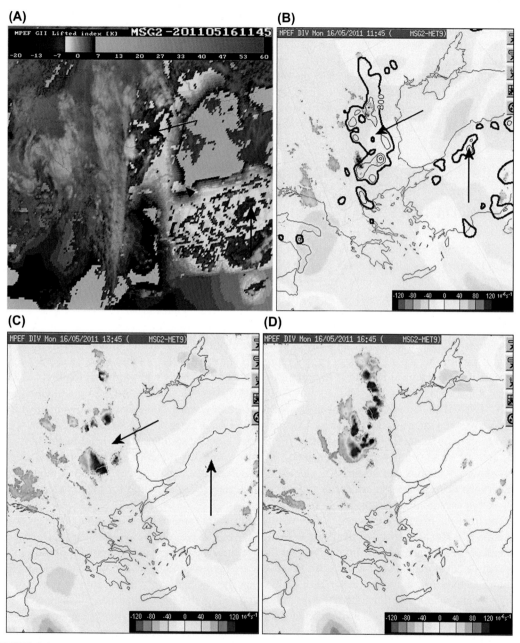

FIGURE 4.4

Joint use of information for tropospheric instability and upper-level divergence on May 16, 2011. (A) MPEF
Global Instability Indices Lifted Index, derived from Meteosat Second Generation (MSG) data for 1145 UTC.
MSG DIV-IR composite of MPEF DIV product and the infrared 10.8 μm channel brightness temperature
(only $<-40°C$); (B) for 1145 UTC, overlaid by nearest ARPEGE analysis of CAPE (black, only ≥ 800, interval
400 J kg^{-1}); (C) for 1345 UTC; and (D) for 1645 UTC. Also shown on (A), (B), and (C) is the color palette for
satellite DIV product.

Georgiev (2013), this threshold (BT = 233 K) has been chosen as a lower limit of cloud visualization in the MSG DIV-IR composite, as this is suitable to depict deep convection over Central Europe and the Mediterranean Sea.

At the same time, a deep convective development occurs in the other area of instability, where preexisting divergence is present (seen by the MPEF DIV product) in the leading part of an upper-level trough (at the black arrow in Fig. 4.4). A large system of deep convective cells with cloud-top brightness temperatures lower than −60°C appeared over this area.

4.3 UPPER-LEVEL DIAGNOSIS OF DEEP CONVECTION
4.3.1 UPPER-LEVEL DYNAMICS FAVORABLE FOR DEEP CONVECTION IN MIDLATITUDES

Most cases of high-impact convective events occur in particular synoptic-scale conditions that are favorable to both produce and bring together the "convective ingredients" described in Doswell et al. (1996). Mesoscale processes may be necessary in some situations to accomplish the final concentration of the ingredients. According to Doswell (1987), convective systems depend primarily on large-scale processes to develop suitable thermodynamic structure, assuming that the primary role of mesoscale processes is to provide the lifting necessary to initiate deep moist convection. At midlatitudes, the forecaster's experience shows that high-impact convective events occur generally in areas where upper-level synoptic disturbance is acting in addition to a conditional instability of the atmosphere. Papers reporting convective case studies show that the role of synoptic scale dynamical forcing is important in the deep convective development (Roberts, 2000; Rabin et al., 2004; Santurette and Georgiev, 2005, 2007; Georgiev, 2013; Stoyanova and Georgiev, 2013). Even if the synoptic vertical motions are relatively weak compared to those involved in convective phenomena as discussed by Doswell (1987) and Doswell and Bosart (2001), the synoptic forcing on ascending motion is often the trigger element to release low-level instability in a majority of strong convective events (see, eg, Browning et al., 2007). Therefore, the upper troposphere dynamics can never be neglected as essential characteristics of the environment that triggers the deep convective ascent or favors small-scale processes acting for the development of convective ascents through the whole troposphere.

To stress the need of upper-level diagnosis in convection forecasting, Figs. 4.5 and 4.6 show an example of a Mesoscale Convective System (MCS) development over the Western Mediterranean at the end of the summer. At this time of the year the Mediterranean region is particularly propitious to convective development (warm sea, specific orography favorable to wind discontinuities producing lifting mechanisms, etc.), and experience shows that the MCSs mostly occur closely linked with upper-level dynamics as illustrated by Figs. 4.6 and 4.7. Fig. 4.6 shows that these V-shape systems develop in areas of upper-level diffluent flow near the left exit of a jet streak, that is, an area of upper air divergence. The typical V-shape is characteristic of a multicellular convective system that regenerates by formation of new cells at the tip of the "V" under a jet-streak exit. This type of MCS is now well known by forecasters due to its potential in producing large accumulated rainfall and has been described by many authors (Fujita, 1978; McCann, 1983; Scofield, 1985; Bader et al., 1995).

(A)   **(B)**

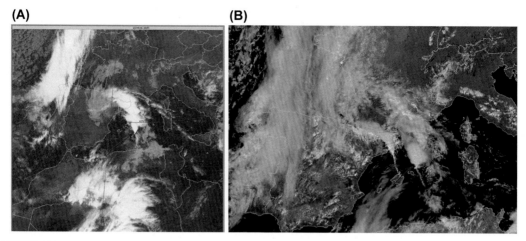

FIGURE 4.5

Meteosat imagery of deep convective systems (V-shape systems, red *arrows*) developing over the Western Mediterranean sea. (A) Infrared image on September 13, 2006, at 0600 UTC; (B) high-resolution visible image at 1245 UTC.

FIGURE 4.6

Infrared image with superimposition of the geopotential (brown, interval 6 dam) and isotachs of the wind (blue, threshold 35 kt) at 300 hPa isobaric surface on September 13, 2006, at 0600 UTC.

In order to confirm the role of upper-level forcing, we can consider a relevant diagnostic output of the DIONYSOS system[1] (see also Pagé et al., 2007) presented in Fig. 4.5. Fig. 4.7A shows the contributions of the main factors to the vertical velocity at the position of the blue cross, indicated on Fig. 4.7B as diagnosed from the omega equation (see, eg, Hoskins et al., 1978). We can see that the vorticity advection (in red) is not negligible in the layer 800—400 hPa; even the latent heat release (in blue) is the most important forcing mechanism.

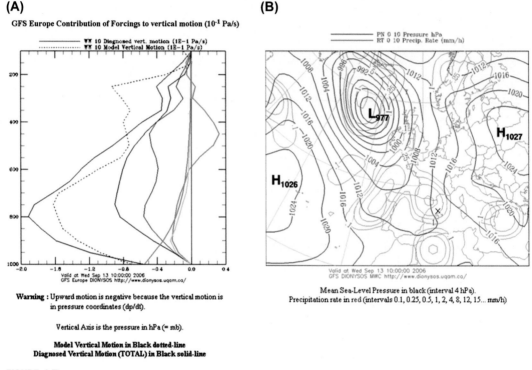

(A)

GFS Europe Contribution of Forcings to vertical motion (10^{-1} Pa/s)

Warning : Upward motion is negative because the vertical motion is in pressure coordinates (dp/dt).

Vertical Axis is the pressure in hPa (= mb).

Model Vertical Motion in Black dotted-line
Diagnosed Vertical Motion (TOTAL) in Black solid-line

(B)

Mean Sea-Level Pressure in black (interval 4 hPa).
Precipitation rate in red (intervals 0.1, 0.25, 0.5, 1, 2, 4, 8, 12, 15... mm/h)

FIGURE 4.7

(A) Diagnostics of contribution of forcing to vertical motion at the position of the blue cross, indicated on (B); (B) mean sea-level pressure and precipitation rate from global forecasting system (GFS) model (the GFS from NOAA, USA; 10 h forecast valid for September 13, 2006, at 1000 UTC). Diagnosed contributions as follows: vorticity advection in red, temperature advection in lime, latent heat release in blue, Sensible heat flux in magenta, friction in orange, orography in green.

Many authors have shown for quite a while that upper-level divergence areas associated with jet streaks were favorable to organized strong convection via their vertical motions and coupled ageostrophic flows (eg, Uccelini and Johnson, 1979; see also Section 4.2.4). However, it is very

[1]DIONYSOS is a diagnostic system specially adapted to standard output from numerical weather prediction models that allows for the interpretation of significant parameters related to weather systems such as development (vorticity, geostrophic vorticity, and height tendencies), vertical motion, divergence, and temperature tendencies.

difficult to detect and follow the crucial features of the upper-level wind fields, that is, the short wave troughs and the jet streak structures' evolution as explained in Section 1.3.2. Following the ideas presented by Hoskins et al. (1985), the passage of a positive (cyclonic) upper-tropospheric PV anomaly (equally a minimum in the height of the constant surface of PV equal to 1.5 PVU) can be accepted as a principal forcing mechanism for convection from upper levels (Browning et al., 2007; see also Section 4.2.2). Based on the PV concept, it has been considered as a good synoptic practice for some years to use maps of the height of the dynamical tropopause in the operational forecasting environment. Many situations in the real atmosphere illustrate the correlation between a dynamical tropopause anomaly interacting with a jet streak and a convective system development.

Using the PV as a key interpretation parameter, Santurette and Argobast (2002) considered the severe convective event, which occurred southeast of France on September 8 and 9, 2002, and led to a catastrophic flash flood. The operational models greatly underestimated the development of this convective weather event, partly due to a deficient simulation of the strong upper-level dynamics. In this context, the quality of the initial states of the global NWP models were assessed by comparing the analyses of the geopotential height of the 1.5 PVU surface with the satellite WV imagery. To diagnose the impact of upper-level dynamics on this convective event over the Western Mediterranean, Santurette and Arbogast (2002) performed numerical experiments with the Météo-France ARPEGE operational forecasting system (Courtier et al., 1991). Appropriate PV modifications were made to the 1.5 PVU surface heights to improve their fit to the satellite WV image and to deepen the area of PV anomaly. The initial three-dimensional distribution of PV in the NWP model was then adjusted to fit these modifications (for details on the technique, see Appendix B).

The modified upper-troposphere PV anomalies were then inverted (in terms of geopotential heights, wind, and temperature fields) and used to directly correct the initial state of the numerical model by using a quasi-geostrophic PV inversion method (Chaigne and Arbogast, 2000) developed at Météo-France. No modifications were made to low levels; the only modifications were those made to the height of the 1.5 PVU surface leading to the new initial state. The new run of the ARPEGE model (which was launched starting from this new initial state) perceptibly improved the forecast. The total rainfall values forecasted by this run for the 12 h period between 0000 and 1200 UTC on September 9 are twice those predicted by the operational run, especially in the most affected area. This experiment (also presented in Santurette and Georgiev, 2005) shows the significance of assessing the impact of the upper-level synoptic situation in the forecasting of such convective events at midlatitudes.

For another case of a Western Mediterranean storm, which produced extremely severe convection on the Algerian coast, Argence et al. (2009) performed experiments with the global ARPEGE model based on PV modifications. They studied the impact of applying PV modifications at initial time on numerical simulations of the precipitation performed with the 10 km model runs of the meso-nonhydrostatic (Meso-NH) model. The model forecast of 6 h accumulated precipitation valid on November 10, 2001, at 1200 UTC was improved, giving more intense precipitation between 50 and 110 mm/6 h in better accordance with the maximum observed values, suggesting that upper-level dynamics do partly control deep convection.

Figs. 4.8 and 4.9 illustrate the correlation between a strong convective system and the atmospheric thermodynamic characteristics in the middle-upper level for a case over the Eastern Mediterranean that developed in a strong blocking regime (see also Section 4.4.1).

FIGURE 4.8

Infrared image (colored only cloud-top BT $<-40°$C) at 1200 UTC on August 6, 2007, overlaid by 18 h forecast from ARPEGE of the 500 h Pa geopotential height (brown, dam) and temperature (blue, every 2°C).

Fig. 4.8 presents an IR image with 500 hPa temperature and geopotential fields superimposed at the mature stage of the blocking system. In this kind of synoptic-scale circulation, deep convective cells develop generally over the region from Greece to Bulgaria and Romania, on the eastern side of the upper-level cut-off low (Fig. 4.8) slowly moving eastward. An experienced observer can recognize some specific features favorable for convective development (strong diffluence on the northeast part of the cut-off associated with cold air and pronounced temperature gradient). However, the use of PV fields shows more clearly the upper-level dynamics of this situation. To illustrate this point, Fig. 4.9 presents the 1.5 PVU surface heights superimposed on infrared imagery; it shows that the convective system (pink arrow in Fig. 4.9) develops on the leading edge of the PV anomaly ("A" in Fig. 4.9A and B), in the area of strong gradient of geopotential of the 1.5 PVU surface.

Fig. 4.9B shows a cross-section of PV (brown), vertical velocity (ascending motions in yellow, descending in blue), and wet-bulb potential temperature (red) along the SW−NE axis indicated in Fig. 4.9A (pink arrow marks the area where the convective process operates); it can be noticed that:

- A strong PV anomaly is present, with folding of the dynamical tropopause (1.5 PVU thick brown line in Fig. 4.9B) down to 500 hPa.
- There are low values of wet-bulb temperature between 500 and 700 hPa marking dry and cold air aloft (green arrow in Fig. 4.9B) accompanying the tropopause folding above a maximum of warm and moist air in low levels that increases the convective instability.
- Strong ascending motions develop in conjunction with these features up to the tropopause (yellow lines in Fig. 4.9B).

As seen in Fig. 4.9C, deep convection develops (at the pink arrow) where the processes of low-level convection and upper-level forcing, acting in parallel, favor a persistent convective development. The role of upper-level blocking circulation will be further considered in Sections 4.3.3 and 4.4.1.

(A)

(B)

(C)

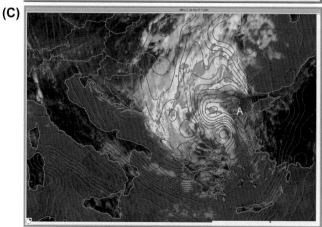

FIGURE 4.9

Infrared image at (A) 1800 UTC on August 5, 2007, and (C) 1200 UTC on August 6, 2007 (colored only cloud-top BT $< -40°C$), overlaid by the corresponding fields of 1.5 PVU surface heights from ARPEGE NWP model (red contours, interval 50 dam). (B) ARPEGE analysis vertical cross-section (along the axis indicated in (A)) of potential vorticity (brown, interval 0.5 PVU), vertical velocity (10^{-2} Pa s^{-1}, ascending in yellow, descending in blue), and wet-bulb potential temperature (red, interval 1°C) at 1800 UTC on August 5, 2007 (see text).

4.3.2 CONVECTION INITIATION AT DEFORMATION ZONES AND UPPER-LEVEL DYNAMICAL DRY FEATURES

In Section 4.3.1 the role of upper-level dynamics in development of deep convection over the mid-latitudes is illustrated by using satellite IR images and various diagnostic parameters. Diagnosing the upper-tropospheric flow characteristics by WV imagery is an important operational practice that allows forecasters a better understanding of the high-impact weather systems. The considerations of Weldon and Holmes (1991) provide a set of examples of how to relate the upper-level wind field properties and the corresponding WV imagery structures in a simple and appropriate way for operational application.

Concerning the diagnosis and forecasting of deep moist convection, it is useful to identify deformation zones (see Martín et al., 1999). A deformation zone is a region in the atmosphere with significant stretching or shearing. Spatial variations in the velocity field between two converging air masses cause a change in the shape of these air masses. Often these deformations result in characteristic cloud and moisture patterns, which can be analyzed in satellite images. WV imagery is the best observational tool to monitor deformation zones, since in the warm season there may be an absence of clouds related to the deformation flow patterns and the IR imagery shows only part of the information from upper levels (see Martín et al., 1999).

A training module on satellite WV imagery interpretation that contains exercises on analysis of various deformation zone patterns has been released by COMET Program (2009). The deformation components and properties of the flow have been described in many texts on atmospheric dynamics, and will only be mentioned here briefly, following the considerations in Martín et al. (1999). By definition, a deformation zone is a cold or hyperbolic region that produces elongated moisture—cloud features along the dilatation and contraction axes. Fig. 4.10 shows a WV image superimposed by the 500 hPa isobaric surface heights and the wind vectors at 300 hPa isobaric surface in such a case over Western Europe. The deformation zone has formed between the couples of mid-level low-pressure (L) and high-pressure (H) systems in Fig. 4.10A. Deep convection developed at the center of the cross-section line in Fig. 4.10B on June 8, 2015. Fig. 4.10A and B shows vertical displacement of the atmospheric circulation to the northwest of the deformation zone, changing from cyclonic at upper level to anticyclonic pattern at low level. The vertical cross-sections in Fig. 4.10C and D show that a baroclinic zone below the middle troposphere to the northwest of the deformation zone (high horizontal gradient of θ_w) as well as orographic effects play certain roles in these convective developments.

The deformation zones are regions where there are no mixing processes and that cannot develop upward (downward) motions or produce (dissipate) clouds or moisture patterns by themselves. They can only modify and elongate the shape of the existing cloud, dry, and moisture systems. Usually, there are sharp imagery contrasts at the edges of the deformation zone. In general, cloud streets, cloud lines, and elongated moisture boundaries are more likely to be parallel to the wind. For deformation zones, this is true as well, particularly when the systems are nearly stationary (slow moving) or lying along the contraction axis. Conversely, for moving systems, the elongated features will not be parallel to the motion with respect to the Earth, and the wind field will be perpendicular to the dilatation axis. Sometimes the deformation zones develop with other wind kinematic components such as rotation (vorticity) centers, jet stream circulation, etc. Based on these considerations, Martín et al. (1999) show that for some flow patterns, it is useful to analyze the atmospheric evolution of the set of the two rotation/vorticity centers (positive and negative) and the deformation zones as a whole. They demonstrated the usefulness of this approach in forecasting deep convection related to nonexplosive situations in the warm season over land. In some types of "weak" convective environments, deep convection appears when the cyclonic vorticity area of a deformation pattern at the upper troposphere overruns lands where solar heating is effective.

Fig. 4.11A shows a WV image superimposed by an absolute vorticity field at 300 hPa in another case over Eastern Europe, where deep convective cells are going to initiate along a deformation zone of

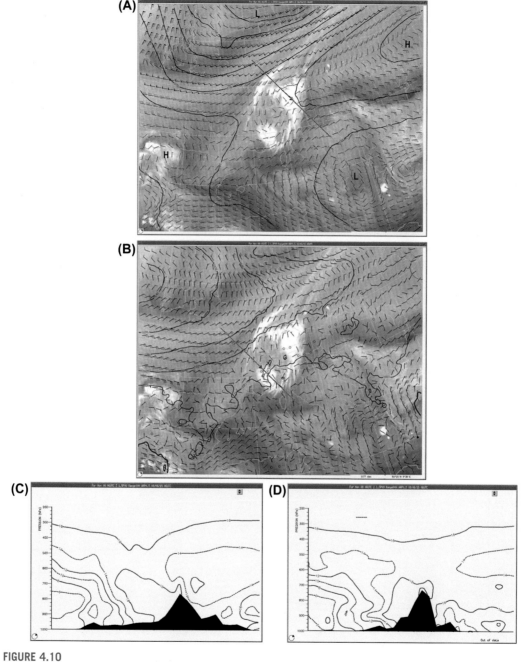

FIGURE 4.10

Deep convection in a deformation zone seen in 6.2 μm Meteosat water vapor images for June 8, 2015, 0600 UTC superimposed by: (A) geopotential of 500 hPa and wind vectors at 300 hPa isobaric surfaces; (B) mean sea-level pressure and wind vectors at 950 hPa isobaric surface. Vertical cross-sections of wet-bulb potential temperature: (C) along the cross-section line in (A); (D) along the cross-section line in (B). The orography is shown in black color in the cross-section.

the upper tropospheric motion field (as seen at the black arrows in Fig. 4.11B). In this case the deformation zone is present between symmetrically distributed cyclonic vorticity features that is a typical deformation flow pattern (see Bader et al., 1995). Fig. 4.11B shows the 6.2 μm WV images overlaid by geopotential and wind vectors at 500 hPa isobaric surface centered at the location of the dilatation deformation axis 3 h after the time of the vorticity field in Fig. 4.11A. This dilatation axis is located along the dry (dark) band in the WV image, elongated to the east of the black arrows in Fig. 4.11B between the circulation of the high-pressure (H) and the low-pressure (L) systems at the middle troposphere.

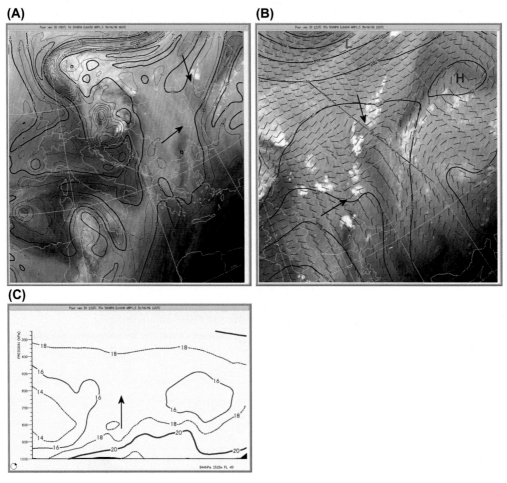

FIGURE 4.11

Deformation zone seen in 6.2 μm water vapor images for June 30, 2006: (A) overlaid by absolute vorticity at 300 hPa isobaric surface (10^{-5} s^{-1}) as a preconvective environment at 0900 UTC and (B) overlaid by geopotential and wind vectors at 500 hPa isobaric surface after the convection initiation at 1200 UTC; (C) vertical cross-section of wet-bulb potential temperature (θ_w) along the line in (B). Also marked in (A) are the areas of cyclonic and anticyclonic circulation patterns associated with mid-level low-pressure "L" and high-pressure "H" systems, respectively.

In certain conditions the horizontal deformation components of the flow may tend to increase and concentrate the temperature (or potential temperature) contrasts (see Martín et al., 1999). Fig. 4.11C shows a vertical cross-section of wet-bulb potential temperature along the line, shown in Fig. 4.11B, where such an increase of the contrasts in the vertical distribution of θ_w at the deformation zone is present on both sides of the blue arrow.

Deep convection may form when an anticyclonic and stable environment disappears from continental areas and an upper-level dynamical structure (cyclonic vorticity center, jet stream, or diffluent flow pattern) approaches the same places. The convective instability was previously capped by the anticyclone influence; normally, this situation is associated with conditions of the absence of cloudiness, and the lapse rate will be enlarged owing to more effective solar heating. In such environments, convective initiation related to various types of deformation zones is possible, for example:

- The slow-moving cyclonic perturbation seen in the WV imagery as a part of a deformation zone (Martín et al., 1999) favors the convective initiation due to a modest and persistent destabilization process, and these synoptic processes are connected with a development of deep convection (see, eg, Doswell, 1987).
- In the case of Fig. 4.11B, the favorable preconvective environment at the upper troposphere is related to the diffluent flow, seen in the wind field at the 500 hPa level along the dilatation axis of deformation zone (the line between the black arrows at 1200 UTC [1400 local time]). Fig. 4.11C shows that the middle–upper troposphere around the center of the deformation zone is unstable (weak vertical θ_w gradient) above the 750 hPa level, at the blue arrow around the center of the cross-section line in Fig. 4.11B). This diffluent pattern is conductive to the growth of deep convective cells under favorable conditions at the low level (warm, moist air and a lifting mechanism due to convergence of the wind, orography).

Different authors have shown that areas of discontinuity in upper-level humidity horizontal distribution seen as moisture boundaries in WV imagery are preferable for convective development (Krennert and Zwatz-Meise, 2003; Ghosh et al., 2008). However, not all moisture boundaries visible in the WV images are favorable for deep convective development, and only very few of them are associated with thermodynamic conditions capable to maintain persistent strong convection (see, eg, Georgiev and Kozinarova, 2009). Fig. 4.12A shows the satellite WV image for the situation of Fig. 4.11B to illustrate that such moisture boundaries are either associated with deformation zones (Fig. 4.12A, at the blue arrow, and Fig. 4.11B, at the black arrows) or related to dynamically active structures (dynamical tropopause anomalies and jets, as at the pink arrow in Fig. 4.12A). Preferable areas of strong deep convective development are moisture boundaries at the leading part of advancing vorticity structures, associated with dynamic dry bands (as defined in Section 3.2.2). A critical point in the forecasting process is to perform a relevant upper-level diagnosis in order to distinguish dynamic dry features (as at the pink arrow in Fig. 4.12) among the variety of gray shade structures on a WV image.

Fig. 4.13A shows WV images for the time of initiation of many convective cells along such moisture boundaries at the areas of favorable surface conditions. Among these cells, the process over Central Europe related to the dynamic WV image feature (at the pink arrow in Fig. 4.12A) is implemented in intense convection, which operates in strong upper-level forcing for vertical motions (see Fig. 4.13B). At the same time, the other moisture boundary (at the blue arrow) was linked to the deformation zone, providing a weaker dynamics in the upper troposphere that is not able to provide strong upper-level enhancement of the vertical motions.

As broadly considered in this chapter, the upper-level dynamic systems cannot explain convective events by themselves, because other factors or ingredients must exist in the same place and at the same time (instability, moisture, low-level lifting mechanisms). The upper-level diagnosis by WV imagery

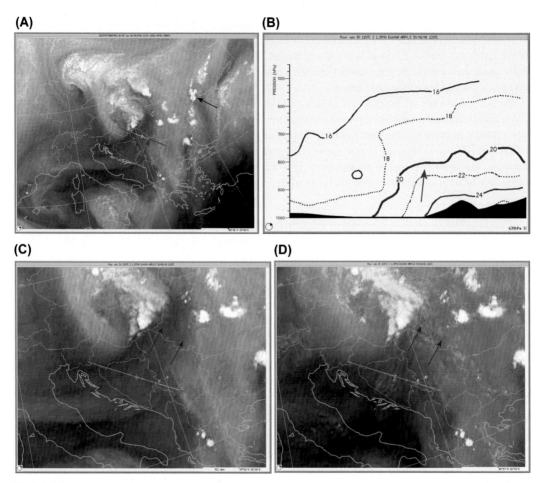

FIGURE 4.12

A preconvective environment (at the pink *arrow*) in the leading part of advancing vorticity structures on June 30, 2006, at 1200 UTC: (A) 6.2 μm water vapor image and (B) vertical cross-section of wet-bulb potential temperature (θ_w) along the line in (A). Also marked in (A) is the deformation zone, at the blue *arrow*. Convective instability seen by the Meteostat Second Generation images in (C) 6.2 and (D) 7.3 μm channels.

analysis could be useful to identify upper-level factors that may create the appropriate setting for convection initiation and development. To illustrate this point, Fig. 4.12B shows a vertical cross-section of wet-bulb potential temperature along the line, shown in Fig. 4.12A. The important atmospheric characteristics, which distinguish the preconvective environment along the cross-section line in Fig. 4.12A from the environment at the deformation zone (Fig. 4.11), can be summarized as follows:

- The θ_w distribution in Fig. 4.12B shows a low-level anomaly of warm and moist air that is much more pronounced than the θ_w anomaly below the deformation zone seen at the position below the blue arrow in Fig. 4.11C.

- The accompanying upper-level structure indicated by the pink arrow in Fig. 4.12 is a dynamic dry zone, connected with a polar strip of cyclonic vorticity at its rear part.
- In addition, the comparison of Fig. 4.12C and D shows that low-level moist air (light-gray shades in the 7.3 μm channel image) is capped by dry intrusion aloft. Such a dry intrusion appears generally in the leading zone of a dynamic dry band (seen in dark-gray shades in the 6.2 μm imagery), connected with a jet stream and a PV anomaly. In the leading part of this area convective instability is present due to buoyancy force of cold/dry air over warm/moist air. As discussed in Section 4.2.3, air masses in which moist low-level air is capped by a deep dry layer are particularly conductive to the rapid growth of deep convective cells. The strong vertical moisture gradient creates a convectively unstable environment and severe convection rapidly develops if a lifting occurs (due to low-level convergence of the wind, orography, upper-level forcing at the jet left exit).

(A)   **(B)**

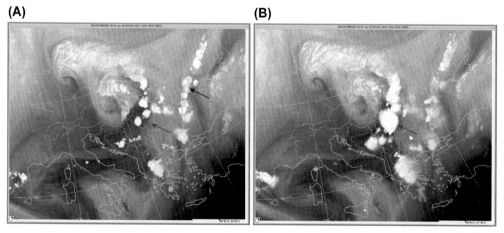

FIGURE 4.13

Convective developments along the dilatation axis of a deformation zone (at the blue *arrow*), and along a dynamic dry zone (at the pink *arrow*), connected with a polar strip of vorticity at its rear part as seen in 6.2 μm water vapor images from MSG on June 30, 2006, at (A) 1400 UTC and (B) 1600 UTC.

This preconvective environment enables strong vorticity advection to take place over a large area where the differential heating and deformation of the flow by the topography can produce surface warm anomaly and associated convergence zone. Such a coupling of these low- and upper-level factors is the driving mechanism for producing an intense convection in a deep tropospheric layer.

4.3.3 CONVECTIVE ENVIRONMENTS OVER THE SUBTROPICAL NORTH PACIFIC

As seen in Section 4.2, deep moist convection is a result of complex mechanisms, which maintain favorable atmospheric environments of elements, mixing at different levels. In some cases favorable conditions dominate at lower levels, while in other situations upper-level forcing is crucial to develop strong convection. Fig. 4.14A shows initiation and development of convective cells over the subtropical North Pacific (around 30°N and 35°N) on September 19, 2015, as seen in a composite imagery of WV image and IR channel brightness temperatures (only cloud tops colder than −40°C) derived from the Himawari satellite of the Japan Meteorological Agency.

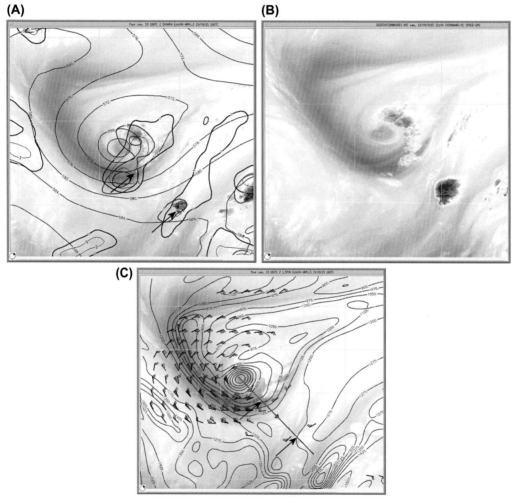

FIGURE 4.14

Lower- and upper-level thermodynamic environments favorable for deep convection. Himawari water vapor images overlaid by the coldest infrared 10.8 μm channel radiance (only cloud-top BT <−40°C) on September 19, 2015: (A) at 1800 UTC superimposed onto ARPEGE model analysis fields of low-level convergence (at 950 hPa, red contours), upper-level divergence (at 200 hPa, blue contours), and geopotential at 500 hPa level as well as (B) at 2100 UTC. (C) Water vapor image at 1800 UTC overlaid by ARPEGE analysis of 1.5 PVU heights (brown contours, interval 75 dam) as well as wind vectors at 300 hPa (red, only ≥60 kt) and 700 hPa (blue, only ≥20 kt).

This location (30°−40°N) at about 177°E is chosen as a suitable area for consideration because of the following special features:

- The transition zone between the subtropics and the midlatitudes is often influenced by two kinds of air masses:
  - Low-level intrusion of warm and moist air coming from low latitudes.
  - Upper-level intrusion of dry and cold air coming from high latitudes.
- The homogeneous ocean surface allows distinguishing the role of different kinds of environments with no impact of orographic effects.
- Satellite imagery is the main observational data source for diagnosis and early forecast of synoptic development over the ocean, where convection can produce bad weather with high impact on the marine activities.

The processes, and hence the conditions, which govern the convective development are acting at each level in different proportions that can strongly vary from case to case. This will be illustrated in Figs. 4.14 and 4.15, considering diagnostic fields from the ARPEGE model (with approximately 25 km grid resolution over this area) overlaid on the composite satellite images and vertical cross-sections: Most of the main atmospheric processes/characteristics of an environment favorable for strong convective development (defined for midlatitudes in Section 4.2) were present in the subtropical area over the North Pacific on September 19, 2015.

Potential instability: The two vertical profiles in Fig. 4.15C and D of temperature (black) and wet-bulb potential temperature (blue) from the ARPEGE model show conditions for increased convective instability in the environments of the two convective situations. However, at the position of the black arrow in Fig. 4.14 the convective instability is much stronger: In Fig. 4.15D the air in the lower-middle troposphere is drier and the atmosphere is more humid at low levels than these in the profile (Fig. 4.15C) to the northwest.

Low-level warm advection: The vertical profiles in Fig. 4.15C and D show rightward wind shear in the lower troposphere from the surface up to the 700 hPa level and moist air over the ocean. However, the area of convective development to the southeast (at the black arrow in Fig. 4.14) is associated with the much stronger baroclinic zone at the low level (Fig. 4.15A, green arrow), due to a distinct warm anomaly and related stronger horizontal gradient of the wet-bulb potential temperature (θ_w).

Presence of a low- to mid-level jet: As usual, such a low-level baroclinic zone originated from subtropical latitudes is associated with a jet stream at the lower−middle troposphere (wind maximum at 700 hPa in Fig. 4.15B). This dynamical feature is seen as a specific moisture boundary on the WV images (at the position of the black arrow in Fig. 4.14A), and especially in the 7.3 μm channel (not presented here). The thermodynamic conditions associated with such a mid-level wind maximum play important roles in producing deep moist convection downstream (as illustrated in Section 3.3.4.1 for the region of Eastern Atlantic and Mediterranean area).

Low-level convergent flow: Fig. 4.14A shows that both convective systems initiated in convergent low-level flow over the ocean.

Upper-level divergence: The convection at the brown arrow as well as at the black arrow in Fig. 4.14A were initiated in local areas of enhanced divergence at the 200 hPa level (blue contours) in the forward diffluent part of a mid-level trough. However, this upper-level favorable factor is more pronounced in the northwestern system where convection occurs near a dynamical tropopause anomaly (Fig. 4.14C).

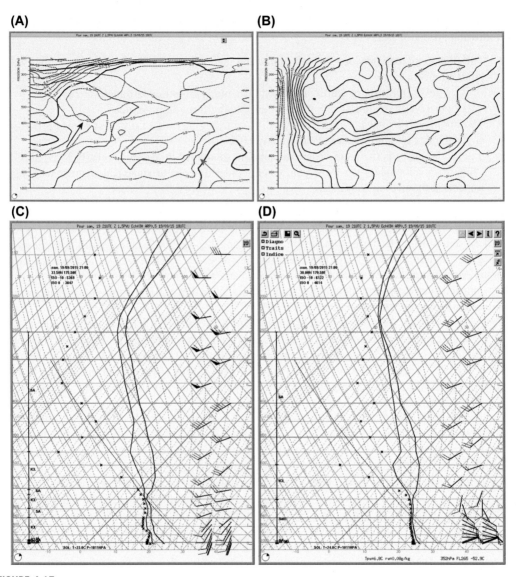

FIGURE 4.15

(A) Vertical cross-sections along the line in Fig. 4.14C of (A) potential vorticity (blue) and θ_w (red) as well as (B) wind speed normal to the cross-section from ARPEGE model. Vertical profiles of temperature (black) and θ_w (blue) from ARPEGE model (C) at the brown and (D) at the black *arrows* in Fig. 4.14.

A cold and dry mid-upper troposphere advection helps to increase instability: There is a line organization of the convective cells at the brown arrow in Fig. 4.14 seen in the satellite images. This is usually a result of the intrusion of cold dry air, which overruns low-level moist air ahead of the PV anomaly (at the magenta arrow in Fig. 4.15A) and leads to a development of an upper cold front (Sections 3.2.2 and 3.5.5).

Vertical wind shear: Fig. 4.15D shows no directional wind shear above the 700 hPa level in the environment of the convective system to the southeast, at the black arrow in Fig. 4.14. The profile in Fig. 4.15C, however, reveals a moderate wind shear from the lower to the upper troposphere due to much more pronounced upper-level dynamics to the northwest at the position of the brown arrow in Fig. 4.14 (see also Fig. 4.15A, magenta arrow, and Fig. 4.15B).

A dynamical tropopause anomaly favors deep convection: The convective cells to the northwest (at the brown arrow in Fig. 4.14) develop in an environment of upper-level forcing, as a result of a tropopause folding (ie, high cyclonic PV down to the middle troposphere; blue contours in Fig. 4.15A), related jet (wind maximum in the layer 400−200 hPa; Fig. 4.15B), and vorticity advection. This is indicated by the existence of a dynamic dry dark zone in the 6.2 μm image (Fig. 4.14).

This case illustrates that the main factors for intense convection, especially in the subtropical latitudes, come from the low-level environment. The upper-level forcing plays a critical role to accelerate the convective development and hence to conduct a vigorous and severe deep convection usually in the middle latitudes. The mid-level trough, which conducts the convective development to the northwest (at the brown arrow in Fig. 4.14C), is a cut-off system. The associated vorticity advection is not significant, since it is not associated with a polar strip of dry intrusion as seen by the WV imagery analysis. There are times, however, when mobile upper-level troughs are intruded in the lower latitudes and the related dynamical tropopause anomalies become important factors in establishment and maintenance of deep convection in the subtropics (see Section 4.4.2).

4.3.4 DEEP CONVECTION IN BLOCKING REGIMES

The main source of heat and moisture in the atmosphere is the Earth's surface. Since heat (and generally moisture) increase equatorward, southerly flow (in the Northern Hemisphere) is usually favorable to supply low levels with warm air and moisture. Extratropical cyclones and associated atmospheric fronts play a reasonably well-understood role in moistening, destabilization, and upper-level forcing on ascending motion. Thunderstorms are much more likely to occur in atmospheric situations of flow from the equator, that is, in the eastern part of upper-level troughs, that produce warm and moist low-level circulation. Therefore, it is usually strong convective events that develop in a blocking regime, which is associated with persistent maintaining of such a synoptic-scale configuration.

A blocking situation is characterized by a stationary low value in the upper-level geopotential fields as well as in the mean sea-level pressure (MSLP) field (see Section 3.3). Due to its persistent nature, the blocking regime plays a major role in surface weather, especially in development of convective systems. The low component of a block has a dynamics, which is very different from that of a midlatitude cyclone. It is difficult to identify fronts in the low of a block. These lows undergo bursts of activity; the most active part tends to rotate around the low with the wind, while "pseudo-fronts" tend to be fixed in the eastern flank of the low. Fig. 4.16 provides a way to schematize the main features involved in a blocking situation that lead to deep convective developments as follows:

- First of all, the persistence of a surface low associated with the negative anomaly of the middle-level isobaric surface height (marked "B" in Fig. 4.16) favors pronounced warm advection on its eastern flank as a result of the durable flow coming from the equator (orange double arrow in Fig. 4.16).
- At the same time, cold flow in the western side of the low pressure (mainly at middle and upper level; green arrow in Fig. 4.16) produces favorable conditions to generate potential instability by the process of differential advection of θ_w (ie, low-θ_w advection aloft above an area of high-θ_w advection; Section 4.2.2).

- Due to the very nature of a block, an area of highly diffluent flow exists on the eastern part of the blocked system (strong diffluent flow symbol at location (1) in Fig. 4.16); such diffluent circulation aloft generates a zone of upper-level divergence that produces ascending motion (double vertical red arrow), which can be pronounced in cases when a tropopause anomaly moving around the center of the upper cut-off low interacts with a jet (black curved line and light brown double arrows at location (3)). Such upper-level dynamical structures producing regions of upper-level forcing on ascending motion (symbolized by vertical red arrow) can displace toward the eastern flank of the low (yellow jet streak structure interacting with tropopause anomaly in gray, at location (2) in Fig. 4.16) and then bring a clear lifting mechanism over the low-level air mass.

Finally, blocking regimes are situations most favorable to set up the process described in Section 4.2.2 and schematized in Fig. 4.1 that produce convective development. In the region contoured by a light green line in Fig. 4.16, on the eastern flank of the low-pressure system, conditions are gathered to generate thunderstorms (as shown in Section 4.2.2, Fig. 4.1), besides in a situation where warm and moist air (which is the "fuel" of thunderstorms) can be pronounced. In summary, this is a very favorable area where strong differential θ_w advection can exist that can be reinforced by a tropopause anomaly moving around the upper low; moreover, according to the nature of the flow around the upper-level low, the dynamical tropopause anomaly can interact with a jet streak bringing forcing on the ascending motion acting on the area of potential instability.

Of course, regional and local conditions are important. Considering Europe, large (and arid) subtropical landmasses are present equatorward of midlatitudes, over North Africa, which is a source region of warm air. Meridional low-level flows in such places transport this warm air over the Mediterranean Sea (which is generally a warm sea) that brings moisture after a more or less long fetch, before it reaches Europe. In the same way, considering North America, the Gulf of Mexico is a source of very pronounced warm and moist environment, which can interact with cold and dry continental air over the American continent in the case of a blocking situation over the United States.

Fig. 4.17A and B presents fields of geopotential (brown lines), and temperature (shading) at 500 hPa isobaric surface and jets at 300 hPa (blue contours) in a case of a wide cut-off low over Europe. They show the large cyclonic circulation in which jet streaks can be identified from the western part up to the southeastern part of the blocked system (blue contours). Fig. 4.17C and D shows the corresponding 6.2 μm WV images where several dark spots can be seen moving along the jet streaks in the cyclonic flow (orange, blue, and yellow arrows).

Fig. 4.18 shows 6.2 and 7.3 μm at 0300 UTC on February 25, 2013. Convective development begins near Greece in the upper-air diffluent flow (that corresponds also to the left exit of a jet streak, Fig. 4.17B), and below a low-tropopause area moving from the east of Sicily up to Greece (orange arrow in Figs. 4.17 and 4.18). Around midday, a distinct V-shape convective system (red arrow in Fig. 4.17D) develops just ahead of a pronounced dark spot moving up in the southwesterly flow toward Greece (blue arrow in Figs. 4.17D and 4.18A).

The upper—middle level diagnosis performed by using satellite imagery in the WV channels shows the following elements of the convective development:

- The reinforcement of moisture flows just ahead of the pronounced dynamical tropopause anomalies (yellow and blue arrows in Figs. 4.17C and 4.18A); these dynamical tropopause anomalies create cyclonic flow in the troposphere and vertical motion (as described in Chapter 1) that increases the southwesterly or southerly component of the flow on their leading edge.

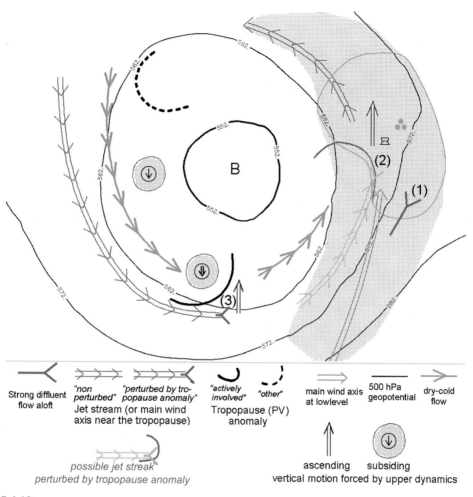

FIGURE 4.16

A schematic representation of main atmospheric features leading to areas very favorable for deep convective development in situations of blocking regime. Red dotted zone is the area favorable for deep convective development (noticed by red cumulonimbus symbol); light green contour area is a privileged region for strong convection (green dotted points indicating strong activity; see text for explanation).

- In the southerly circulation on the eastern flank of the block, a reinforcement of the moisture flow marked by maximum of wind at 600 hPa that can be diagnosed by the moisture boundary on the 7.3 μm WV image (between the green arrows in Fig. 4.18B); near Greece (at the northern green arrow and around the pink arrow in Fig. 4.18B) the moisture increases particularly ahead of the low tropopause area (marked by blue arrow in Fig. 4.18A). This is a sign that vertical motion has begun downstream of the MLJ (red vectors of maximum wind, Fig. 4.18B).

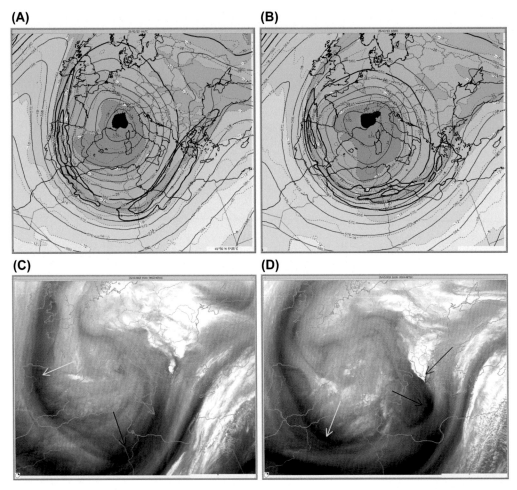

FIGURE 4.17

Convective development in blocking regime on February 25, 2013. Geopotential (brown lines) and temperature (shading) at 500 hPa isobaric surface, with 300 hPa jet superimposed (blue lines, threshold 100 kt) for (A) 0000 UTC and (B) 1200 UTC. Water vapor 6.2 μm images for (C) 0000 UTC and (D) 1200 UTC of a V-shape convective system (red *arrow* in (D)) ahead of a dark spot (at the blue *arrow*). The yellow *arrow* indicates the approaching dynamic dark band, associated with a dynamical tropopause anomaly.

At the southwestern side of the large blocking system (corresponding to location (3) in Fig. 4.16), reinforcement of the moisture flow (light blue arrow in Fig. 4.18B) also occurs on the leading edge of the pronounced tropopause anomaly moving over Spain (yellow arrow in Fig. 4.17C and light blue arrow in Fig. 4.18B). This area, however, is not as favorable for strong convection as regions (1) and (2) in Fig. 4.16 are, as discussed earlier.

Fig. 4.19 presents a cross-section along the A—B axis indicated in Fig. 4.18b. The area around the pink arrow (Figs. 4.18B and 4.19) is a zone of very humid low-level air and increased mid-level

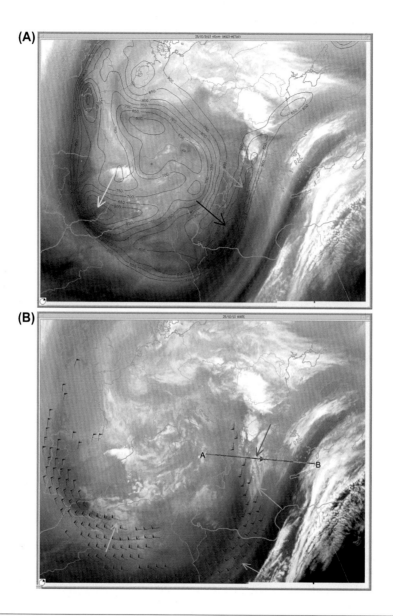

FIGURE 4.18

(A) 6.2 μm water vapor image with geopotential of the 1.5 PVU surface superimposed (ARPEGE 3 h forecast, interval 50 dam, threshold 900 dam), and (B) 7.3 μm water vapor image with 600 hPa maximum wind superimposed (red, threshold 50 kt) at 0300 UTC on February 25, 2013. Also shown on (B): the A—B axis of the cross-section in Fig. 4.19.

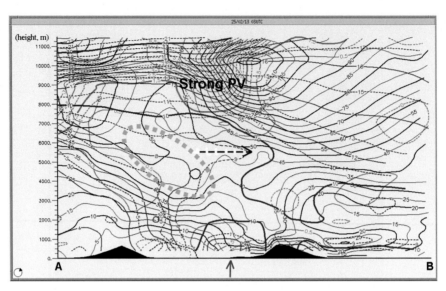

FIGURE 4.19

Vertical cross-section along the A–B axis indicated in Fig. 4.18B of ARPEGE 3 h forecast valid at 0300 UTC on February 25, 2013, of potential vorticity (green, every 0.5 PVU); wind speed perpendicular to the cross-section (black, every 5 kt); wet-bulb potential temperature (red, every 1°C).

moisture seen on the 7.3 µm WV image (Fig. 4.18B) where vertical motion has begun and convective clouds are going to develop. The numerical model data shown in Fig. 4.19 confirms the conclusions based on WV imagery analysis regarding the upper-middle level diagnosis presented above, regarding the main elements of the convective situation as follows:

- An area of instability (at the pink arrow in Fig. 4.19) with high wet-bulb potential temperature in the low levels (from the surface up to about 4000 m) and intrusion of dry cold air (low wet-bulb temperature) at the upper level (dashed blue arrow in Fig. 4.19). In this region the 7.3 µm WV shows a light-gray feature.
- An MLJ exit; this jet is marked by a maximum speed of wind component perpendicular to the cross-section (light blue oval in Fig. 4.19). We can notice this jet corresponds to a strong temperature gradient at low levels and to the gray shade gradient in the 7.3 µm WV image (Fig. 4.18B).
- A strong PV intrusion aloft (indicated in Fig. 4.19) corresponding to dark-gray features in Figs. 4.17C and 4.18A (orange and blue arrows).

Analysis based on images in the two WV channels with the NWP model fields helps to understand and to monitor the evolution of the main features associated with a blocking system that can lead to convective development.

4.3.5 UPPER-LEVEL DYNAMICS AND DEEP CONVECTION IN TROPICAL AREAS

There is a lack of systematic studies about the role of larger-scale processes in tropical convection. The reason for this low interest is that the dominated thinking about the interaction between deep moist convection and the environment in the tropical area is the statistical view of a quasi-equilibrium state of the radiation–convection adjustment process of tropical weather systems. Studies, however, suggest that the tropical atmosphere is far from a convective quasi-equilibrium, with much weaker warming in the middle and upper troposphere (Lin et al., 2015) that is possibly caused by the ubiquitous existence of shallow convection and stratiform precipitation. In this interpretation approach the full troposphere convective quasi-equilibrium is basically a two-phase first-vertical-mode view of the tropical atmosphere with the atmosphere switching between two phases: deep convection and no convection (or weak-deep convection). The result of Lin et al. (2015) could be a limitation for global climate/weather prediction and regional models, relying on the parameterization of convection, which is dependent on the notion of statistical equilibrium. At the same time the statistical equilibrium theory made several concrete predictions about tropical phenomena that were later verified by observations (Emanuel, 2007).

As discussed in Emanuel (2007), the cornerstone of the contemporary statistical equilibrium theory of convection was first stated by Arakawa and Schubert (1974), who postulated that moist convection consumes potential energy at the rate it is provided by larger-scale processes. For quantitative considerations the existence of convective quasi-equilibrium in the tropics is a suitable viewpoint to explain the differences in the processes conductive for convection over tropical and midlatitude areas.

- Due to a convective near-equilibrium in the tropics, convection maintains a nearly moist adiabatic lapse rate from the cloud base to the tropopause. The potential energy, as quantified, for example, by CAPE, is not accumulated in the atmosphere but is released by convection as fast as it is produced. In case of instability, there is an immediate release of the CAPE.
- In midlatitudes, the instability may be inhibited for many hours, even days (in blocking systems, to be shown in Section 4.4.1). Such a CIN acts as a "pressure cooker," producing a lot of high pressure.
- Wind shear and vorticity advection are more frequent and stronger in midlatitudes than in the tropics.

The upper-level dynamics is fully responsible for wind shear and vorticity advection and it partially governs CIN formation. Therefore, there is much difference between the role of the atmospheric environment of deep convective systems in the midlatitudes and in the tropics. It is well explained by Emanuel (2007) that compared to the extratropical atmosphere, the dynamics of the tropical atmosphere are poorly understood. The cornerstone of dynamic meteorology—quasi-geostrophic theory and its contemporary encapsulation in PV thinking—works well in middle and high latitudes, where the motion is quasi-balanced over a large range of scales; also diabatic and frictional effects are usually small there and can often be neglected on short time scales. For this reason, the dynamics of fundamental processes such as Rossby wave propagation and baroclinic instability are well understood. By contrast, much of what occurs in the tropical atmosphere is neither quasi-balanced nor adiabatic, and thus the tools that have served well to study atmospheric processes in the middle and high latitudes are poorly suited to understanding the meteorological processes in the tropics.

In this section it is shown that considering the structure and characteristics of the dynamical tropopause has proved to be also an effective means to study the upper-troposphere dynamics in tropical areas. Experimentations have been carried out by forecasters from tropical centers of Météo-France (La Réunion, New Caledonia, French Polynesia, French Indies) on convective situations (excluding hurricanes) and have shown the efficiency of such an approach. However, it has been found that the 1.5 PVU surface is not a good level to represent the tropopause for these latitudes. Fig. 1.4

(Section 1.2.3) shows that for latitudes below 30° the PV in the upper troposphere is on average lower than in the higher latitudes: PV values between 0.5 and 1 PVU are climatologically present between 400 and 200 hPa in the tropics. In other words, the strong vertical PV gradient around the tropopause begins with values of about 1 PVU and not about 1.5 PVU as in midlatitudes. Fig. 1.5 (Section 1.2.3) confirms that the lower stratosphere, as seen through the ozone concentration (to diagnose the lower stratosphere), remains substantially stronger below the 1.5 PVU surface in tropical areas. Finally, after testing by the forecasting services of Météo-France, the surface of PV equal to 0.7 PVU was chosen as representative for the dynamical tropopause in the tropics.

Several authors (Morwenna et al., 2000; Funatsu and Waugh, 2008) have discussed the effects of destabilizing the lower troposphere (especially in moist air) by the presence of pronounced potential vorticity maximum aloft. In many cases of strong convection development in the tropics a low dynamical tropopause height (maximum of cyclonic PV) is present in the upper troposphere. Fig. 4.20 shows IR satellite images overlaid by the analysis of the geopotential of the 0.7 PVU surface in the South Indian Ocean region. There is a good correlation between the main cloud masses and the structures of the dynamical tropopause:

- In the bottom right corner of the images (black arrows in Fig. 4.20A and B), the cloud mass is closely related to an area of low height of the dynamical tropopause, the 0.7 PVU surface (that is to say, to a maximum of cyclonic PV in the upper troposphere): Its western edge is well correlated with the region of strong gradient of the dynamical tropopause height.
- Southeast of Madagascar, the V-shape convective system (yellow arrows in Fig. 4.20A and B) developed in conjunction with the arrival of a minimum of height of the dynamical tropopause (noted as "A" in Fig. 4.20A and B) and moves toward the southeast in association with the movement of this tropopause minimum height (Fig. 4.20B).

Fig. 4.20D presents a vertical cross-section of wet-bulb potential temperature (red contours) and PV (green) through the area where the convective system develops below the leading edge of strong PV. It can be noticed that:

- There is a maximum of wet-bulb potential temperature (warm moist air) near the surface (pink arrow in Fig. 4.20D).
- Above the low-level warm and moist air, a PV intrusion is present aloft (strong PV noticed in Fig. 4.20D); this strong PV intrusion aloft is associated with dry and cold air (blue arrow in Fig. 4.20D) penetrating at mid-levels (which contributes to destabilization of the low-level air).

Therefore, PV fields and related concepts can be effective operational tools to understand the dynamical structures of the upper troposphere in analysis and very short-range forecasting, even in tropical latitudes. There is a lack of systematic studies on the potential influences of upper-level dynamics on the tropical deep convection.

In summary, deep convection in the tropics is primarily a consequence of the moist and unstable warm air convergence, but the upper troposphere conditions often play a significant role in the more or less pronounced activity as well as in the life cycle of mesoscale convective systems. Weldon and Holmes (1991) have provided some guidance to interpret basic large-scale WV imagery features regarding the forecasting applications of deep convective weather at tropical locations. In certain cases, WV imagery analysis can be performed in terms of PV concept for upper-level diagnosis of deep convection in tropical areas.

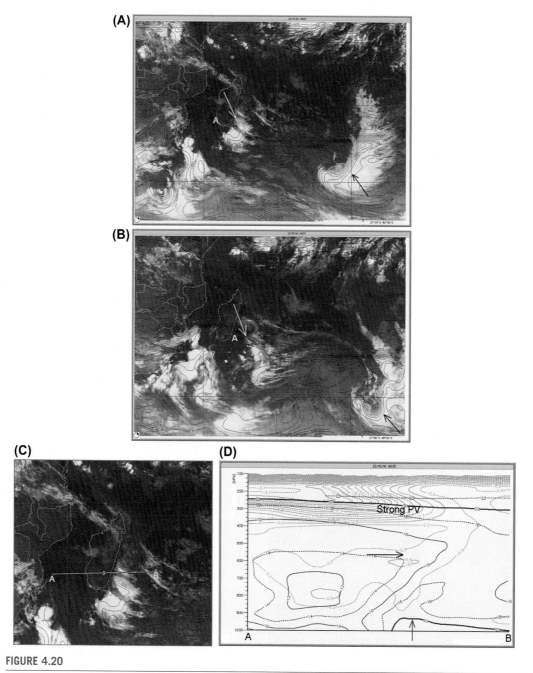

FIGURE 4.20

(A) Infrared image with superimposition of 0.7 PVU surface heights from ARPEGE analysis (red contours every 50 dam): (C) on May 18, 2006, at 1800 UTC, and (B) on May 19, 2006, at 0600 UTC; (D) vertical cross-section of wet-bulb potential temperature (red contours every 2°C) and potential vorticity (green contours every 0.2 VU) along the A—B axis in (C).

4.4 USE OF DATA FROM WATER VAPOR CHANNELS IN DIAGNOSING PRECONVECTIVE ENVIRONMENTS

4.4.1 UPPER-LEVEL FORCING/INHIBITION IN THE ENVIRONMENT OF MOIST CONVECTION

An important forecast issue is determining when and where severe deep moist convection may develop that is especially complicated in situations of blocking regimes. The satellite WV imagery is a powerful tool for diagnosis and predicting the evolution of upper-level circulation and has an essential operational role in diagnosis of a convective environment favorable for strong convection. Considering a case of widespread, persistent convective development over Southeastern Europe on August 6, 2007, this section focuses on diagnosis of large-scale processes, which may produce inhibition and acceleration of the convective developments.

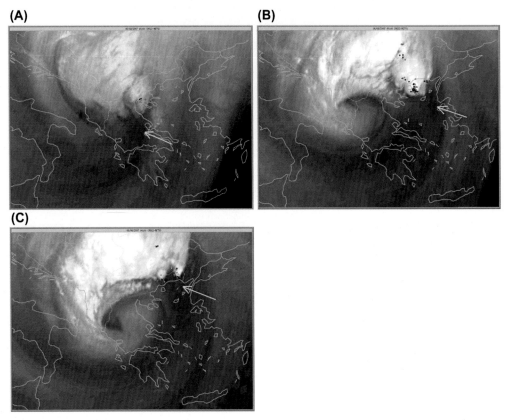

FIGURE 4.21

Water vapor image in 6.2 μm channel with superimposition of the last 15 min lightening (colored dots) at (A) 1800 UTC on August 5, 2007; (B) 0000 UTC on August 6, 2007; and (C) 0400 UTC on August 6, 2007.

Fig. 4.21 presents 6.2 μm images with superimposition of the last 15 min lightening (colored dots). As a result of the cyclonic flow around the upper-cut-off low, an upper-air dry zone (warm brightness temperature related to dark-gray shades in the WV image, yellow arrow in Fig. 4.21) is moving in the south and then in the east part of the low. The deep convection begins and then increases at the leading edge of the dynamic dry feature (the darkest image gray shades), which is related to a critical upper-level dynamical structure involved in the acceleration of convective development. The associated tropopause anomaly is identifiable as a dark spot on the 6.2 μm WV image (in the night from August 4 to 5, yellow arrow in Fig. 4.22A) in the westerly part of the cut-off low. The fields in Fig. 4.22A and B

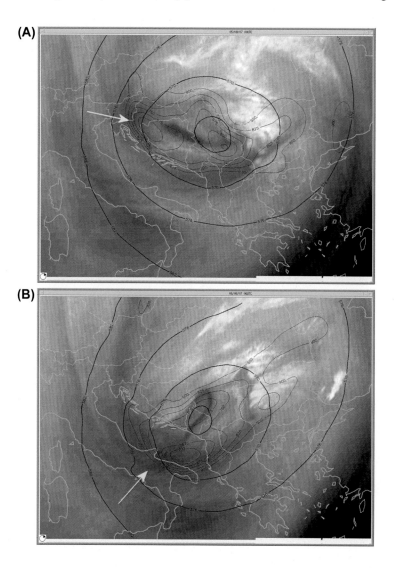

FIGURE 4.22

Water vapor image in 6.2 μm channel at (A) 0300 UTC and (B) 0900 UTC on August 5, 2007, with superimposition of geopotential at 500 hPa (brown, every 60 dam) and geopotential of the 1.5 PVU surface (red, ≤900 hPa every 75 dam).

confirm that it is clearly associated with an isolated minimum of height of the dynamical tropopause (upper PV maximum) turning round persistently within the upper-level low.

Fig. 4.23A shows the MSG WV image for 1200 UTC overlaid by the fields of the maximum wind (black vectors) and convergence (red contours) at the 950 hPa level as well as the maximum wet-bulb potential temperature (θ_w, yellow contours) at the 850 hPa level. These diagnostic fields show that the

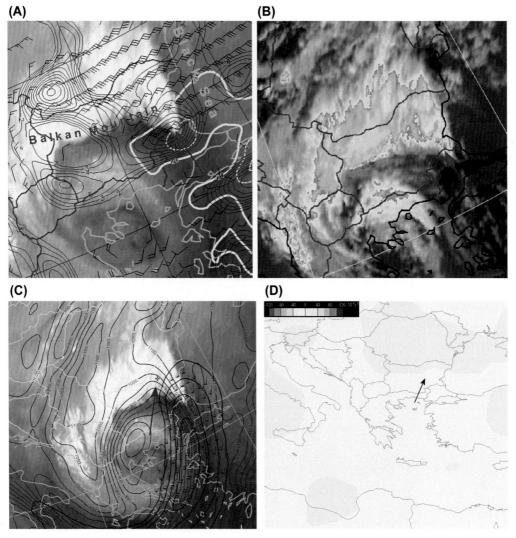

FIGURE 4.23

Meteosat water vapor 6.2 μm image on August 6, 2007, for 1200 UTC superimposed by the corresponding ARPEGE analysis of (A) wind vectors (black contours, only ≥10 kt) and convergence (red contours, only ≤−20 × 10^{-5} s^{-1}) at 950 hPa level as well as the 850 hPa wet-bulb potential temperature (yellow contours, only ≥20°C); (C) the constant surface of PV = 1.5 PVU heights (brown), the wind vectors (red) and wind speed (black contours) at 300 hPa (only ≥60 kt); (B) enhanced infrared 10.8 μm image (colored only brightness temperature <−40°C) for 1215 UTC; (D) EUMETSAT DIV product for 1145 UTC.

low-level conditions of the preconvective environment were critical for the potential of the atmosphere to produce deep moist convection, as follows:

- Persistent surface flow of moist air from the Black Sea.
- Convergence of the low-level flow.
- A low-level anomaly of warm and moist air near the maximum of low-level convergence (at the pink arrow in Fig. 4.23A). The enhanced IR image in Fig. 4.23B shows that strong convective development has initiated to the north of this area.

To focus on the role of upper-level dynamics in determining where deep moist convection may develop, Fig. 4.23C shows the WV image at the time of Fig. 4.9C in Section 4.3.1 overlaid by ARPEGE model analyses of the height of the 1.5 PVU surface (brown contours) as well as the maximum wind vectors (red) and wind speed (black contours) at 300 hPa. It is seen that the process developed in the leading part of the PV anomaly is related to a deep (down to 600 dam) folding of the dynamical tropopause. In this case, two areas of deep convection can be discriminated by the occurrence of low-level convective initiation and upper-level forcing in the leading diffluent part of the PV anomaly.

- A primary area is present to the north of the pink arrow in Fig. 4.23A and B, where the processes of low-level convection and upper-level forcing, acting in parallel, govern the convective development.
- A secondary area is located around the blue arrows in the enhanced IR imagery on Fig. 4.24A−C. In this area the low-level convection interacts with the favorable upper-level dynamics at a later stage of its initiation, after its movement into an environment of upper-level divergent flow, associated with ascent in the upper troposphere.

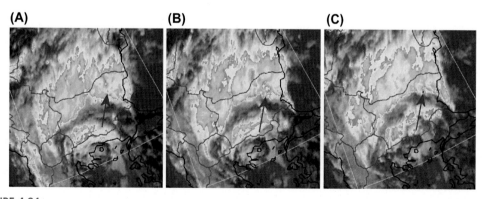

FIGURE 4.24

Meteosat enhanced infrared 10.8 μm image (colored only brightness temperature <−40°C) on August 6, 2007, for (A) 1330 UTC, (B) 1445 UTC, and (C) 1600 UTC.

Regarding the primary area, as seen in the enhanced Meteosat IR image (Fig. 4.23B), deep convective cells develop near the sea coast close to concentrating maxima of convergence and θ_w in the low levels. Large CAPE (2370 J kg^{-1}) was predicted in the 3 h forecast of ARPEGE valid for 0900 UTC over this region, together with strong vertical wind shear from the surface to the 300 hPa level. Fig. 4.23A also shows a potential for lifting due to the interaction of the low-level flow (about 15 m s^{-1} as analyzed by the ARPEGE model) with the topography of the Balkan Mountains downstream (to the north and northwest of the pink arrow).

The enhanced IR imagery (Figs. 4.23B and 4.24, as well as other Meteosat images, not presented) shows that persistent development of thunderstorms occurred over the primary area, while the secondary area was affected by abrupt development (at the position of the blue arrows) of several convective cells that move to the north followed by the development of a subsequent cell. Although the primary area was affected by extremely heavy precipitation (more than 120 mm, measured for 24 h), the highest severity of this convective event occurred in this secondary area.

Fig. 4.23C and D shows that the intense convection in the "secondary" area developed in synoptic-scale upper-troposphere diffluent (divergent) flow at the left exit of the jet in the leading part of the PV anomaly. The presence of the upper-level PV anomaly indicates coherent areas of descending and ascending motions (see Fig. 4.1). The possible upper-level impact on the convective development initiated at low levels is determined by the position of the growing convective cell relative to the upper-level ascent areas (contributing to deep convective development) and descent areas (contributing to convective inhibition). It is also seen in Fig. 4.23C that the persistence of such a favorable environment is related to a strong blocking regime of the upper-level circulation in the leading part of the PV anomaly. In such cases of convective development, the use of WV imagery is essential in helping to distinguish between the environments of upper-air descent (resulting in dark image gray shades) and environments of ascending motion at upper-troposphere levels (light image gray shades). Depending on the environment in which the convection occurs, a wide range of horizontal shifts in the location between the low-level initiation and the subsequent possible upper-level forcing may occur.

Regarding the secondary area, a horizontal shift (about 50 km) appears in the location between the low-level convection initiation and the upper-level forcing (in the environment of southeasterly flow, 15−20 m s^{-1} from 700 hPa up to 400 hPa levels as analyzed/predicted by ARPEGE model). Large CAPE was predicted for this region in the 3 h ARPEGE forecast valid for 0900 UTC (940 J kg^{-1}) and 1500 UTC (2786 J kg^{-1}), together with strong vertical wind shear from the surface to the 500 hPa level. The most severe weather was a result of several convective developments to the north of the WV image moisture boundary (at the blue arrow in Fig. 4.23C), downstream of the area of upper-level subsidence marked by dark-gray shades on the WV image. The dry area on the WV image indicates descending motions in the upper troposphere, and hence the initiation of deep convection was a result of a low-level forcing (which could come from any appropriate mesoscale mechanism). Downstream of the moisture boundary seen in the image, deep moist convective cells suddenly developed into the middle and upper troposphere, influenced by the interaction between mesoscale (low-level) and large-scale (low- as well as upper-level) processes, which governed the intensity and evolution of the resulting convection.

Therefore, it appears that after the convection initiated and developed to the middle troposphere, it was then inhibited due to upper-level descending motion, confirmed by the persistent dark dry zone in the WV imagery. Then, the convective cells intensified abruptly (growing above the middle troposphere) after their approach to the WV moisture boundary and entering the secondary area, where they met ascending motion in the upper troposphere related to divergent flow aloft.

Although the low-level convection operates in a ubiquitous favorable upper-level environment only in the primary area of convection, the precipitation was much more intense and heavier in the secondary one. Assuming the presence of similar conditions conducive for deep convection at low levels, the essential difference was that the upper-level dynamics influenced the convective events in the secondary area in two ways:

- For a certain time after its onset the low-level convection operated beneath upper-level descending motion (diagnosed by the presence of a dry zone in the WV imagery). Such a descent acts as a lid that tends to inhibit the development of deep convection above the middle troposphere allowing warm, moist air to be bottled up at lower–middle levels, thereby increasing CAPE (Browning et al., 2007). The diagnosis of downward motions can be made by using EUMETSAT DIV product. As seen in Fig. 4.23D, to the south of the secondary area (at the black arrow), upper-level convergence is present (a sign for descent). At the same time, the dry air aloft is acting to increase the convective instability in the environment of the convective development.
- Then, moving with the flow, the convective cells move under the influence of the upper-level forcing mechanism (upper-level divergent flow, green shades in Fig. 4.23D), which produces an environment favorable for strong convective evolution.

To conclude, when the convection initiation process operates in an upper-level environment favorable to deep moist convection, the result is most likely to be widespread, intense thunderstorms. On the other hand, mesoscale processes may at times be sufficient to initiate deep convection in environments that may be only marginally favorable for (or even inhibitive to) convection (Doswell, 1987). The result in this case is usually short-lived thunderstorms or storms that remain isolated (although possibly intense) rather than growing into a widespread area of convection. Such an evolution may be at least partly due to preceding neutral (neither divergent nor convergent) or negative (convergent) upper-level conditions. In many cases, it is possible to have significant divergence of the upper-level flow without any convective development at all that may occur in the absence of some required ingredients, as described in Section 4.2.1.

In the midlatitudes, high-impact convective events often occur in areas where instability and low-level forcing meet the synoptic-scale upper-level forcing mechanisms, which produce upward motion in the upper troposphere related to divergent flow aloft. As discussed in Doswell (1987), the relative rarity of widespread, extremely violent convective weather can be understood in terms of the rarity of the coincidence of extremely favorable large-scale environments with the appropriate mesoscale processes to initiate (and, perhaps, enhance) the large-scale potential. The example of the event of August 6, 2007, serves to illustrate how important the upper-level context can be in midlatitudes and shows how the upper-troposphere circulation may govern the low-level convection as well as the possibilities for mesoscale features to release their potential in a strong convective environment.

4.4.2 UPPER-LEVEL FORCING AND CONVECTIVE INSTABILITY IN SUBTROPICAL AREAS: MIDDLE EAST CASE STUDY, DECEMBER 22, 2009

This section shows the applicability of imagery and products based on satellite data in the 6.2 μm WV channel in diagnosing upper-level forcing of convection over the lower subtropical latitudes down to about 20°. The case study example presents a strong convective development in December 2009 that followed the Red Sea coast of Saudi Arabia from its initiation close to the Gulf of Aqaba to its slow dissipation near the Yemen border. The convection was aligned with some low-level convergence over the Red Sea (Al Badi et al., 2009). The mesoscale and small-scale convective ingredients in the low levels over this specific region acted in tandem with preexisting upper-level divergent flow for convection initiation. During the convective development, advection of PV anomaly played a critical role for intensification of the process (as discussed in Section 4.4.1). As a result, the storm moved south-southeastward at an average speed of 25 km h^{-1} over a period of 15 h, and had a continuous outflow at the cloud top in a northeasterly direction (Al Badi et al., 2009).

4.4.2.1 Diagnosis of Upper-Level Preconvective Environments

Fig. 4.25A shows WV 6.2 μm images superimposed by color-enhanced IR 10.8 μm brightness temperatures of cloud-top heights (only colder than −40°C) and ARPEGE analysis of 500 hPa heights just before the appearance of deep convective clouds over the Northeastern Red Sea coast (at the position of the black arrow). The upper-level flow structure as diagnosed by EUMETSAT DIV product is shown in Fig. 4.26. For the purposes of operational forecasting, the following diagnostic considerations regarding the upper-level preconvective environment are relevant:

- The convection was going to develop in the leading diffluent part of a rapidly deepening Mediterranean trough, which is propagating down across the Red Sea (Fig. 4.25A).
- Six hours prior to the convection initiation, strong divergence of the upper-level flow is present over this area (Fig. 4.26A−C, at the black arrows). This is a sign for existence of persistent synoptic-scale vertical motions at the upper troposphere that can strongly enhance the depth of convective updraft from the low level.
- As seen in Fig. 4.25B, deep convective clouds have appeared around 0000 and 0100 UTC on December 22 at the moisture boundary, seen in the 6.2 μm channel image elongated across the southern border of Egypt to the northeastern coast of the Red Sea. This is a sign for the presence of a jet in the upper-level divergent flow and forcing for vertical motion, especially at its left exit (at the black arrow).

(A)   **(B)**

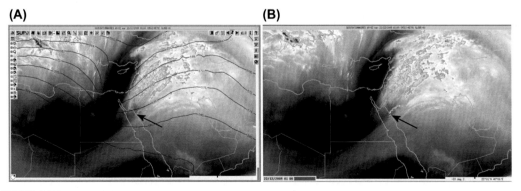

FIGURE 4.25

Meteosat water vapor 6.2 μm images superimposed by color-enhanced infrared 10.8 μm brightness temperatures (only ≤−40°C) on December 22, 2009, at (A) 0000 UTC superimposed by ARPEGE analysis of 500 hPa heights and (B) 0100 UTC.

As in the others considered in Sections 4.2.4 and 4.4.1, in this case the advanced EUMETSAT DIV product (based on data from WV 6.2 μm channel) performed well as a tool for the identification of divergent flow (related to upper-level vertical motion as a positive factor for deep convective evolution). However, there is no one-to-one correspondence between the upper-level divergence derived by satellite data and the areas of convective development. Deep moist convection occurs where favorable conditions at low and upper levels coincide. That imposes a limitation on using the upper-level divergence as a single parameter in diagnosing the convective environment and forecasting deep convective evolution.

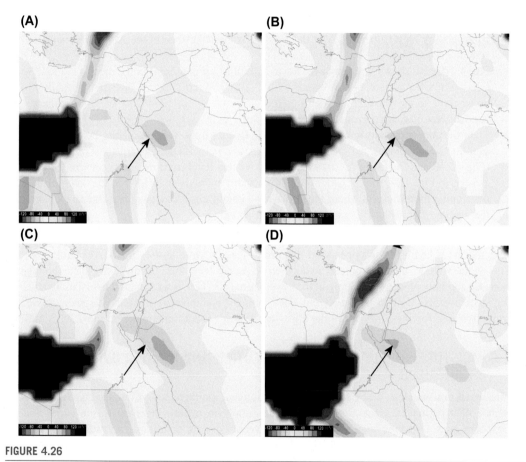

FIGURE 4.26

EUMETSAT DIV product (based on data from water vapor 6.2 μm channel) for December 21, 2009, at (A) 1945 UTC, (B) 2145 UTC, and (C) 2345 UTC; (D) for December 22, 2009, at 0445 UTC.

4.4.2.2 Diagnosing Thermodynamic Context of Convective Development

The sequence of WV images in Fig. 4.27 shows the propagation of moisture boundary related to the left jet exit along the eastern coast of the Red Sea, where the low-level convective ingredients are present. In addition, strong upper-level forcing can be diagnosed through analysis of WV imagery superimposed by relevant thermodynamic fields, as follows:

- The gray shade pattern seen in the WV image in Fig. 4.27A confirms that the development of deep convection is closely attached to the left exit of a moderate upper-level jet. The associated upper-level forcing is conductive for development of a strong convective complex being closely connected to the distinct jet moisture boundary in the WV 6.2 μm images (Fig. 4.27A and B).
- The upper-level jet is located along the boundary between the dry (dark) band, which indicates high PV, and the light warm part of the upper-level jet. This dynamic dark band was linked to stratospheric intrusion of dry air into the upper troposphere on the cold side of the jet. This is confirmed by high values of PV at the 300 hPa isobaric surface in the ARPEGE analysis (Fig. 4.27B).
- Fig. 4.27B also shows that low-level warm and moist air is present in the area of the Red Sea (wet-bulb potential temperature values of 18−20°C). Accordingly, convective instability is present due to the buoyancy force of cold/dry air over warm/moist air that is particularly conductive to the rapid growth of deep convective cells (see Section 4.3.2).
- Then the convective storm develops closely in line with the surge moisture boundary in the satellite 6.2 μm images and specifically at its cross-point with the Eastern coast of the Red Sea (Fig. 4.27C and D). Over this area the convective low-level environment is present, although, as seen in Fig. 4.27, the maximum low-level warm anomaly is located to the south of the developing thunderstorm where the upper-level conditions are not as favorable as they are in the leading part of the PV anomaly at the left jet exit.

This case is a manifestation of the relation between the development of deep moist convective storms and the strength of the upper-level dynamics. It is shown that the upper-level forcing mechanisms work well in the subtropical area when an upper-level trough and associated dynamical tropopause anomaly propagates deeply in the lower latitudes. There is an apparent association between upper-level divergent flow at the left jet exit, a PV anomaly, and the convective development related to a strong dynamic forcing of ascent in the upper troposphere (Santurette and Georgiev, 2005, 2007).

4.4.3 MOISTURE SUPPLY FOR DEEP CONVECTION AND RELATED DYNAMICAL STRUCTURES

As discussed in Chapter 3, a key factor for generation and maintenance of intense convection is the existence of a permanent synoptic-scale moisture supply in a deep layer up to the upper-middle troposphere over a specific area where the low-level mesoscale conditions as well as upper-level dynamics are also favorable. Such a moisture supply can come from large-scale convergence and evaporation from the surface as well as via strong large-scale horizontal transport of moist air in the lower and middle troposphere (see Section 3.3.4.2). Bao et al. (2006) showed that some enhanced bands of vertically integrated water vapor (IWV) in the central and eastern Pacific are associated with direct poleward transport of tropical moisture. For the purposes of satellite imagery interpretation, it calls for distinction between moisture fluxes, associated with dynamical structures at different tropospheric levels that will be considered in this section.

(A)

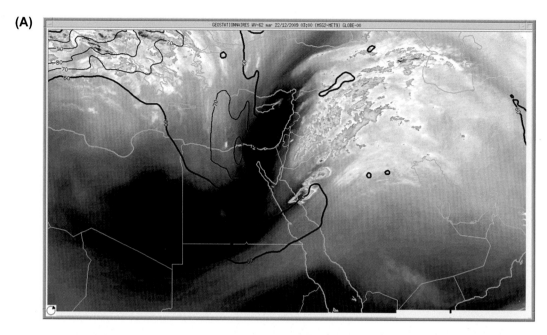

(B)

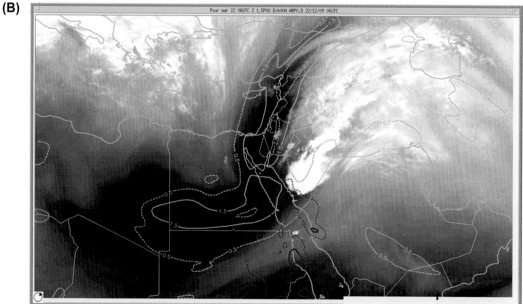

FIGURE 4.27

Meteosat water vapor 6.2 μm images superimposed by color-enhanced infrared 10.8 μm brightness temperatures (only ≤−40°C) on December 22, 2009, superimposed by ARPEGE analysis of thermodynamic fields: (A) at 0300 UTC superimposed by 300 hPa wind speed (only ≥60 kt), (C) at 1500 UTC superimposed by the potential vorticity field at 300 hPa isobaric surface, and (D) at 1800 UTC. (B) Water vapor 6.2 μm image at 0600 UTC superimposed by the potential vorticity at 300 hPa isobaric surface (green) and wet-bulb potential temperature (red, only ≥18°C).

(C)

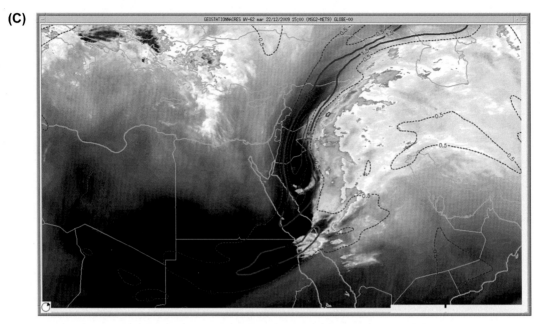

(D)

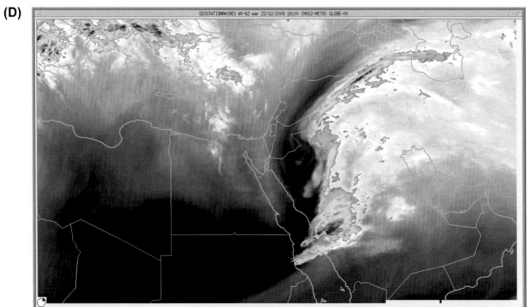

FIGURE 4.27 Cont'd

4.4.3.1 Atmospheric Rivers

It is well known that the precold-frontal low-level jet (LLJ) streams in the extratropical cyclones are characterized by warm temperatures, weak stratification, large WV content, and strong low-altitude winds (eg, Ralph et al., 2005). In various situations, such LLJs are conducive to the production of heavy rainfall through orographic forcing (eg, when the LLJ strikes the coastal mountains). These LLJs are associated with deep, narrow corridors of concentrated WV transport referred to as atmospheric rivers (ARs), reached from the subtropics to high latitudes. The type of LLJ that is the focus of the studies devoted to ARs is part of a broader region of generally poleward heat transport within extratropical cyclones that is referred to as the warm conveyor belt (WCB); see, for example, Carlson (1980). The ARs studied over South America, to the south of Australia, and over the Northeastern Pacific were found to be located to the east of a midlatitude cyclone, and their formation may be independent of the genesis and development of baroclinic cyclones (Zhu and Newell, 1998).

Ralph et al. (2004) show that the deep prefrontal moisture plume in conjunction with the strong front-parallel flow ahead of a cold front in lower latitudes (around $33°-35°N$) resulted in an AR of enhanced horizontal WV transport. Using flight-level observations and airborne radar data, they reveal that deep convection was triggered in a line within this narrow feature, where a convective updraft and maximum cloud liquid water was also observed. The ARs are found to be a critical factor for extreme precipitation during the winter in the east coasts of the Northeastern Pacific (Ralph et al., 2005) and Atlantic (Sodemann and Stohl, 2013).

4.4.3.2 Moist Conveyor Belts

Other events of enhanced moisture flux to the midlatitudes are rooted in the tropics and may be related to the tropical moisture conveyor belts (MCBs) in which most of the moisture transport takes place above the boundary layer (Bao et al., 2006; Knippertz and Martin, 2007). These are events of horizontal transport of moist air as with the case in Fig. 3.20D, Section 3.3.4.2. Knippertz and Martin (2007) studied this MCB associated with a single precipitation event in the Southern United States. They show that the majority of air parcels associated with significant WV fluxes originate from the mid-level (700–800 hPa) tropical easterlies and then slowly circulate northeastward toward North America. This is considered by Knippertz and Martin (2007) as a very marked difference to the classical WCB, where air parcels originate in the lower troposphere, and justifies the usage of the term "MCB" for this coherent trajectory ensemble.

Contrary to the atmospheric rivers and WCBs, the tropical moisture conveyor belt is a band of poleward moisture transport that originates from the tropics, above the boundary layer, and usually can be easily identified by satellite WV imagery, especially by images in the 7.3 μm channel.

4.4.3.3 Axes of Maximum Winds at Middle-Upper Troposphere and Related Movements of Moisture

The interaction of the synoptic airflow in the upper-lower troposphere and deep convection have been a focus of much research for a long time (eg, Uccellini and Johnson, 1979). The studies presented earlier support the view in Section 3.3.4.2 that moisture, transported from the (sub)tropics to the midlatitudes in distinct, individual plume-like events of moisture movements can be associated with MLJs and seen as moisture boundaries in the satellite 7.3 μm images (Georgiev and Santurette, 2009). As discussed in Section 3.3.4.1, for geographical regions influenced by the effects of a latent heat source, such MLJs are related to a thermodynamic environment favorable for convection, as they are associated with enhancement of advection of low-level warm subtropical air masses.

On the other hand, deep convection may play a key role in developing and intensifying upper-tropospheric jet streaks and also initiating midtropospheric dynamical features that aid in severe weather development. Hamilton et al. (1998) discuss how upstream convection drives an MLJ streak from diabatically forced geostrophic adjustment that produces a rapid unbalanced response and further accelerates parcels downstream controlling the removal of mass in the right exit region of this jet. This MLJ is maintained through convection being driven continuously, and the latent heat release forces the ageostrophic response ultimately culminating in geostrophic adjustments. Hamilton et al. (1998) implemented a primitive equation model with a prescribed thermal forcing to examine the development of the right-flank MLJ. Different sensitivity tests have been completed using latent heat release and barotropic as well as baroclinic flows. From both model results, they confirm that latent heating plays a primary role in the development of the MLJ. Transverse vertical circulations favor ascent on the right flank of the wind maximum at mid-level. It is shown that such unbalanced mid-tropospheric dynamical features could be critical processes that aid in severe weather development.

The findings of Hamilton et al. (1998) are consistent with the mechanism of moisture movements seen in the 7.3 μm WV images that operate through ascending motions and evaporation from the moist surface as well as by horizontal transport of moist air. As seen in Fig. 3.18, Section 3.3.4.2, the associated MLJ feature is produced and is reinforced by the continual growth and reformation of the updraft in a low-level baroclinic zone associated with a distinct positive anomaly of wet-bulb potential temperature. This results in a continuous acceleration of the winds near the level of maximum heating and the eventual enhancement of the unbalanced MLJ. Unbalanced refers to the direction of the transverse ageostrophic circulation, which is reversed from that known to preserve thermal wind balance in the upper-level jet entrance and exit regions (Hamilton et al., 1998). Such ageostrophic circulation enabled a nearly continuous acceleration downstream, as well as an increase in mid-level mass divergence on the right flank of the MLJ. This takes effect in the evacuation of mass from the underlying column on the right flank of the MLJ as it propagated northeast. Hamilton et al. (1998) concluded that this integrated mass loss enhanced the formation of a surface mesolow, the development of a low-level jet, as well as the eventual formation of an environment that was capable of producing and supporting the simulated deep convection.

Many authors have shown that the transport of warm moist air carried by jets in the lower and middle troposphere provides the latent heat required to drive deep convective updrafts and enhances convection and precipitation. This mechanism is particularly efficient to bring convective ingredients in the subtropics (eg, Insel et al., 2010) and the adjacent midlatitude areas such as the Southern United States, Mediterranean, and East Asia (Thiao et al., 1993; Georgiev and Santurette, 2009; Yasunari and Miwa, 2006). Among the other environment characteristics of the mean location for genesis and initiation of mesoscale convective complexes are areas of maximized warm, moist air transport via a low-level jet, and upper-level divergence superimposed low-level convergence (Durkee and Mote, 2010).

There are times when deep convective cells develop in association with a flow structure in the upper and middle troposphere that consists of a very wide zone of strong winds. In some of those cases there may be multiple axes of maximum winds and related movements of moisture at different levels within the wide zone. Such a case over the Mediterranean is shown in Fig. 4.28A with three parallel jet zones in the field of wind speed at 300 hPa (red, only ≥60 kt) located at the warm side of corresponding boundaries in the WV image. The cross-section of wind normal to the cross-section plane shows that the wind maximum on the left side is associated with the jet core at the 250 hPa level in the rear side of a trough of polar origin. The role of this jet and associated PV anomaly will be considered in Section 4.4.4.

(A)   **(B)**

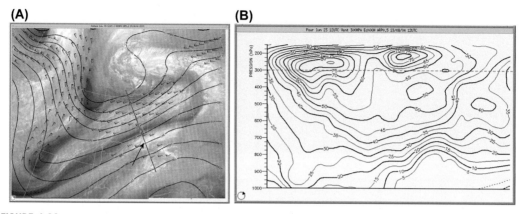

FIGURE 4.28

(A) Water vapor image of Meteosat-8 in 6.2 μm channel overlaid by 500 hPa heights and wind vectors at 300 hPa (red, only ≥60 kt) for August 23, 2004, at 1200 UTC. (B) Vertical cross-section of wind normal to the cross-section plane indicated by the green line on (A).

4.4.3.4 Diagnosis of Dynamical Moisture Structures

The imagery analysis based on the two WV channels, 6.2 and 7.3 μm, provides a more complete view of the upper-level dynamics and moisture movements related to specific large-scale flow patterns and can be performed for diagnosis of dynamical moisture structures. In order to analyze the origin and structure of the moisture flux, associated with the jets on the southeastern side of the cross-section in Fig. 4.28, WV images upstream of the wide zone of maximum winds, that is, over the southwest part of Fig. 4.28A, are shown in Fig. 4.29A and B. Fig. 4.29C and D shows cross-sections of wet-bulb potential temperature (θ_w, red contours) and relative humidity (pink, 90% filled) as well as wind speed normal to the cross-section (black contours) along the axis, shown in the images. The diagnosis based on the imagery analysis shows that:

- The jet in the middle of the zone (at the black arrow in Fig. 4.28A), located at about the 200 hPa level (Fig. 4.28B), originates from midlatitude upper-level dynamics. This jet produces an upper-level moisture boundary, which is distinctly seen in the 6.2 μm channel image in Fig. 4.29A (at the red arrow).
- The axis of maximum winds at the middle troposphere (at the blue arrows in Fig. 4.29B and D) is associated with large-scale subtropical moisture flux toward the Iberian Peninsula along the corresponding moisture boundary on the 7.3 μm channel image.
- The analysis of WV image gray shade patterns in the 7.3 μm image on Fig. 4.29B and the superimposed field of wet-bulb potential temperature at 600 hPa confirms that these two moisture movements have quite different origins: The first one is rooted in the midlatitudes to the west, while the second one comes from the subtropical latitudes.

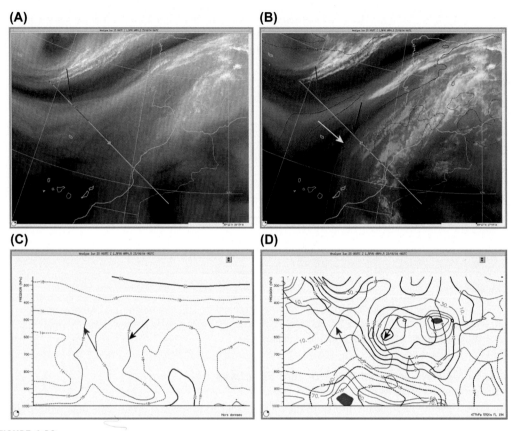

FIGURE 4.29

Meteosat-8 water vapor images for August 23, 2004, at 0006 UTC in (A) 6.2 μm and (B) 7.3 μm channel, overlaid by wet-bulb potential temperature at 600 hPa. Vertical cross-sections along the axis depicted in (A) and (B) of: (C) θ_w (red, 20°C thick) and (D) relative humidity (pink, 90% filled) as well as the wind normal to the cross-section (black contours).

The cross-sections in Fig. 4.29C and 4.29D show that the 7.3 μm channel image in Fig. 4.29B is sensitive to distinguish the differences of humidity at the mid-level (at the blue and the red arrows, respectively). The changes of θ_w and relative humidity along the cross-section line at 600 hPa in Fig. 4.29C and D correspond to the observed moisture features in the 7.3 μm channel image in Fig. 4.29B and fit with the θ_w contours at the blue and the red arrows, respectively.

In order to better analyze and understand the complicated low-level moisture field, Fig. 4.30A shows the corresponding cross-sections of wind speed normal to the cross-section plane (black contours) and total wind vectors (red). The image in the IR 10.8 μm channel in Fig. 4.30B shows that the cloud-free areas (nearly black), low-level clouds (dark gray, at the pink arrow), and medium-level clouds (light gray, at the yellow arrow) are present along the cross-section axis. Concerning the two

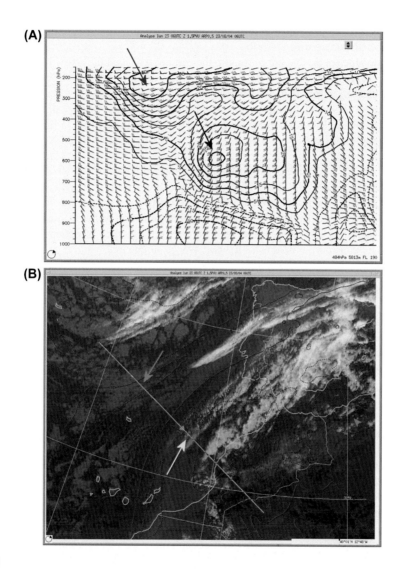

FIGURE 4.30

(A) Vertical cross-sections of wind speed normal to the cross-section plane (black contours) and total wind vectors (red) along the axis depicted in (B) Meteosat-8 IR (10.8 μm) image for August 23, 2004, at 0006 UTC overlaid by θ_w at 600 hPa.

quite different moisture movements, the following related changes in θ_w and relative humidity along the cross-section line are observed:

- The higher low-level θ_w values at the surface below the red arrows in Fig. 4.29C and D are due to the presence of very humid air and stratocumulus clouds in an area of low-level northerly winds seen in the cross-section of wind in Fig. 4.30A. Cold advection is present in the lower troposphere over this area (strong leftward shift of the wind from the surface up to 600 hPa level). In effect, stratocumulus clouds have been formed, as seen at the position of the pink arrow in the IR image in Fig. 4.30B.

- The highest values of θ_w that are present to the southeast of the blue arrow in Fig. 4.29B and C result from very warm low-level air due to the warm advection (right shift of the wind in the low levels as seen in Fig. 4.30A). Over this area, the relative humidity is significant near the surface capped by warm dry air in low levels. Moisture transport takes place above, in the middle troposphere, seen as a moisture boundary at the yellow arrow in Fig. 4.29B rooted in the subtropical latitudes.

4.4.3.5 Diagnosis of Large-Scale Confluent Moisture Movements by 6.2 μm and 7.3 μm Images

During the warm seasons when high moisture content is available, deep moist convection can develop in association with dynamical moisture structures at the middle and upper troposphere. Diagnosis of convergent large-scale moisture movements of different origin in such a preconvective environment can be performed by considering images of the two WV channels in 7.3 and 6.2 μm bands, as seen in Fig. 4.29A and B:

- The axis of the upper jet stream at the position of the red arrow in Fig. 4.30A is located along the moisture boundary at the red arrow in the 6.2 μm image in Fig. 4.29A.
- The MLJ stream at the position of the blue arrow in Fig. 4.30A is indicated by the moisture boundary at the yellow arrow in the 7.3 μm image in Fig. 4.29B. This mid-level moisture boundary cannot be distinguished in the 6.2 μm images in Fig. 4.29A because it is related to a large-scale WV movement, on its equatorward side, located at the 600 hPa level, and the sensitivity range of the 6.2 μm channel is quite narrow at this level (see Chapter 2 and Appendix A).
- A zone of two confluent moisture movements (one at the middle troposphere and another one at the upper troposphere) is shown by the difference between the directions of the red and yellow arrows, which are normal to the moist boundary associated with the jets in Fig. 4.29B. Correspondingly, westerly total winds are related to the jet with the WV movement at the upper level (Fig. 4.30A, upper left), and southerly/southeasterly winds in the MLJ (Fig. 4.30A, center) are responsible for the mid-level movement of moisture.
- The two moisture boundaries, which can be considered for diagnosis of the jet features at different troposphere layers, tend to approach each other downstream, and the two associated large-scale WV movements seem to parallel each other over the Iberian Peninsula.

A sequence of 6.2 μm images from 15 to 21 UTC showing the initiation and development of the deep convective systems is presented in Fig. 4.31A–C. The 7.3 μm image for 18 UTC is also shown in Fig. 4.31D. The convection initiated just to the northeast of the axis of the cross-sections in Fig. 4.32 and developed in close association with the large-scale moisture movements. Orographic effects are of significance concerning the initiation of some convective cells near the Pyrenees. As a result, two lines of mesoscale convective systems appear downstream (forward to the red and the blue arrows, respectively):

- An MCS organized in a line to the left of the cross-section axis center (looking downstream) where a dynamical upper-level forcing takes place after the upper-level jet axis get to a position above the MLJ axis, seen in the wind cross-section at the black arrow in Fig. 4.32A. The low-level conditions are also favorable in association with the local maximum of $\theta_w = 20°C$ at the red arrow in Fig. 4.32A.

- An MCS organized in a line to the right of the cross-section axis center (looking downstream) in the area of the most favorable low-level conditions with a large area of warm surface air of θ_w higher than 20°C, at the blue arrow in Fig. 4.32A. This MCS develops also in association with a local mid-level wind maximum aloft. This local jet is not visible as a moisture boundary in the 7.3 μm image because it is capped by the convective cloud feature at the blue arrow in Fig. 4.32B, top right.

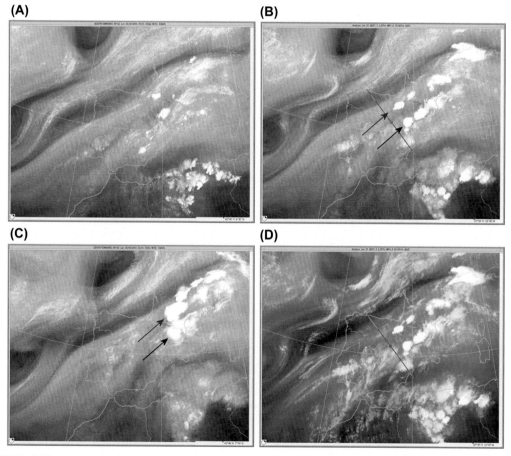

FIGURE 4.31

A sequence of 6.2 μm images from Meteosat-8 that shows the development of deep convective cells on August 23, 2004, at (A) 1500 UTC, (B) 1800 UTC, and (C) 2100 UTC. (D) 7.3 μm channel image for 1800 UTC.

The diagnosis by WV imagery analysis shows that large-scale mid-level WV movements take place in a deep layer over a wide area around the cross-section line center, depicted by the blue arrows in Fig. 4.32B, and provide moist air (relative humidity above 90%), feeding the maintenance of the strong deep convection.

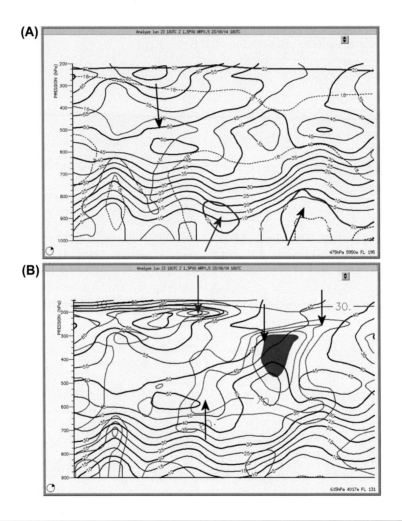

FIGURE 4.32

Vertical cross-sections across the axis depicted in Fig. 4.31C and D at the entrance region of the area of deep convection on August 23, 2004, at 1800 UTC. (A) Wet-bulb potential temperature (red, 20°C thick) and normal wind speed (black); (B) relative humidity (pink, 90% filled) and normal wind speed (black).

4.4.4 REINFORCEMENT OF CONVECTIVE DEVELOPMENT THROUGH A POTENTIAL VORTICITY ANOMALY ADVECTION: A SURGE MOISTURE BOUNDARY IN THE WATER VAPOR IMAGES

Severe convection may occur after reinforcement of an intense convective system, produced by an interaction with a dynamical tropopause anomaly. This section is aimed to further elucidate the mechanism of such a reinforcement applying upper-level diagnosis in the convective environment by using 6.2 μm images.

The WV 6.2 μm images in Fig. 4.33 show that finally the convective clouds, which have developed in two lines (as shown in Figs. 4.31 and 4.32) associated with the large-scale moisture movements in the upper and middle troposphere have embed in a large strong convective system over Southeastern France. Also overlaid on the images in Fig. 4.33 are the fields of 1.5 PVU surface heights and wind speed (blue contours) at 300 hPa (only ≥60 kt). These compositions show that the convection intensification was produced by interaction of the two jets along the southeastern part of a polar low within the wide zone of strong winds considered in Fig. 4.28, which involves large-scale confluent moisture movements, seen in Fig. 4.29B.

As usual, the jet stream axis on the polar side of the wide zone of maximum wind is associated with a strong PV anomaly as in Fig. 4.33A that will play a critical role in a further reinforcement of the convective process. The advection of PV anomaly is associated with the upper polar trough and can be followed by the evolution of the surge boundary on the WV image (at the red arrow in Fig. 4.33A).

The surge boundaries are defined by Weldon and Holmes (1991), who use the term "surge" to describe this type of significant but temporary motion of a WV imagery pattern with a dark dry region

(A) **(B)**

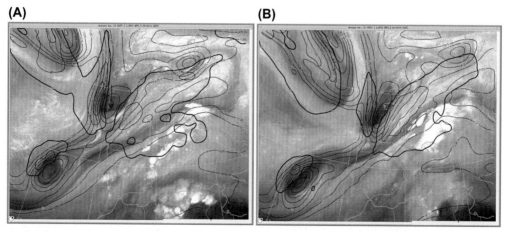

FIGURE 4.33

Convection intensification produced by interaction of two jets along southeastern part of a polar low. Water vapor images of Meteosat-8 in 6.2 μm channel overlaid by 1.5 PVU surface heights (brown contours, only ≤1200 dam) and wind speed at 300 hPa (blue contours, only ≥ 60 kt) for (A) August 23, 2004, at 1800 UTC and (B) August 24, 2004, at 0000 UTC.

at its upstream side. The feature accelerates and then decelerates significantly as it changes position and shape. The boundaries are usually very distinct, since high clouds and abundant high-level moisture are commonly present on the forward sides adjacent to the wide dark areas. Darkening of the 6.2 μm channel image in a specific area reflects the drying aloft from the sinking air due to the tropopause folding. From an operational perspective it is important to follow the surge boundaries, since the darkening surge process in the high amplitude trough circulation pattern conveys upper-level conditions, which are favorable for convection.

As shown with the consideration concerning Fig. 4.33A, a dual jet stream structure is observed along the southeastern side of the low over Ireland. By the second image in Fig. 4.33B, the double jet structure is still identifiable in the wind speed and PV fields. However, during the next 6 h between Fig. 4.34A and C, a fast approaching of the surge boundary occurred. As a result, the northern and the

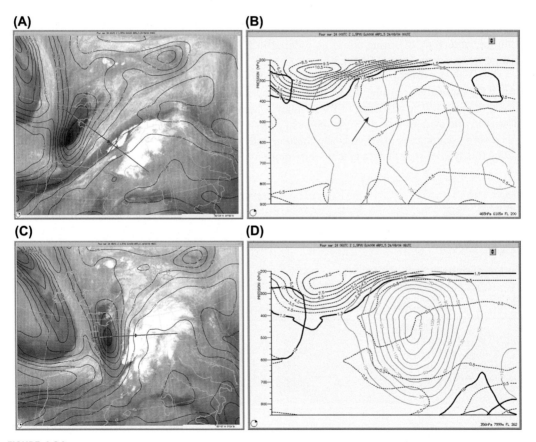

FIGURE 4.34

Reinforcement of convective development by interaction of the jet system with a dynamical tropopause anomaly on August 24, 2004. Meteosat water vapor images in 6.2 μm channel at (A) 0000 UTC and (C) 0600 UTC overlaid by 1.5 PVU surface heights. (B) and (D) the corresponding cross-section of potential vorticity (brown, only ≥1.5 PVU) and vertical velocity (10^{-2} hPa/s, ascent in orange, descent in blue) along the lines in (A) and (C), respectively, at the upstream side of the developing convective system.

southern jets tend to merge with each other, and a long jet stream system has formed in association with the tropopause folding along the dry surge structure (in Fig. 4.34C). This is evident in the merger of the corresponding moisture patterns associated with the jets. The interaction of the PV anomaly with the complex jet system contributed to formation of a large area of intense convection over France along the eastern side of the diffluent trough.

Fig. 4.34B and D shows the PV anomaly advection toward the convective system and associated forcing coming from the upper-level (enhanced ascent at the red arrow in Fig. 4.34B). This produces extremely strong ascent (170×10^{-2} hPa s^{-1}) in the next 6 h at the forward side of the PV anomaly maximum seen in Fig. 4.34D.

4.4.5 EARLY FORECAST OF UPPER-LEVEL FORCING FOR INTENSE CONVECTION

The dry surge WV imagery features considered in Section 4.4.4 are conservative dynamical structures associated with upper-level jets and dynamical tropopause anomalies. A critical nowcasting issue is to assess the potential of this type of dynamic structure to approach an area, where convective ingredients are present, and to produce upper-level forcing for intense convective development. Fig. 4.35 illustrates such a case in the leading part of the blocking circulation over the Mediterranean that exhibits a very diffluent upper-level flow pattern.

4.4.5.1 Diagnosing the Strength of the Upper-Level Dynamics

Various studies have shown the relation between the development of deep moist convective storms and the strength of the upper-level dynamics, which can be analyzed by WV imagery (Rabin et al., 2004; Santurette and Georgiev, 2005; Ghosh et al., 2008; Georgiev and Kozinarova, 2009; Georgiev, 2013). Extrapolating the evolution of dynamic dry bands and moisture boundaries in the 6.2 μm channel imagery is a way of early forecasting a potential strengthening of the upper-level dynamics over a specific area. Fig. 4.35A and B shows WV images overlaid by 500 hPa heights 24 h before severe convection developed over such an area of interest (at the black arrows). This process caused a catastrophic flash flood on the afternoon of August 5, 2005, and produced enormous damages of property as well as 10 fatalities over Bulgaria (Southeastern Europe). In order to diagnose the dynamical rate of the upper-level circulation over the area of interest, we may consider the following three types of existing moisture boundaries in the WV imagery pattern:

- The dry band at the position of the pink arrow, south of the area, where convective clouds develop on August 4 in the leading part of the blocking system. The images in Fig. 4.35A and B do not show a darkening process within the blocking circulation toward the area of interest (from the pink to the black arrow). This is a sign for a weak upper-level dynamics related to this upper-level feature.
- The moisture boundary at the blue arrow is a surge boundary, which (moving through the southern periphery of the blocking system) tends to increase its dynamical rate associated with a darkening process seen in Fig. 4.35A and B. At the time of Fig. 4.35A and B, this surge boundary is far from the area of interest, but the WV imagery analysis shows a potential for its propagation to this region moving within the circulation of the blocking system.

- The inside moisture boundary (indicated by the red arrow), which is related to the deformation zone on the equatorward side of a newly formed upper-level ridge (seen in Fig. 4.35A–C). The formation of this ridge leads to strengthening the blocking regime as a result from anticyclo-genesis (see Section 4.4).

In order to further illustrate and explain the different dynamical rates of these WV imagery features, Fig. 4.36 shows the corresponding WV images, overlaid by the contours of the geopotential (only $\leq$850 dam) and wind vectors (only $\geq$65 kt) at the 1.5 PVU surface. Using this diagnostic tool in the view of the considerations in Section 3.2.2, the following conclusions can be made:

- The dry band at the position of the pink arrow in Fig. 4.36A is not related to a distinct PV anomaly: Low values of the dynamical tropopause heights are not seen in Fig. 4.36B. This dry band tends to move away from the area of interest in accordance with the upper-level wind maximum south of the blocking system (at the position of the green arrow). No forcing of convection occurred in the area of interest, because the upper-level circulation south of this area tends to progressively reduce diffluence of the flow in the leading part of the blocking system.
- The surge boundary at the position of the blue arrow in Fig. 4.35A is associated with a jet stream and a PV anomaly (low heights of the 1.5 PVU surface heights), as seen in Fig. 4.36B. Strong subsidence occurred in its leading part, reflected by a darkening process in the WV imagery (Fig. 4.35A and B). This is a sign for further significant development and the potential for this dynamic feature to play a critical role in producing upper-level forcing.
- To diagnose strengthening of the blocking regime in the northwestern part of the cut-off system, the evolution of the WV imagery pattern can be considered. Fig. 4.35 shows that the inside boundary (indicated by the red arrows) is becoming more distinct and moving inward to the cut-off circulation with time. The northerly flow strengthens and a polar strip of dry air with high PV is intruded into the rear side of the blocking system that is seen by surging of the corresponding dry feature in the 6.2 μm WV imagery. The evolution of the PV anomaly and related dynamic dry band (indicated by the blue arrows in the WV imagery) is associated with the corresponding deformation of the wind field seen in Fig. 4.36. The blocking regime, in which the surge boundary (related to a PV anomaly) propagates, conveys in its northeastern part mid- to upper-level divergence to the area of interest. This is confirmed by increasing diffluence of the 500 hPa height contours downstream from the surging moisture boundary. The highly divergent flow pattern is shown by the diffluent imagery moisture features around the yellow arrow in Fig. 4.36D.

4.4.5.2 Early Forecast of Upper-Level Forcing for Convective Development

Fig. 4.37 illustrates that following a surge boundary associated with a dynamical tropopause anomaly, the forecasters may anticipate the process of upper-level forcing for vertical motion over a low-level area favorable for convection. Fig. 4.37A and B shows 18 h forecast of low-level thermodynamic fields (at 925 hPa level) valid for 12 UTC on August 5 from the ARPEGE operational numerical model. They are overlaid by WV imagery 12 h (Fig. 4.37A) and 6 h (Fig. 4.37B) prior to the time of the forecast fields. The following features of this process can be inferred through WV imagery analysis:

- The dry WV image signature (indicated by the pink arrows) is surging as a conservative dynamic structure before its intrusion in the northeastern part of the blocking system, where the low-level environment is conducive for convection initiation and development (Fig. 4.37).

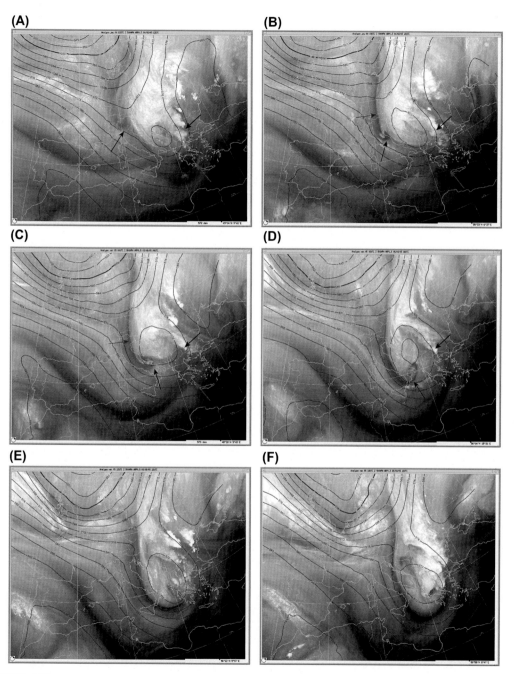

FIGURE 4.35

Strengthening of the inside boundary in sequence of water vapor 6.2 images, overlaid by 500 hPa surface heights on August 4, 2005, at (A) 1200 UTC and (B) 1800 UTC; August 5, 2005, at (C) 0000 UTC, (D) 0600 UTC, (E) 1200 UTC, and (F) 1800 UTC.

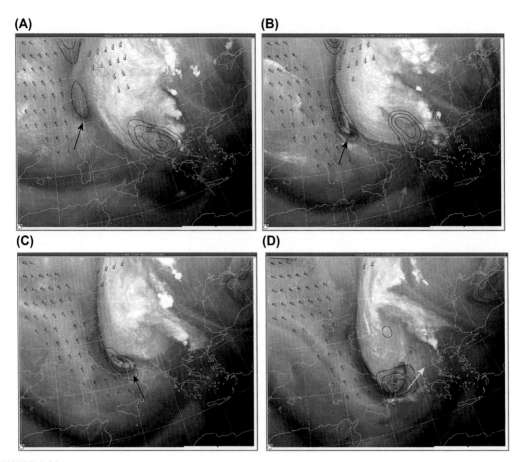

FIGURE 4.36

Serge boundary evolution in the southern part of the cut-off seen in 6.2 μm images, overlaid by the geopotential (only ≤850 dam) and wind vectors (only ≥65 kt) of the 1.5 PVU surface on August 4, 2005, at (A) 1200 UTC and (B) 1800 UTC; August 5, 2005, at (C) 0000 UTC and (D) 0600 UTC.

- Following the propagation of the surge moisture boundary in the WV imagery (Fig. 4.37A and B), an experienced observer is able to anticipate the dynamical tropopause anomaly advection over the area where the NWP model has forecasted a low-level environment favorable for convection:
  - Diurnal heating
  - Positive θ_w anomaly of warm/moist air
  - Convergence of the flow
- Strengthening of the jet takes place and, as a result, strengthening and propagation of the PV anomaly southeastward is observed (at the position of the pink arrows in Fig. 4.37A and B). At 06 UTC on 5 August (Fig. 4.37B) the surge boundary and related PV anomaly is in the southern part of the blocking circulation.

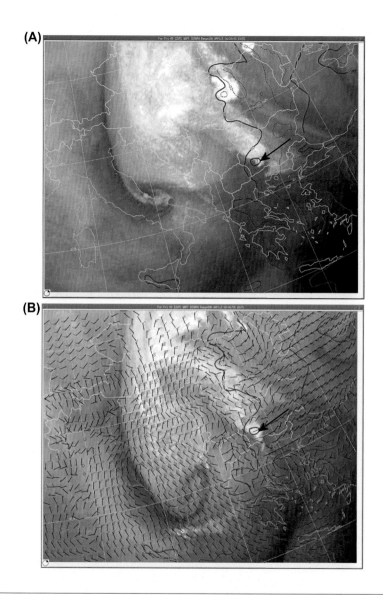

FIGURE 4.37

Advection of a dynamical tropopause anomaly, seen as a water vapor image surge boundary over a low-level environment favorable for convection on August 5, 2005. Water vapor images in 6.2 μm channel (A) at 0000 UTC overlaid by 18 h forecast of 925 hPa wet-bulb potential temperature (θ_w, red contours, only $\geq$20°C) valid for 1200 UTC and (B) at 0600 overlaid by ARPEGE 18 h forecast of θ_w and wind (blue *arrows*) at 925 hPa level valid for 1200 UTC.

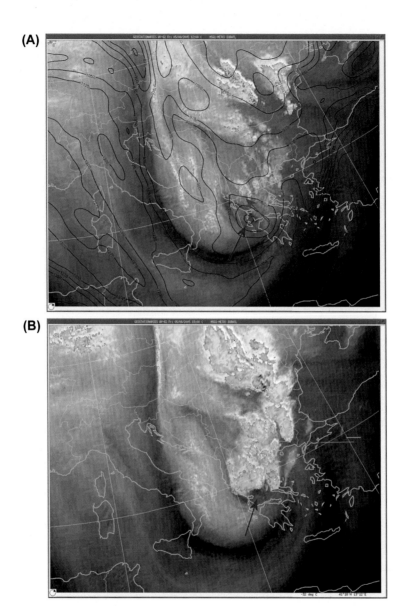

FIGURE 4.38

Meteosat Second Generation water vapor 6.2 μm images superimposed by color-enhanced infrared 10.8 μm brightness temperatures (only ≤−40°C) on August 5, 2005: (A) overlaid by the ARPEGE 12 h forecast of the 1.5 PVU surface heights valid for 1200 UTC and (B) at 1500 UTC.

Then, from 12 to 15 UTC (as seen in Fig. 4.38), the upper-level PV anomaly (red arrow) propagated further to the northeast in the blocking circulation and strongly reinforced the development over the area favorable for convection beneath a highly diffluent upper-level flow (seen by the imagery moisture features at the pink arrows in Fig. 4.38A). This process was implemented in a deep moist convection over the area of interest strongly intensified by the influence of upper-level dynamics.

4.5 SUMMARY OF THE CONCLUSIONS

A way to conduct efficient operational synoptic analysis and forecasting is to interpret the available information in the view of the appearance and development of key thermodynamic structures, relevant to the processes, which dominate over the area of interest. Section 4.2 presents such a framework regarding the processes of intense deep moist convection with a specific focus over the midlatitudes.

In Section 4.3, the synoptic conditions are diagnosed in terms of typical circulation patterns that may be clearly identified by imagery in WV channels. These are jet stream patterns at the mid- and upper levels, dynamical tropopause anomaly, blocking regime, advection and convergence of moisture in a deep tropospheric layer, and vertical displacement of air mass. Section 4.3.3 illustrates that the main factors for intense convection, especially in the subtropical latitudes, come from the low-level environment. The upper-level forcing plays a critical role, most often in the middle latitudes, to accelerate the convective development and hence to conduct a vigorous and severe deep convection.

Section 4.4 focuses on the interpretation of WV images in studying the potential for initiation and development of deep convection that can be performed from the perspective of the presence of three groups of conditions, as follows:

1. Large-scale thermodynamic structures of the convective environment related to dynamical imagery features, some of them being defined in Chapter 3.
2. Coincidence of these favorable large-scale flow/moisture patterns indicated in the WV imagery with relevant subsynoptic-scale conditions responsible for initiation of convection (associated with topography, diurnal heating, mesoscale vertical displacement of air masses, convective instability, and low-level moisture convergence).
3. Synoptic development of the flow structures responsible for upper-level forcing, maintenance, and reinforcement of intense convection.

The group (1) conditions are systematized in Table 4.1, along with a set of corresponding characteristic features of the thermodynamic environment of intense convection that may be indicated and followed in the images of the 6.2 and 7.3 μm channels in the WV absorption band.

The two WV channels of MSG, Himawary, and other geostationary satellites can be used as alternative or complementary information to assess mesoscale ingredients of convection such as those mentioned in the group (2) conditions for the purposes of issuing early warnings of intense convective developments as shown in Sections 4.2.4 and 4.4.5.

To illustrate the role of group (3) conditions, several types of convective developments in midlatitude, subtropical, and tropical areas are presented in Section 4.3 in the view of the proposed dynamical concept for interpretation of the system evolution in complex situations. Section 4.4

Table 4.1 Characteristic Features of the Thermodynamic Environment of Deep Convection That May Be Indicated and Followed in the Images of 6.2 and 7.3 μm Water Vapor Channels

Synoptic Conditions	Water Vapor Imagery Features
Upper-level forcing by advection of a dynamical tropopause anomaly	Dynamic dry dark zone associated with dry surge boundary feature in water vapor 6.2 μm imagery (Sections 4.3.2, 4.4.4 and 4.4.5)
Left exit region of an upper-level jet	At the eastern poleward side of the moist surge boundary in 6.2 μm (Sections 4.4.1−4.4.3)
Leading diffluent part of blocking systems	Leading poleward side of blocking systems seen in 6.2 μm imagery (Sections 4.3.4, 4.4.1, and 4.4.5)
Mid-level jets associated with strengthening of θ_w gradient in a low-level baroclinic zone	Moisture boundary in 7.3 μm image, which may not be distinct in the 6.2 μm image (Section 4.4.3)
Vertical displacement of air masses leading to increasing and release of convective instability	Moist pattern in 7.3 μm overlaid by dry surge in 6.2 μm (Sections 4.3.2 and 4.4.3)
Mid- to upper-level moisture supply	Water vapor movement (moist surge) in 6.2 μm (Section 4.4.3)
Low- to mid-level moisture supply	Water vapor movement (moist surge) in 7.3 μm (Section 4.4.3)
Confluence of large-scale moisture movements at mid-upper level	Confluence of moisture boundaries in 6.2 and 7.3 μm (Section 4.4.3)
Upper-level deformation flow pattern	Diffluent and vorticity moisture features (Section 4.3.2)

provides guidance for operational meteorologists in diagnosing and forecasting crucial elements of the strong convective development leading to severe weather as follows:

- Diagnosis of upper-level forcing/inhibition in the environment of moist convection (Section 4.4.1).
- Diagnosis of upper-level preconvective environment in subtropical areas (Section 4.4.2).
- Diagnosing the thermodynamic context of convective development (Section 4.4.2).
- Diagnosis of large-scale confluent moisture movements (Section 4.4.3).
- Diagnosing the strength of the upper-level dynamics (Sections 4.4.4 and 4.4.5).
- Early forecast of upper-level forcing for convective development (Section 4.4.5).

In diagnosing preconvective environments the weather analysis can be efficiently performed by using WV imagery, advanced satellite products, and other relevant information to apply dynamical concept in weather forecasting as well as to identify areas where the upper- and low-level positive factors for strong convection are coupled (see Sections 4.2.4, 4.4.1, 4.4.2, and 4.4.5) as follows:

- NWP upper-level wind and PV fields as well as EUMETSAT MPEF Divergence product.
- NWP model forecasts of low-level thermodynamic fields and air mass instability parameters.
- Advanced air mass instability parameters from hyperspectral satellite soundings such as the EUMETSAT MPEF Global and Regional Instability Indexes and/or upper-air sounding data.

USE OF WATER VAPOR IMAGERY TO ASSESS NUMERICAL WEATHER PREDICTION MODEL BEHAVIOR AND TO IMPROVE FORECASTS

5

CHAPTER OUTLINE

Weather Analysis and Forecasting. http://dx.doi.org/10.1016/B978-0-12-800194-3.00005-4

5.1 OPERATIONAL USE OF THE RELATIONSHIP BETWEEN POTENTIAL VORTICITY FIELDS AND WATER VAPOR IMAGERY

5.1.1 NATURE AND USEFULNESS OF THE RELATIONSHIP

As discussed in previous chapters, a joint interpretation of water vapor (WV) imagery and upper-level potential vorticity (PV) fields may provide valuable information in the process of operational weather forecasting, as follows:

- The satellite WV image is representative of the upper-level motion field (see Section 3.2).
- The PV concept (see Section 1.2) may be used to gain quick and direct insight into the upper-level dynamics.
- The evolution of the dynamical tropopause (in midlatitudes the 1.5 or 2.0 PVU surface) gives a good representation of the upper-level PV anomalies and associated perturbations in the upper-air dynamics.

Dynamically active regions in the upper-troposphere circulation are associated with several significant processes that are apparent in the WV imagery:

1. They are associated with a dynamical tropopause anomaly (cyclonic PV anomalies) and strengthening of the jet that produces areas of upper-level convergence and divergence in their close vicinity.
2. In such a region, the jet stream is characterized by a strong dark/light gradient on the WV imagery with dry air on the polar side.
3. Subsidence associated with the convergence occurs and, as a consequence, dry air of stratospheric origin is intruded in a particular area of the upper troposphere. The moisture content of that area decreases, and therefore the satellite WV image becomes darker.

4. The divergence is associated with ascending motions, which moisten the upper troposphere and lighten the WV imagery.

These four processes are responsible for the PV—WV relationship. As the dynamics become more active, the PV—WV relationship becomes more meaningful and can be used as a basis to validate NWP output by comparing the imagery with model-derived PV fields. Two general points of view are taken:

- In a dry (dark) feature PV—WV comparison, PV patterns are compared with dark-gray features in the imagery (see Fig. 5.1, red arrows). The dry comparison is applied in dynamically active regions to establish any mismatches between WV image dark features and cyclonic PV anomalies (dynamical tropopause anomalies). In Fig. 5.1B, for example, the dry comparison shows disagreement between the darkest parts of the image and the maxima in the PV field at 400 hPa.
- In a moist (light) feature PV—WV comparison, the PV—WV relationship is analyzed by considering light patterns on the imagery. The purpose of the moist comparison is to establish any mismatches at the moist side of the jet associated with high geopotential of the dynamical tropopause and light image gray shades. When the numerical weather prediction (NWP) output is correct, the moist convex patterns and cloud heads on the imagery correspond well to ridges in the field of dynamical tropopause heights. In Fig. 5.1A, areas of agreement are shown by blue arrows. However, to the west-southwest of the blue arrows, parts of the light cloudy features in the imagery are associated with troughs in the dynamical tropopause height because of shortcomings in numerical model performance.

A dry comparison allows identification and monitoring of WV dark zones associated with upper-level forcing of cyclogenesis. As shown in Section 3.2.2, dry intrusion consists of very dry air, which often descends to low levels near cyclones and produces dark areas on the WV imagery associated with cyclonic PV anomalies.

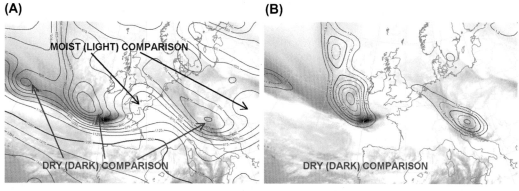

FIGURE 5.1

Water vapor image of December 27, 1999, at 1200 UTC superimposed on numerical weather prediction output fields: (A) heights (dam) of 1.5 PVU surface and (B) potential vorticity (only contours ≥1.2 PVU, the 2.0 PVU contour thick) at the 400 hPa isobaric surface. The *red and blue arrows* in (A) indicate dynamically active regions to make dry and moist comparison, respectively.

However, the PV—WV relationship is not always clear, and a forecaster must apply some knowledge of its nature when using such an approach. In regard to the dry features comparison, the point is that after strong subsidence has formed a dry streak to the rear of a cyclone, horizontal transport plays a major role in producing variability of the upper-troposphere flow. As a result, some dry air that had originally subsided also moved horizontally or even rose. Thus, although dark-gray shades of a WV image are associated with dry intrusion, they do not equate with instantaneous subsidence patterns but reflect a long history of parcel motion that can include ascent as well (see Section 3.2.1). In the next section, we discuss some complications that may arise when monitoring dry intrusions by superimposing WV images and PV fields.

Performing a moist feature comparison may be an effective operational forecast tool, especially during very early stages of disturbance development, before the expanding dry zone can be distinguished in the imagery. However, as discussed in Chapter 2, the WV channel's sensitivity to detect differences of humidity depends on altitude. Moreover, the level of maximum sensitivity varies with humidity, becoming lower with drier air (see Appendix A). As a result of these effects, a moist feature comparison is not always efficient. Some mismatches in the PV–WV relationship are possible because changes in the height of the dynamical tropopause surface lead to variable sensitivity of the WV channel. A dry feature comparison is usually more precise because it is made only in the areas where the PV–WV relationship works best. On the polar side of the dynamically active regions the tropopause lowers down to the middle troposphere. In this case, the WV channel exhibits its maximum sensitivity at mid-levels because of very dry upper-troposphere conditions.

The application of PV–WV comparisons of dark and light imagery features will be further considered in Section 5.4, where the focus is on assessing NWP model behavior in real situations.

5.1.2 INFORMATION CONTENT OF VORTICITY FIELDS RELATED TO WATER VAPOR IMAGERY

The WV imagery represents the dynamics near the tropopause or often some layers below the tropopause. Since PV anomalies associated with stratospheric dry intrusions have vertical depth, it is necessary to examine them on a range of vertical levels as well as to look at cross-sections. For that purpose, when interpreting the PV–WV image relationship it is convenient to superimpose various fields of PV onto the imagery (as shown in Chapters 3 and 4) or absolute vorticity. The most useful fields are the following:

- The height of the dynamical tropopause for midlatitudes, most often the surfaces of PV = 1.5 PVU (or PV = 2 PVU in the higher latitudes) and PV = 0.7 PVU in the tropics
- The cyclonic PV anomalies at upper-level isobaric surfaces (for 400 hPa most often the contours of PV $\geq$ 1.0 PVU are representative)
- The absolute vorticity (see Carroll, 1997b)

These fields may be used in various conceptual models for synoptic-scale analysis because they are associated with different aspects of the troposphere dynamics. The constant-PV surfaces of 1.5 or 2.0 PVU are chosen to lie between tropospheric and stratospheric values of PV and reflect characteristics of the dynamical tropopause in the middle latitudes. From these surfaces, the lower one (1.5 PVU) is usually more representative of the dynamical tropopause anomalies in the lower-middle latitudes, and it is more efficient to make dry feature comparisons. The PV anomalies at the 400 hPa level are closely associated with the dry air intrusion to the rear of the cyclones, since this level is near the level of maximum sensitivity of the 6.2 μm WV channel. Although the indicated vorticity fields are reflected by the gray shades in the WV images, they all show some disadvantages and limitations for use in a forecasting environment. For example, when comparing them with the imagery, any mismatch might be present in old cyclonic systems associated with any kind of blocking regime. It is also useful to use cross-sections for better viewing of the related structure of the tropopause folding. The field of PV anomalies at the 400 hPa isobaric surface is especially efficient when applying a dark-features comparison, and the fields of dynamical tropopause heights/absolute vorticity are quite useful for both dark and light comparisons.

5.1.3 COMPLEXITY OF THE RELATIONSHIP BETWEEN DRY INTRUSION AND POTENTIAL VORTICITY ANOMALIES

A number of studies have focused on describing the relationship between PV and WV channel radiance; however, the relationship is complex and far from being easily described. Complications can arise from various factors:

- Stratospheric intrusions occur only in the vicinity of the jet, so the relation between PV and the imagery dark areas is valid only in this vicinity, on the polar side of it.
- Mid-level PV anomalies may not be associated with dry zones on WV images in the leading part of diffluent troughs.
- The relationship can depend on the temperature profile in the path of WV channel radiation, on latitude, and on the season.

Therefore, there is no simple one-to-one correspondence between high cyclonic PV and high radiance, and there are features of the PV—WV relationship that can cause confusion and misinterpretation. In this section we elucidate some of the problems.

WV channels are sensitive to WV temperature (see Section 5.2), which varies with the altitude of the moisture layer, with changes in season and in the circulation system, and with differences of air masses at different latitudes. Whereas the temperature range in the troposphere may be near $100°C$ in summer, it is much smaller during winter. In a cold season, especially on the poleward side of the jet stream or within an upper-air trough environment, the warmest air in the vertical column may be near $-30°C$ and would produce light-gray shades on the imagery. Therefore, during winter seasons at high latitudes, or at middle latitudes during cold weather regimes, dark-gray shades on the WV images may not appear. Although the atmosphere can be very dry, the cold temperatures contribute to light-gray shades. Elsewhere, and during other seasons, dark areas and features are common, ranging from single-pixel sizes at the lee of the mountains or at the rear of convective clouds to very large regions over subtropical latitudes.

The PV—WV image relationship depends on the synoptic situation as well. In general, the PV maximums are associated with dry features, but:

- Dark zones in the WV images are not always connected with PV anomalies.
- The correspondence does not hold everywhere.

The relationship works best at midlatitudes, in cases associated with cyclogenesis, and close to, or poleward of, fronts (see Mansfield, 1996) and frontal zones at lower latitudes, such as over the Mediterranean. However, depending on the synoptic situation and the position of the area where the comparison is made, the dry areas may not correspond to PV anomalies and vice versa. Some important caveats must be considered:

- The relationship is meaningful only in connection with developing cyclonic systems. It is not valid, for example, in connection with decaying cyclones, since by that stage of development the moist air and dry air have already mixed, nor is it meaningful for anticyclonic systems, in general.

- As shown in Georgiev (1999) and Santurette and Georgiev (2005), the latitude effect should be taken into account when interpreting WV imagery; that is, the same darkening process on the imagery can denote larger PV anomalies in the higher latitudes or smaller PV anomalies in the lower latitudes.

Together with the latitude dependence of the PV—WV image relationship, another source of complexity in interpreting WV imagery to monitor dry intrusions can be disagreement between dry zones and regions of high cyclonic PV. Usually, such disagreement happens in specific synoptic situations and the interpretation problem may be considerably simplified by superimposing WV imagery onto different PV and diagnostic fields. As pointed out by Demirtas and Thorpe (1999), this relationship is not reliable in a variety of situations, either because the radiances are affected by factors other than WV (such as clouds), or because variations in WV content are not related to upper-tropospheric dynamics. In particular, over subtropical regions the air may be warm enough to appear as dark regions in WV images without a relationship with PV dynamics. This is illustrated for the case over the area to the Northeast of New Zealand in Fig. 5.2, where thermodynamic fields are superimposed on the WV image from MTSAT.

Analyses of the dynamical fields and wet-bulb potential (θ_w) at 850 hPa (see Fig. 5.2A) show warm air (high θ_w values) corresponding to a very dark area (at the green arrow). At the same time a folding of the dynamical tropopause at the black arrow in Fig. 5.2B is related to a much less dark dry zone in the imagery. In a moist feature comparison, the 1.5 PVU surface height shows a rising of the dynamical tropopause (at the blue arrow in Fig. 5.2B) that is related to white image shades to the west but to darker shades to the east of the blue arrow.

Other cases of poor relationship that can appear within upper-level cut-off lows and in the leading part of a diffluent trough are presented in Santurette and Georgiev (2005). Although the relationship between the PV distribution and WV imagery is complex, the visualization of an animation of WV images helps to monitor the evolution of the upper-level dynamics and to control the behavior of NWP models. The usefulness of the WV imagery analysis for operational forecasting will be considered in Sections 5.4 and 5.5.

5.2 SYNTHETIC (PSEUDO) WATER VAPOR IMAGES

Synthetic WV images, which are simulated as analysis and forecast output fields of some NWP models, are used as additional information in the operational forecasting environment. These are generated through any radiation transfer algorithm (eg, Schmetz and Turpeinen, 1988; Eyre, 1991) that processes NWP model—derived vertical temperature and humidity profiles. Fig. 5.3 shows comparisons between a Meteosat WV image and the corresponding synthetic (pseudo) WV image derived for a 6 h (T+6) forecast by the ARPEGE model and projected onto a grid of 0.1 degrees of horizontal resolution as a 6.2 μm channel image of Meteosat Second Generation (MSG).

As seen in Fig. 5.3, the two pictures exhibit the same general large-scale moist and dry features in gray shades. However, some differences are present as well (at the red arrows). Such differences may be associated with areas of mesoscale convective clouds due to the lower resolution of the pseudo WV image, which depends on the numerical model resolution and is less than the resolution

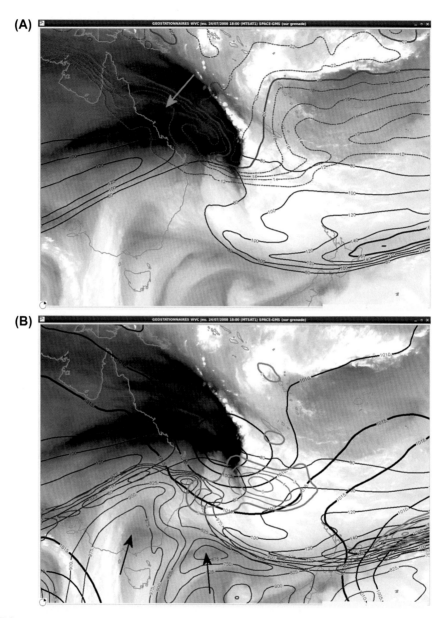

FIGURE 5.2

Satellite water vapor image superimposed on isotachs of the maximum wind on the 300 hPa isobaric surface (*blue lines*, every 20 kt) and (A) wet-bulb potential temperature at the 850 hPa surface (*red lines*, values above 12°C) and (B) mean sea-level pressure (*black contour*, 5 hPa intervals, values ≤1015 hPa), 1.5 PVU surface heights (brown, every 75 dam, values ≤975 dam), and 200 hPa divergence (green, >2×10^{-5}s^{-1}).

(A)   **(B)**

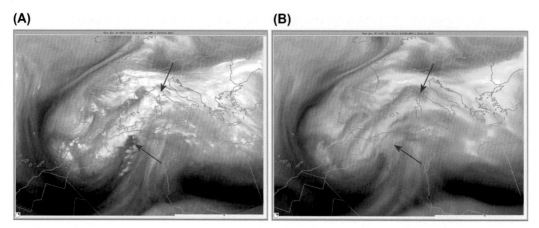

FIGURE 5.3

Comparison between Meteosat water vapor (WV) image and synthetic WV image for 1800 UTC on November 30, 2014, in (A) Meteosat WV image and (B) synthetic WV image derived at 6 h forecast by ARPEGE model. Also marked are features of disagreement (*red arrows*).

of the MSG WV images (3 km at the subsatellite point). Significant disagreement may be due to inadequacy of the radiation transfer algorithm, difficulties of large-scale numerical models to correctly predict convective clouds, or errors in the NWP model output (as will be discussed in Section 5.3).

Synoptic-scale differences between the satellite and the synthetic images are significant since they may be due to poor numerical model output. At locations indicated by red arrows, there are disagreements between the real (satellite) and the synthetic WV images, derived at the T+6 forecast. Such disagreements related to poor performance of the model output will be discussed in Sections 5.3 and 5.4.

5.3 COMPARING POTENTIAL VORTICITY FIELDS, WATER VAPOR IMAGERY, AND SYNTHETIC WATER VAPOR IMAGES

It is clear from the discussion in Section 5.1 that some care is required when interpreting the PV—WV relationship for operational purposes. Although mismatches between a WV dark zone and a PV anomaly might be present, the poor correspondence may result from the specific synoptic situation rather than from any NWP errors, and the comparison between model PV fields and WV imagery cannot tell us whether the numerical model is correct. In such cases, a powerful approach can be to compare the PV fields with synthetic (pseudo) WV images derived by the same NWP model as well as with satellite WV images, thereby allowing comparison of PV fields with moisture distribution from two different data sources.

5.3.1 CONCEPT OF VALIDATING NUMERICAL WEATHER PREDICTION OUTPUT BY APPLYING A WATER VAPOR—POTENTIAL VORTICITY—PSEUDO WATER VAPOR COMPARISON

The concept to use a comparison between satellite WV images, PV fields, and pseudo water vapor (PWV) images (WV—PV—PWV comparison) to validate and adjust NWP output is illustrated in Fig. 5.4. The data sources and products, both from satellites and NWP models, are shown in rectangular boxes, and the processes responsible for generating and using the products are shown in ellipses. A solid ellipse denotes an objective process based on theoretical considerations and computational techniques; a dashed ellipse denotes a subjective process based on theoretical assumptions but dependent on the experience of the human interpreter.

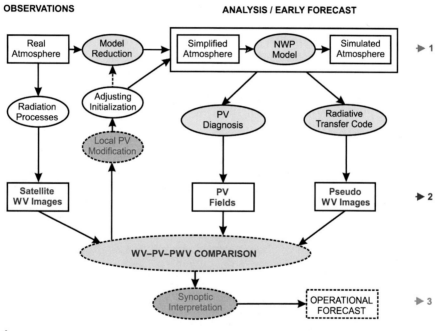

FIGURE 5.4

Concept of validating numerical weather prediction output at analysis and early forecasts by using the relationship between satellite water vapor (WV) imagery and model-generated potential vorticity fields and synthetic (pseudo) WV images.

The conceptual scheme shown in Fig. 5.4 reveals three levels of the weather forecasting process:

1. The first level (at the top of the scheme, marked "1") represents the numerical model run. The process of model reduction (analysis and initialization) of the real state generates a simplified atmosphere. Then, the simplified atmosphere is used as an initial state by the model, which predicts the evolution and produces a simulated atmosphere for the forecast.

2. The second, middle level (marked "2" in Fig. 5.4) represents the three kinds of products for comparison that are generated by satellite measurements or as output of the numerical model. The following two forward chains (downward arrows between level 1 and level 2) illustrate the processes of extracting the level 2 products:

   a. The left forward chain depicts the extraction of WV images from radiation measurements by the satellite instrument, which sees the real atmosphere via the radiation processes in the WV absorption band.

   b. The right forward chain represents the extraction of analysis and forecast products at level 2. The process of generating analysis/forecast PV fields is referred to as PV diagnosis, which means a process of calculating various PV fields from NWP output parameters (temperature and wind). The second branch of this chain represents the process of synthetic PWV image generation for analysis/forecast from NWP output parameters (temperature and humidity) by means of a radiative transfer code.

3. Level 3 (at the bottom of the scheme in Fig. 5.4) represents the final step of the procedure. At this level, a synoptic interpretation of the WV−PV−PWV comparison is performed and, on this basis, NWP output can be validated and operational forecasts can be improved.

The essential point is the WV−PV−PWV comparison, depicted between levels 2 and 3 in the scheme in Fig. 5.4. If any significant disagreement among satellite WV images, model-generated PV fields, and synthetic PWV images is detected, it may be useful in two ways:

1. It may be interpreted from a synoptic point of view (forward procedure, downward arrows between level 2 and level 3) to help the validation of NWP output and to adjust operational forecasts.

2. A backward procedure of a local modification of PV (upward arrows between level 3 and level 2) can be applied to adjust initial conditions in the NWP model. After adjusting the PV field to improve its fit with the satellite WV image, several methods have been tested and the following have appeared in the literature:

   a. The PV inversion-derived winds and temperatures can be assimilated into the model as bogus observations (Demirtas and Thorpe, 1999). This approach is indicated by the dashed upward arrow, which connects the processes of adjusting initialization and model reduction.

   b. Wind and temperature increments (as derived by PV modification and PV inversion) can be inserted directly into the model's initial fields (Swarbrick, 2001; Arbogast et al., 2008). Accordingly, the final step of the procedure links adjustments in initialization to a simplified atmosphere (solid upward arrow to the left).

An efficient method has been developed in Météo-France and is briefly described in Appendix B. Let us focus on the process of WV−PV−PWV comparison for analysis and early forecasts, which is the critical point of the concept presented in Fig. 5.4.

- Usually, the fastest check of NWP analyses and very short period forecasts is accomplished by comparing real (derived by satellite) and PWV imagery.
- However, PV fields should also be compared with the imagery because, when a real error is detected, this comparison may provide knowledge of the error in the PV distribution and allow potential adjustment of NWP initial fields by PV modification.
- It is also useful to compare synthetic WV images with PV fields generated by the same NWP model that produces the pseudo image. Because the PWV images are synthetic products of the model, they can be used to indicate whether any mismatches correspond to real NWP model errors.

A PV—PWV comparison must always be performed, since mismatch between the PV field and satellite WV image may be due to NWP errors or to one of the following:

- An inappropriate PV diagnosis resulting from errors or assumptions made in the PV calculation program, including the approximation of the Rossby-Ertel PV, the model resolution, the coordinate system, and so on.
- A lack of a useful PV—WV image relationship because of a special type of synoptic situation (see Section 5.1) that depends on the reliability of the PV fields to reflect any diagnosed aspect of the atmosphere (eg, to reflect dynamical processes in the troposphere, such as upper-level convergence/divergence associated with low-level descending/ascending motions).

In other words, validating NWP output calls for comparing all three different products at level 2 in Fig. 5.4. Since the satellite WV image is an observation of the real atmosphere, it is always assumed to be correct.

5.3.2 TYPICAL INSTANCES OF WATER VAPOR—POTENTIAL VORTICITY—PSEUDO WATER VAPOR COMPARISON

Applying the approach presented in Section 5.3.1, we can distinguish five possible cases of agreement/disagreement, each indicating a different kind of validity/inaccuracy along the chains of products and processes depicted schematically in Fig. 5.4. For operational applications, it is helpful to classify the possible results of the comparison into three main groups, depicted schematically in Fig. 5.5.

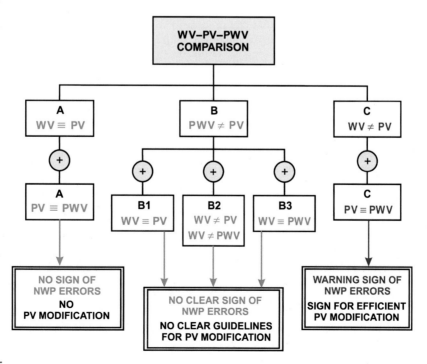

FIGURE 5.5

Instances of agreement/disagreement in the comparison among water vapor images, potential vorticity fields, and pseudo water vapor images (WV—PV—PWV comparison).

Also shown in Fig. 5.5 is the final decision, as suggested from the resulting comparison, and the relevance of applying PV modification to adjust the NWP initial conditions to reduce any model errors.

A. *Agreement between the WV image and the PV field as well as between the PV field and the synthetic image (WV $\equiv$ PV $\equiv$ PWV): no sign of NWP errors.*
This instance indicates no sign of NWP errors, and an assessment of the processes and products depicted in Fig. 5.5 would involve:
- An appropriate PV diagnosis (because PV $\equiv$ PWV)
- A useful PV–WV image relationship (because PV $\equiv$ PWV)
- An adequate radiative transfer code (because WV $\equiv$ PWV)
- An atmosphere well simulated and simplified by the NWP model (because PV $\equiv$ WV $\equiv$ PWV)

B. *Instances of mismatch between the synthetic image and the PV field (PWV $\neq$ PV): no clear sign of NWP errors.*
There are three possible subinstances that may indicate NWP errors, but the nature of the errors is unclear.

B1. Agreement between the WV image and the PV field (WV $\equiv$ PV).
The instance can indicate:
- The lack of a useful PV–WV image relationship (because PV $\neq$ PWV), or otherwise, an appropriate PV diagnosis and an adequate dynamical behavior of the NWP model (because WV $\equiv$ PV)
- Errors in the NWP model-derived moisture distribution or any inadequacy of the radiation transfer algorithm that produces PWV images (because PWV $\neq$ WV)
- No potential for efficient PV modification to adjust NWP initial fields

B2. Mismatches between the WV image and the PV field as well as between the WV image and the synthetic image (WV $\neq$ PV and WV $\neq$ PWV).
The instance can indicate:
- An inadequate PV diagnosis (because PV $\neq$ PWV)
- The lack of a useful PV–WV image relationship (because PV $\neq$ PWV)
- Errors in the NWP model-derived moisture distribution or any inadequacy of the radiation transfer algorithm that produces PWV images (because PWV $\neq$ WV)
- A lack of suggestions about the NWP model behavior
- A lack of clear guidelines for PV modification

B3. Agreement between the WV image and the synthetic image (WV $\equiv$ PWV).
The instance can indicate:
- An inadequate PV diagnosis (because PV $\neq$ PWV)
- A lack of a useful PV–WV image relationship (because PV $\neq$ PWV)
- An adequate radiative transfer code and NWP model-derived moisture distribution (because WV $\equiv$ PWV)
- A lack of suggestions about the NWP model behavior
- A lack of clear guidelines for PV modification

C. *Mismatches between the WV image and the PV field (WV $\neq$ PV) as well as agreement between the PV field and the synthetic image (PV $\equiv$ PWV): a warning sign of NWP errors.*

This instance indicates some sign of NWP errors and can indicate the following:

- An appropriate PV diagnosis (because PV ≡ PWV)
- A useful PV−WV image relationship (because PV ≡ PWV)
- Atmospheric dynamics poorly simulated or simplified by the NWP model (because WV ≠ PV)
- A potential to validate NWP output and adjusting NWP initial fields
- Clear guidelines for PV modification

In the first (**A**) instance the WV−PV−PWV comparison indicates "No Errors" (green), and the last (**C**) instance is a sign for "Existence of Errors" (red) in the dynamical performance of the numerical model. In the other three B instances, model errors are possible (orange), but it is unclear how to relate mismatches seen in the WV−PV−PWV comparison to the PV distribution. In some cases, a synoptic interpretation of the moisture distribution can be applied to gain knowledge of the errors.

Of course, when making a WV−PV−PWV comparison, instances of agreement or mismatch do not concern the image as a whole. They concern only specific details associated with important dry and moist features seen in the dynamical fields.

In cases of disagreement between WV and PWV images or if synthetic images are unavailable, the final conclusion of the PV−WV image comparison may be adjusted by considering the vertical distribution of humidity and vertical velocity derived by the NWP model.

The case when the WV image, the PV field, and the synthetic WV image all agree (WV ≡ PV ≡ PWV) suggests a general validity of the NWP model. However, it is rare to find a situation of complete coincidence of features on the fields/images being compared. When making the WV−PV−PWV comparison, it is important to identify whether there is a general agreement associated with synoptic-scale dynamical features. Smaller patterns of disagreement can be considered as insignificant. Some examples of applying this approach are presented in this section below.

Fig. 5.6A and B shows ARPEGE analyses of 1.5 PVU surface heights and winds at the 300 hPa isobaric surface associated with the jet superimposed onto the satellite WV image Fig. 5.6A and the synthetic WV image Fig. 5.6B of the ARPEGE model with a grid of 0.5 × 0.5 degrees. The important agreement among the three different products (WV, PV, and PWV) may be established by moist comparison (indicated "A") associated with the light-gray shade feature, which is located to the south of the green arrows. Such a moist pattern appears as a result of a secondary wave development close to the polar side of a strong jet.

Fig. 5.7A depicts a further undulation associated with the secondary development 6 h after the appearance of instance A in the WV−PV−PWV comparison. Also seen in Fig. 5.7A is excellent PV−WV image agreement associated with the moist pattern marked "M," which is a signature of cyclogenesis. Since the synthetic WV image was not available at 1800 UTC, the model analysis of humidity distribution may be used instead to check the validity of the PV−WV relationship with regard to this moist feature comparison. Fig. 5.7C shows a vertical cross-section of the ARPEGE analysis of relative humidity along the line A−B in Fig. 5.7B that intersects the moisture feature of interest M. Relative humidity of 60−70% is present at the 400 hPa level where the Meteosat WV channel 6.2 μm exhibits its maximum sensitivity.

Inspection of the humidity distribution in Fig. 5.7C shows that the appearance of the M feature in Fig. 5.7A results from a mid-tropospheric layer of moisture (between 350 and 450 hPa) with dry air

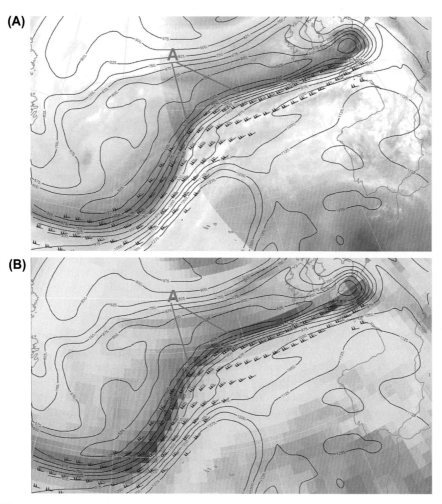

FIGURE 5.6

ARPEGE analysis of 1.5 PVU surface heights (brown, every 75 dam) and strong wind vectors (red, threshold 100 kt) at 300 hPa on March 18, 2002, at 1200 UTC, superimposed on corresponding (A) satellite water vapor image and (B) synthetic water vapor image derived at the 12 h forecast.

above, which is similar to case 4 shown in Fig. 2.9A. However, the depth of the moisture (about 90%) has risen to 700 hPa in Fig. 5.7C instead of to the 800 hPa level as in case 4 in Fig. 2.9A. Therefore, the brightness temperatures produced by the moisture distribution in Fig. 5.7C have to be slightly lower than those produced in case 4 of Fig. 2.9A because of the colder temperature of the low-level moisture layer. This takes an effect in the medium/light image gray shades seen in Fig. 5.7C and confirms that the vertical moisture distribution given by the model is relevant according to the radiative transfer theory for the satellite 6.2 μm WV channel (see Appendix A). This means that the ARPEGE model has correctly analyzed the upper-level moist feature M that appears in the WV image as a medium/light-gray shade feature associated with the secondary cyclone development.

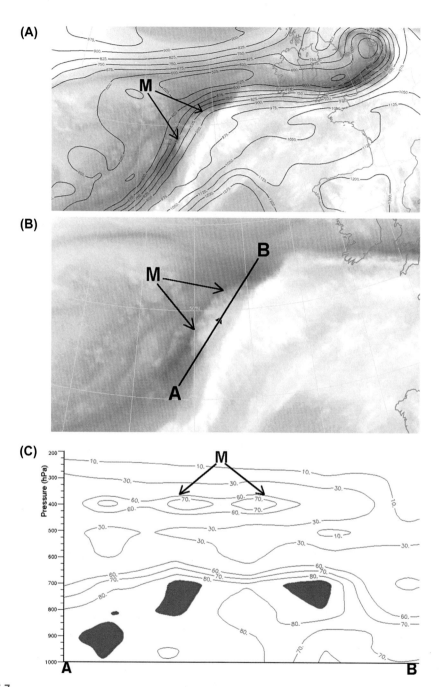

FIGURE 5.7

Properties of the moist pattern (M) associated with the secondary development, seen on March 18, 2002, at 1800 UTC. (A) Water vapor image overlaid by ARPEGE analysis of 1.5 PVU heights, interval 75 dam; (B) water vapor image showing the axis location (A–B) of the cross-section in (C); (C) cross-section of relative humidity (shading for 90%) along the axis A–B given in (B).

The cases of disagreement between synthetic images and PV fields (PV ≠ PWV; B instances in Fig. 5.5) include significant mismatches among the three different data sources. They may be seen in WV−PV−PWV comparisons for specific situations where the NWP model might be in error and/or the PV−WV image relationship lacks meaning as well. Examples for each of the instances, defined in Fig. 5.5, are presented in Santurette and Georgiev (2005).

Both the agreement in the PV−WV moist comparison at location M in Fig. 5.7A and the appearance of the high-level moisture feature M in the vertical cross-section in Fig. 5.7C confirm the validity of the dynamical performance of the model.

Fig. 5.8A and B shows a Meteosat WV image and the corresponding synthetic image derived at the 12 h forecast of the ARPEGE model, with a grid of 0.1 × 0.1 degrees over Europe. The same satellite and synthetic WV images with the field of geopotential of the constant surface of PV = 1.5 PVU superimposed are shown in Fig. 5.9A and B.

The comparison between the three data sources WV−PWV−PV shows the presence of the following instances:

- **A** at the yellow arrow: Relatively low dynamical tropopause that corresponds to dry air and dark-gray shades in both the satellite and the synthetic WV images.
- **B1** at the blue arrow: Relatively high dynamical tropopause that corresponds to moist air and light/white-gray shades in the satellite WV image, but the synthetic image shows a much drier (darker) zone there.
- **C** at the black arrow: The location of the tropopause folding (ie, local minimum of the 1.5 PVU surface heights) given by the NWP model corresponds well to the darkest dynamic dry zone on the model-derived synthetic image but it is related to a light-gray shade structure on the satellite images.
- **C** at the red arrow: At an area of relatively dark-gray shades on the satellite images the NWP model forecast gives wrongly a light-gray shade structure on the synthetic image and local maximum of the 1.5 PVU surface.

An important element of instance **B1** is the mismatch between the satellite and the synthetic WV images. This suggests errors in the moisture distribution analyzed or predicted by the NWP model. Such errors may be due to poor humidity data assimilation, which is a process independent of the dynamical performance of the model. Consequently, the model behavior as well as the surface pressure forecast may be quite accurate, even if mismatch between the satellite and synthetic images exists. The WV−PWV mismatch also may result from poor model-derived temperature fields or from any inadequacy in the radiation transfer algorithm used to produce synthetic or WV images. So, in Fig. 5.5, instance **B1** is shown to give the conclusion "no clear sign of errors."

Numerical model errors may be associated with mismatches of instance **B2** (the location of any dynamical feature on the synthetic image is shifted from its real position seen in the satellite image, and they both do not well correspond to the field of 1.5 PVU surface heights). However, if PV modification is made, it is uncertain that the result will be positive for the following reasons:

- The PV diagnosis may not be a relevant approach in such a kind of synoptic situation, which results in disagreement of the PV field to both synthetic WV product from the NWP model as well as the satellite WV image.

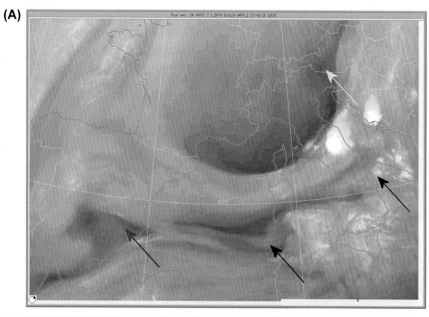

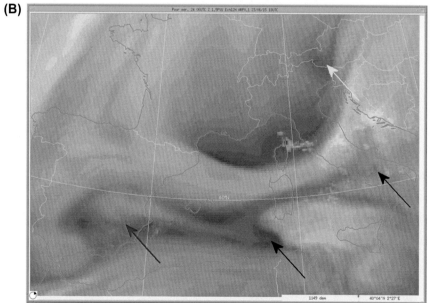

FIGURE 5.8

(A) Meteosat water vapor image for June 24, 2015, at 0600 UTC and (B) the corresponding synthetic image derived at the 12 h forecast of the ARPEGE model, with a grid of 0.1 × 0.1 degrees over Europe.

(A)

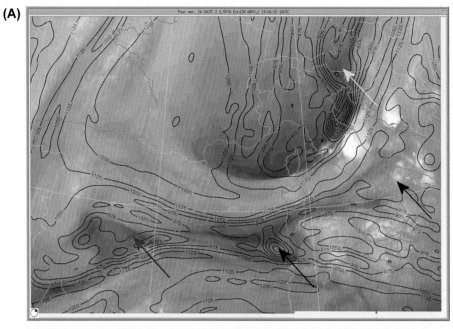

(B)

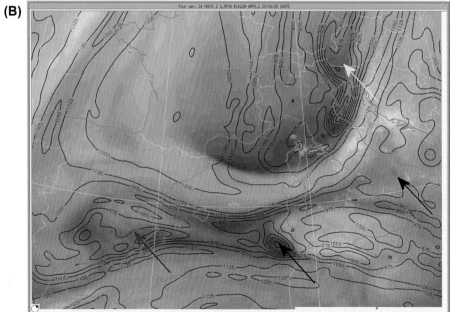

FIGURE 5.9

(A) Meteosat water vapor image for June 24, 2015, at 0600 UTC and (B) the corresponding synthetic image derived at the 12 h forecast of the ARPEGE model, with a grid of 0.1 × 0.1 degrees over Europe, overlaid by the 1.5 PVU surface heights from the 12 h forecast of the ARPEGE model.

- The mismatch between the satellite and the synthetic WV images may result from an inadequate simulation by the radiative transfer code or from inaccuracy of the NWP model moisture distribution that leads to a disagreement of the synthetic WV image to the real satellite observation of the atmosphere.

Therefore, in such cases, adjusting the PV field to the satellite WV image according to the PV concept (ie, to try to fit dark/light features to areas of low/high tropopause heights) is not recommended.

The disagreement of instance **B3** (the synthetic WV image fits the satellite image but there is a mismatch as to the PV field) often happens in old cyclonic vortexes associated with the presence of a blocking regime. Such cases of disagreement are usually not significant since the PV−WV image relationship might not be useful depending on the specific nature of the synoptic situation (Santurette and Georgiev, 2005). Generally instance **B3** happens in regions of low dynamical activity of the synoptic-scale processes. The reasons for such disagreements may result from inadequacy of PV diagnosis because (1) the PV concept may not be relevant to describe the upper-level motion field over that area or (2) the PV calculation procedure may suffer from errors in the model that simulate the PV distribution or from interpolation errors when changing coordinate systems. This kind of disagreement between the PV field and the imagery often occurs in a decaying cyclone where the dry and moist air are too mixed. The circulation is not dynamically active in such a decaying cyclone and does not exhibit a pronounced upper-level anomaly at dynamical tropopause heights. In synoptic situations that exhibit instance **B3** of the PV−WV−PWV comparison, the PV−WV image relationship is not an efficient tool for assessing dynamical behavior of the NWP model. From an operational point of view, the PV−WV relationship is useful only at the dynamically active regions.

5.4 SITUATIONS OF MISMATCH BETWEEN WATER VAPOR IMAGE AND POTENTIAL VORTICITY FIELDS AS A WARNING SIGN OF NUMERICAL WEATHER PREDICTION ERRORS

The most important instance of the WV−PV−PWV comparison is a mismatch between the WV image and the PV field as well as agreement between the PV field and the synthetic image/NWP moisture distribution (C in Fig. 5.5). It gives a warning sign of an NWP error and knowledge of this error in the PV distribution. The forecaster then can make the best estimate of how to subjectively adjust the subsequent forecast. We will present this approach by considering various cases of typical situations in which the numerical models can be in error in dynamically active regions, where the PV−WV relationship is useful.

5.4.1 CYCLONE DEVELOPMENT WITHIN A CUT-OFF LOW SYSTEM

Fig. 5.10 show the 6 h forecast field of 1.5 PVU surface heights superimposed onto the corresponding satellite and synthetic WV images for a case of a large cut-off low system located over the Eastern Atlantic and Western Mediterranean. A process of strengthening the upper-level dynamics in the southeastern part of the cut-off circulation is under way that finally results in cyclogenesis over the Western Mediterranean. The disagreements in the WV−PV−PWV comparison (at the blue and the red arrows) illustrates that in such cases the NWP models often are in failure to distribute correctly the total vorticity among dynamically active regions of the blocking system (see also Georgiev and Martín, 2001).

(A)

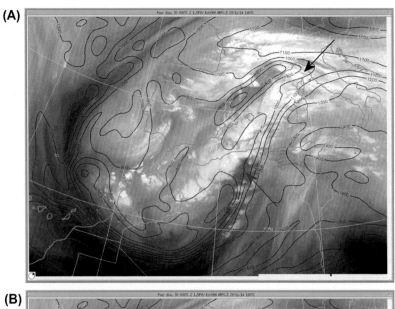

(B)

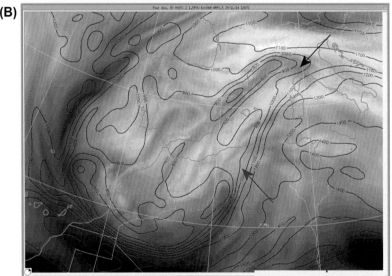

FIGURE 5.10

Numerical weather prediction errors seen in the WV—PV—PWV comparison at 0000 UTC on November 30, 2014. (A) Satellite water vapor (WV) image and (B) synthetic WV image from the 6 h forecast of ARPEGE overlaid by the 6 h forecast field of 1.5 PVU surface heights. Also marked: features of disagreements at the *red* and the *blue arrows.*

A significant error in the model performance is seen at 0000 UTC on November 30 by the PV−WV comparison in Fig. 5.10A, which shows that the correspondence between PV field and WV imagery is rather poor over the most dynamically active area of cyclogenesis over the Western Mediterranean:

- The dark dry slot in the satellite image at the red arrow in Fig. 5.10A does not correspond to minimum in the geopotential of the 1.5 PVU surface in the 6 h forecast field from ARPEGE.
- The tropopause folding (dynamical tropopause heights lower than 700 dam) given by the model is wrongly located significantly to the southwest of the pronounced dark zone of the WV image (at the red arrow in Fig. 5.10A).

The nature of this disagreement is analyzed through the comparison with the synthetic WV image forecasted by the same NWP model, shown in Fig. 5.10B. In the PV−PWV comparison, we have the following important observations:

- In Fig. 5.10B, to the southwest of the red arrow, the area of minimum of 1.5 PVU surface heights corresponds to a dry zone on the synthetic WV image derived from the 6 h forecast of ARPEGE.
- Consequently, the darkening process was handled by the model, but in the analysis fields, the disturbance is located southwestward (behind) of the real situation seen in the Meteosat WV image in Fig. 5.10A.
- There is a quite large disagreement between the satellite WV image and the synthetic WV image simulated by the ARPEGE forecast over the area of interest (at the red arrows) shown in Fig. 5.10A and B.

These results of the analysis confirm that the WV−PV−PWV comparison gives a warning sign of a model error (instance C according to the scheme in Fig. 5.5). The failure of the model to simulate correctly the advancing upper-level dry zone is a signal that an underestimation of the strength of vorticity advection and the related cyclogenesis would be expected in the NWP model performance. The error of ARPEGE can be seen in the T+18 forecasts of 500 hPa surface heights (Fig. 5.11A) and mean sea-level pressure (MSLP; Fig. 5.11B), which were overestimated by 3 dam and 2 hPa, respectively. Fig. 5.11B also shows errors regarding the position and the gradient of the surface low.

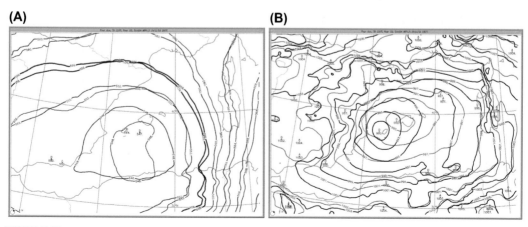

FIGURE 5.11

ARPEGE numerical weather prediction model error seen in (A) 500 hPa surface heights at T+18 forecasts (blue) and analysis (brown), and (B) MSLP at T+18 forecasts (red) and analysis (black) valid for 1200 UTC on November 30, 2014.

Knowledge of the bad model performance can be gleaned by comparing the fields in the satellite infrared (IR) image and the synthetic image from the 12 h forecast of ARPEGE derived through the same radiation transfer algorithm, which produces synthetic WV images, but applied for the MSG channel 10.8 μm. This kind of comparison based on Fig. 5.12A and B shows a wrong forecast of the cloud system of the cyclone developing over the Balearic Islands that is well seen in the location of the red arrows.

(A)   **(B)**

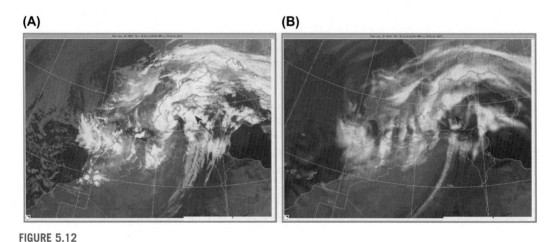

FIGURE 5.12

ARPEGE numerical weather prediction model error seen in (A) satellite infrared (IR) image and (B) synthetic IR images at T+12 forecasts valid for 0600 UTC on November 30, 2014.

5.4.2 CYCLOGENESIS WITH UPPER-LEVEL PRECURSOR IN STRONG ZONAL FLOW OVER THE NORTHEASTERN ATLANTIC COAST OF AMERICA

This section illustrates a case study example of an NWP error in the simulation of cyclogenesis with upper-level precursor over Northeastern America. Fig. 5.13 shows WV images of GOES-E satellite superimposed onto ARPEGE fields of geopotential and wind vectors (only ≥80 kt) at the 300 hPa isobaric surface on June 16, 2015, 1800 UTC. At this stage, upper-level precursors for cyclogenesis are not distinct in the strong zonal flow over Northeastern America. Six hours later the process is evident in the WV image and the field of 1.5 PVU surface heights shown in Fig. 5.13B together with vectors of strong winds at 300 hPa. By using WV imagery and analysis of vertical cross-sections in Fig. 5.13C of PV (blue) and wet-bulb potential temperature, θ_w (red), the following features of the process can be distinguished:

- A dynamic dry band has appeared in the WV image at the polar side of the cross-section line in Fig. 5.13B that is a signature of a PV anomaly and precursor for upper-level forcing of cyclogenesis.
- This is confirmed by folding of the 1.5 PVU surface down to the 450 hPa level seen in Fig. 5.13C (blue contours).

- The pronounced tropopause folding is closely in phase with a baroclinic zone in the lower-middle troposphere (strong horizontal gradient of θ_w in the lower troposphere seen in Fig. 5.13C). This is a signature for rapid cyclogenesis on the pressure field.

At the onset of cyclogenesis, there is a good agreement in the PV–WV comparison as follows:

- The dry band of the WV image at the polar side of the cross-section line in Fig. 5.13B is related to the area of low tropopause heights (brown contours).
- The jet stream is undulated and the NWP model well locates it on the equatorward side of the moisture boundary in the WV image.

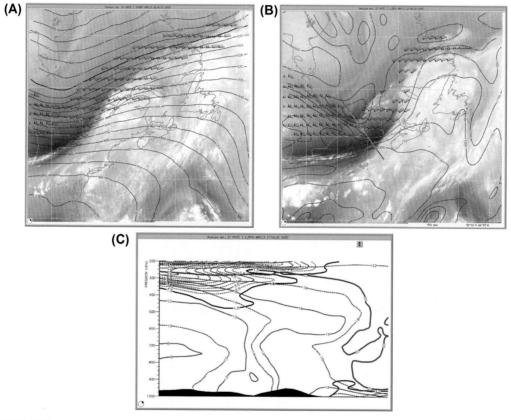

FIGURE 5.13

Water vapor images of GOES-E superimposed by ARPEGE fields: (A) 300 hPa heights and wind vectors on June 16, 2015, at 1800 UTC as well as (B) vectors of strong wind (threshold 80 kt) at 300 hPa and 1.5 PVU surface heights on June 17, 2015, at 0000 UTC; (C) vertical cross-section (along the line on (B)) of potential vorticity (blue, only ≥1.5 PVU) and with wet-bulb potential temperature (red).

Fig. 5.14A provides a way of checking on the validity of the NWP model performance in the next 6 h. The composition combines the satellite WV image and the 6 h forecast of the 1.5 PVU surface heights valid for June 17 at 0600 UTC.

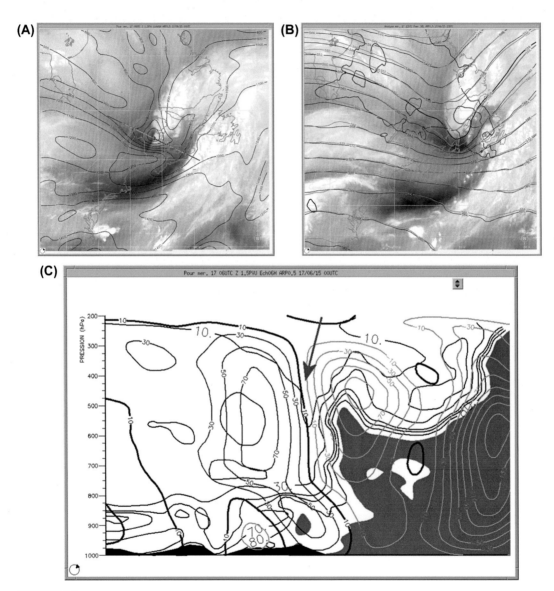

FIGURE 5.14

Water vapor images of GOES-E on June 17, 2015, superimposed by corresponding ARPEGE fields: (A) 6 h forecast valid for 0600 UTC of 1.5 PVU surface heights; (B) 500 hPa heights at analysis (brown) and 12 h forecast (green) valid for 1200 UTC as well as differences (*red contours*) between the MSLP fields from the same analysis and forecast; (C) vertical cross-section (along the line on (A)) of vertical velocity (10^{-2} hPa s^{-1}, ascending in orange, descending in blue) and relative humidity (*black contours*, above 90% *pink dashed*) from ARPEGE 6 h forecast valid for 0600 UTC.

The following conclusions of the PV−WV imagery analysis are of importance:

- The low heights of 1.5 PVU surface are well correlated with the dark areas upstream of the geopotential minimum, but there is a disagreement downstream.
- The area of minimum tropopause height (which normally corresponds to a dark dry zone in the WV imagery) now partially covers the area of deep clouds in the developing storm.
- This is a sign that the NWP model has overestimated the speed of the system propagation downstream. It seems to be a consequence of underestimation of the undulation of the zonal jet stream that leads to slight underestimation of the strength of cyclogenesis, which took place during the next few hours.

Since synthetic WV images are not available, the model-derived distribution of relative humidity associated with the developing cyclonic feature may provide additional insight into the model behavior. The cross-section in Fig. 5.14C along the vertical plane of the tropopause folding shows additional information for the NWP performance, as follows:

- There are discrepancies in the moisture distribution simulated by the model, which forecasted less than 10% relative humidity (at the red arrow in Fig. 5.14C) where clouds are present in the satellite image (Fig. 5.14A).
- On the other hand, the vertical distribution of vertical velocity is well predicted by the model having the transition between the strong ascending/descending motions at the rear side of the surge boundary in the WV image (at the center of the cross-section line in Fig. 5.14C).
- Therefore, the operational model well handled the areas of subsidence and ascendance. This means that the model behavior tends to follow the dynamics of cyclone development, which is the reason for a comparatively small error in the surface pressure.

The result of this model shortcoming is seen in Fig. 5.14B where the ARPEGE fields of 500 hPa heights at analysis (brown) and 12 h forecast (green) valid for 1200 UTC are presented. The model was wrong, underestimating the mid-level trough sharpening with 2 dam that was related to the under-estimation of the strength of jet undulation. In the same plot, the field of differences (red contours) between the MSLP at analysis and 12 h forecast valid for June 17 at 1200 UTC shows that the model has underestimated the low deepening with 2−3 hPa.

The case study example presented here is associated with a dynamical situation of a rapidly deepening cyclone. In such cases, PV fields and related concepts are beneficial in validating dynamical evaluation of model analysis and improving very short-range forecasts. Comparing upper-level PV fields and WV imagery helps to reveal errors in the numerical model performance. It was shown that, where synthetic WV images from model-analyzed or forecast fields are not available, cross-sections of humidity and vertical motions can be used to provide a more complete and accurate view of model behavior.

5.4.3 MOIST ASCENT AT CUT-OFF UPPER-LEVEL FLOW OVER THE NORTHEASTERN ATLANTIC

Dynamically active regions of ascending moist air in the middle and upper troposphere appear as distinct features of light image gray shades in the WV imagery and usually are associated with surface cyclogenesis (see, eg, Georgiev and Martín, 2001; Santurette and Georgiev, 2005). Fig. 5.15A shows the ARPEGE 9 h forecast of 1.5 PVU surface heights (brown contours) valid for 0900 UTC on June 17, 2015, superimposed by the corresponding WV image. The area of upper-level moist ascent (white image gray shades at the black arrow) has appeared in the previous 6 h on the polar side of a cut-off low over the Northeastern Atlantic. It is seen that the agreement between the NWP PV field and the satellite observations is poor: The upper-level updraft at the cut-off center is not well reflected in the PV field, which exhibits too slight elevation of the dynamical tropopause (Fig. 5.15A, along the black arrow). This area of mismatch corresponds to the center of the line of the cross-sections, presented in Fig. 5.15C and D, where dry air and weak ascending motions in middle-upper troposphere are simulated by the NWP model.

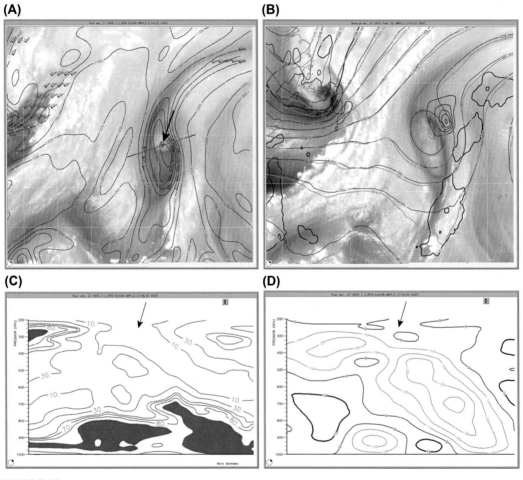

(A) **(B)** **(C)** **(D)**

FIGURE 5.15

Water vapor images from GOES-E on June 17, 2015, superimposed by ARPEGE fields: (A) 9 h forecast valid for 0900 UTC of 300 hPa wind vectors (in red, only $\geq$ 70 kt) and 1.5 PVU surface heights (*brown contours*); (B) 500 hPa heights at analysis (brown) and 18 h forecast (green) valid for 1800 UTC as well as differences (*red contours*) between the MSLP fields from the same analysis and forecast. Vertical cross-sections (along the line on (A)) of (C) relative humidity (above 90% dashed) and (D) vertical velocity (10^{-2} hPa s^{-1}, descending in blue, ascending in orange) from ARPEGE 9 h forecast valid for 0900 UTC.

In order to assess the model behavior regarding the simulation of low-level thermodynamic conditions, Fig. 5.16 shows the WV image overlaid by the field of wet-bulb potential temperature at the 850 hPa isobaric surface at analysis (black) and 12 h forecast (red) valid for 1200 UTC. It is seen that at this time

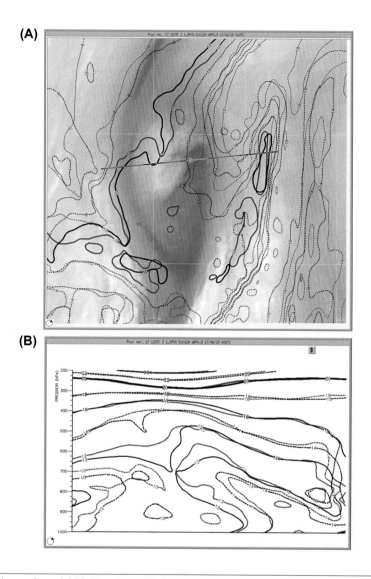

FIGURE 5.16

(A) Water vapor image from GOES-E on June 17, 2015, superimposed by corresponding ARPEGE fields of wet-bulb potential temperature (θ_w) at analysis (black) and 12 h forecast (red) valid for 1200 UTC; (B) vertical cross-section (along the line on (A)) of θ_w at analysis (black) and 12 h forecast (red).

the area of upper-level moist ascent has been developed as a hook cloud head seen in the WV image with a dark slot inside that is a sign for surface low deepening. The NWP model has not caught this development although it has well simulated the thermodynamic evolution as seen by the θ_w distribution:

- The θ_w field at the 850 hPa isobaric surface is overall well handled at the ARPEGE 12 h forecast (red contours in Fig. 5.16A), showing a good agreement with the analysis (black contours).
- The cross-section in Fig. 5.16B shows that the low-level distribution of θ_w from the surface up to the 700 hPa levels was correctly predicted (red contours) compared with the analysis (black contours in Fig. 5.16B).

Therefore, the ARPEGE error in the 18 h forecast seen in Fig. 5.15B is mainly due to wrong simulation of upper-level dynamics, which is seen in the analysis based on the satellite WV image and PV field 9 h earlier (Fig. 5.15A).

5.4.4 UPPER-LEVEL INFLUENCE ON DEEP CONVECTION WITHIN A CUT-OFF LOW SYSTEM

The considerations in Chapter 4 illustrate that in addition to its usefulness in analysis of synoptic-scale processes, WV imagery can provide a more realistic understanding of the convective system evolution and thus enable forecasting such extreme events in more detail. Following the dynamical upper-level structures, a forecaster may assess coupling of dynamically active regions at upper levels with a favorable preconvective environment in the lower troposphere. Thus the WV imagery and PV analysis can be used to assess the NWP model behavior and adjust early forecasts of strong convection. This section illustrates the application of this methodology in a convective situation, which produced a flood event on June 23, 2006, over the Balkans in Southeastern Europe that was poorly forecasted by the operational NWP model (the ARPEGE global model and the ALADIN mesoscale model, which is run through initial fields of ARPEGE). A detailed synoptic description of the event based on operational materials and evidences for the numerical model errors are presented in Georgiev and Kozinarova (2009). Low-level moist air, diurnal heating, and a terrain-induced lifting are existing factors favorable for the initiation of this convective process, but there is no evidence for a typical low-level convergence of the flow. This severe convective development was primary driven by PV advection over the region where low- to mid-level moist air is present, but the ALADIN operational NWP model had not predicted the intensive precipitation, which occurred there on June 23, 2006.

5.4.4.1 Dry Feature Comparison to Validate Numerical Weather Prediction Simulation of the Upper-Level Descent

Fig. 5.17A shows the error in the NWP forecast valid for 0600 UTC on June 23, 2006, as seen by a WV imagery analysis in terms of PV concept. By examination of differences between the dynamic dry dark patterns in the 6.2 μm WV image and the geopotential field at the dynamical tropopause (the surface of constant PV = 1.5 PVU), the following features of the model performance to simulate mid- to upper-tropospheric ascent are noted:

- The minimum of the tropopause height given by the model (at the blue arrow) is shifted about 200 km to the southwest from the real vorticity center seen as a dynamic dark zone in the satellite image (south of the black arrow).
- The area of the tropopause folding (975 contour of 1.5 PVU surface height, indicated by the black arrow) does not mirror the surge moisture boundary in the WV image that is a sign of the PV-gradient underestimation by the model simulation.

5.4.4.2 Reliability of Potential Vorticity–Water Vapor Relationship

If synthetic WV images are not available, the reliability of the PV–WV relationship can be analyzed by the vertical distribution of relative humidity across the area of upper-level descent (over the dark area in the WV image). The analysis of the cross-section in Fig. 5.17B confirms the conclusion (based on Fig. 5.17A) that the NWP model failed to catch the drying of the upper troposphere related to the vorticity feature: The relative humidity distribution in the middle and upper troposphere forecasted by the NWP model simulates values in the range 30–60% where the satellite observations show the presence of very dry air (at the black arrow).

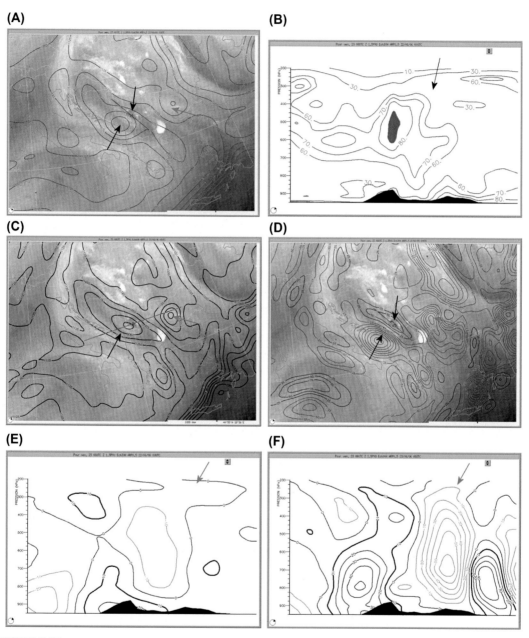

FIGURE 5.17

Meteosat water vapor image on June 23, 2006, superimposed by ARPEGE fields valid for 0600 UTC: 1.5 PVU surface heights in (A) 30 h forecast and (C) analysis as well as (D) the differences between the fields in (A) and (C). (B) Vertical cross-sections of relative humidity (above 90% dashed) along the line on (A) in 30 h model forecast. Cross-sections of vertical velocity (10^{-2} hPa s^{-1}, descending in blue, ascending in orange) along the line on (C): (E) 36 h forecast and (F) analysis.

The NWP analysis on June 23 at 1200 UTC, presented in Fig. 5.17C as a reference, shows a good correspondence to the satellite image: The vorticity feature fits much better to the WV image dark zone. Fig. 5.17D shows the differences between the 30 h ARPEGE forecast and the analysis in Fig. 5.17A and C, respectively. It is seen that the cross-section in Fig. 5.17B is performed along the dark zone in the WV image, where the tropopause folding was underestimated in the 30 h forecast of the NWP model with 75 dam (Fig. 5.17D, at the black arrow).

5.4.4.3 Moist Feature Comparison to Validate Numerical Weather Prediction Simulation of Upper-Level Ascent

Fig. 5.17E and F shows cross-sections of vertical velocity (along the line in Fig. 5.1C) in the 30 h forecast and the analysis, respectively. In the WV−PV analysis, we note the following features of the numerical simulation of vertical velocity:

- The area of ascending motion in the middle and upper troposphere seen in light-gray shades of the satellite WV image does not correspond to significant rising of the dynamical tropopause (at the green arrow in Fig. 5.17A).
- At the analysis field (Fig. 5.17C) this area corresponds to an area of much higher tropopause heights. Fig. 5.17D shows that in this light zone in the WV image the rising of the tropopause was highly underestimated by the NWP model 30 h forecast with 175 dam.
- The cross-sections of vertical velocity in Fig. 5.17E confirm that the operational model has wrongly forecasted descending motions over the area of image light shades (at the green arrows in Fig. 5.17A and E). This confirms the reliability of WV−PV relationship and the model failure in its 30 h forecast. The better correspondence in WV imagery and PV field (Fig. 5.17C) leads to much better simulation of vertical velocity by the NWP model analysis seen in Fig. 5.17F at the green arrow.

5.4.4.4 Nature of the Poor Numerical Weather Pattern Forecast

By comparing the ARPEGE analysis and forecast fields of 1.5 PVU surface heights, Fig. 5.18 shows that in the next hours the underestimation of the strength of the upper-level dynamics by the NWP model simulation has increased. The following considerations are noteworthy:

- The upper-level PV gradient in the forecast field (pink contours) in Fig. 5.18A is much weaker than in the analysis (black contours).
- The elongated dry (dark) vorticity feature seen in the WV image (at the blue arrow in Fig. 5.18A) is not well correlated with the shape and position of the minimum of the 1.5 PVU surface height in the forecasted field (pink contours).
- The difference in the geopotential of the 1.5 PVU surface between the forecast and analysis fields in Fig. 5.18B shows that the tropopause folding was underestimated in the 36 h forecast of the NWP model with 175 dam (at the black arrow).

In this section we have considered an NWP model performance in the simulation of upper-level dynamics of a convective environment by using satellite WV imagery and PV analysis. Applying this practice can help the forecasters to assess the validity of NWP dynamical fields and correctly diagnose the severity of the convective situation in the presence of low-level favorable synoptic factors. As shown in Georgiev and Kozinarova (2009), in the presented case study the poorly predicted

(A)

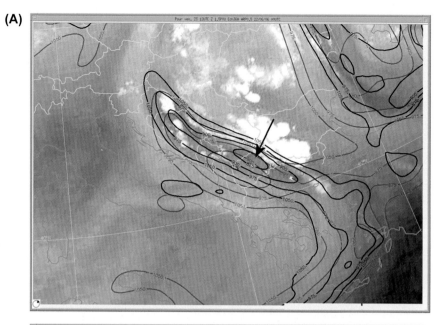

(B)

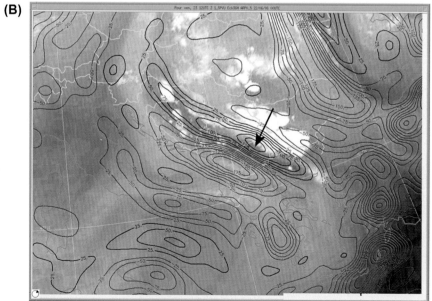

FIGURE 5.18

Meteosat water vapor image on June 23, 2006, superimposed onto corresponding ARPEGE fields: (A) 1.5 PVU surface heights (only ≤1050 dam) from 36 h forecast (pink) and analysis (black) valid for 1200 UTC as well as (B) onto differences between the fields in (A).

convective storm was related to the underestimation of the interaction between the upper-level positive PV anomaly and a low-level warm anomaly. As a result the ARPEGE model underestimated both the upper-level forcing of the vertical motion and the strength of the low-level warm anomaly, which produced a baroclinic zone (strong θ_w gradient).

5.4.5 RAPID BAROCLINIC CYCLOGENESIS IN A STRONG ZONAL ATLANTIC FLOW

This section focuses on a secondary cyclone development in the polar side of a strong jet where a large mismatch between PV fields and the satellite WV imagery is present. Interaction between the jet stream and a dynamical tropopause anomaly is involved in this process that plays a critical role in promoting cyclogenesis by strengthening the jet streak (see Section 3.3.2). The cyclone that developed on December 27, 1999, exhibits just this sort of development, but none of the operational models available were able to correctly predict the deepening of the surface low, which moved very fast over France, producing windstorms and catastrophic damage. This case is a manifestation of the importance for operational weather forecasting to perform WV imagery analysis in terms of "PV thinking."

Fig. 5.19A shows a satellite WV image for December 27 at 0000 UTC overlaid by an ARPEGE analysis of dynamical tropopause heights; Fig. 5.20A is the same picture superimposed onto the PV anomalies at 400 hPa (blue) and wind vectors (red). The cyclogenesis begins as an interaction between the zonal jet stream and the strong PV anomaly to the northwest of the jet (see Fig. 5.20A). The high PV values of about 4 PVU associated with this dynamical tropopause anomaly indicate that the numerical model has analyzed stratospheric air at 400 hPa as a result of the tropopause folding. The zonal jet stream is seen in the WV image as a high dark/bright gradient area, with the dry air on the polar side (see Section 3.2). The beginning of cyclogenesis is associated with undulation of the jet at location L

(A)   **(B)**

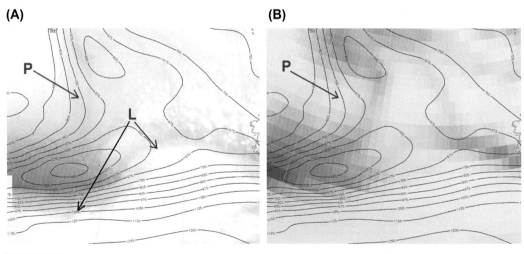

FIGURE 5.19

Field of 1.5 PVU surface height (dam) ARPEGE analysis superimposed with the corresponding satellite and synthetic water vapor (WV) images for December 27, 1999, at 0000 UTC. Also marked: "P," the dark strip of a polar potential vorticity anomaly, and "L," the undulated jet stream feature in the WV image. (A) Meteosat WV image; (B) synthetic WV image derived from ARPEGE output analysis.

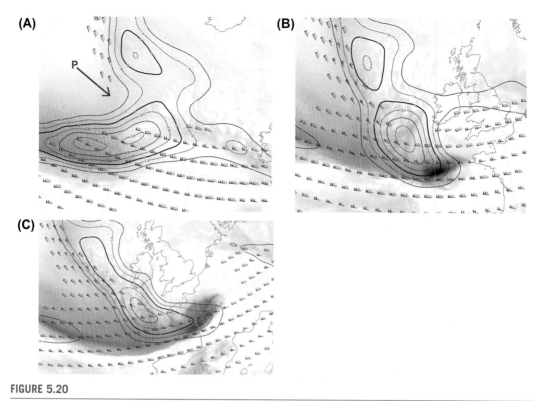

FIGURE 5.20

Water vapor image overlaid by ARPEGE fields of potential vorticity (PV) anomalies at 400 hPa (only contours ≥1.0 PVU, the 2.0 PVU thick) and wind vectors at 300 hPa (red, threshold 80 kt) for December 27, 1999. Also indicated is "P," the dark strip of a cyclonic PV anomaly: (A) at 0000 UTC, analysis; (B) at 1200 UTC, analysis; (C) at 1800 UTC, 6 h forecast.

with cyclonic curvature at the blue arrow and anticyclonic curvature at the red arrow in Fig. 5.19A. However, the wave-shaped feature of the jet is not reflected in the corresponding dynamical fields, nor in the dynamical tropopause heights or in the 300 hPa winds (see Fig. 5.20A). The convex part of the WV-image jet feature in Fig. 5.19A (at location L, red arrow) corresponds to relatively low dynamical tropopause heights. This means that at 0000 UTC, the operational numerical model did not catch the ridging (associated with the moist ascent in the circulation of the wave) and was late in analyzing correctly the process of rapid cyclogenesis.

Together with the major pattern of interaction between the zonal jet stream and the strong dynamical tropopause anomaly a northern-latitude PV anomaly is present northward of location P in Figs. 5.19A and 5.20A. Inspection of the satellite WV image in Fig. 5.19A shows that a dark zone directly connects these two dynamical objects, allowing the major PV anomaly (which interacts with the zonal jet) to be enforced by a polar intrusion of high PV. However, in the synthetic image of the ARPEGE model with a grid of 0.5 × 0.5 degrees, this PV anomaly is disconnected from the polar intrusion by a light area, at location P in Fig. 5.19B. Therefore, the model has analyzed a false area of ascending motions instead of the polar strip of dry intrusion.

Fig. 5.21C and D shows vertical cross-sections along the line N—S in Fig. 5.21A and B. It is evident that the N—S line crosses different gray-shade patterns on the two images. Within the real area of dry air that produces medium-gray shades at location P in Fig. 5.21A, the numerical model sees a moisture pattern of 80% relative humidity at 500 hPa (see Fig. 5.21C) that gives the synthetic image a light appearance. This false area of ascent is associated with a false rising of the dynamical tropopause at location P in Fig. 5.21D. It is obvious that this ARPEGE error leads to a false disconnection of the dynamical tropopause anomaly at location S (interacting with the zonal jet) from the polar strip of PV anomaly advection at N.

After onset of rapid cyclogenesis, the behavior of the model tends to follow the real evolution of the cyclonic PV anomaly: In Fig. 5.20B and C, the two areas of PV = 2.0 PVU at 400 hPa (thick blue contours) are connected. The northern area of high PV is associated with the polar dry strip P, and the southern area of high PV (northward of the blue arrow in Fig. 5.19A) produces a dry slot to the rear of the developing vortex (Fig. 5.20B and C). Although the model performance was not perfect, this development corresponds much more accurately to the pattern in the satellite WV imagery than the situation in Fig. 5.20A does. Therefore, at 1200 (analysis) and 1800 UTC (6 h forecast) the model tends to attenuate the error and to directly connect the polar anomaly to the zonal jet streak, as seen in Fig. 5.20B and C. This behavior of ARPEGE confirms that interaction between the two dynamical features occurs during the cyclogenetic process.

In summary, comparison between the satellite and synthetic WV images overlaid by dynamical tropopause heights at ARPEGE analysis on December 27 at 0000 UTC (see Fig. 5.19A and B) shows

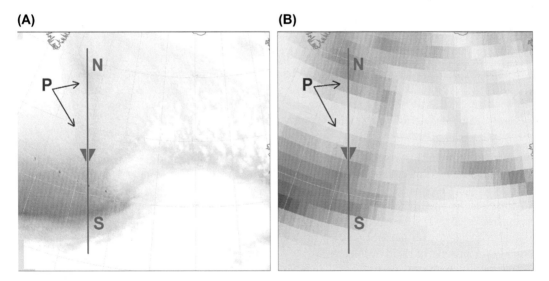

(A)　　　　　　　　　　　　　　**(B)**

FIGURE 5.21

Satellite and synthetic water vapor images in (A) and (B), showing the N—S line of the vertical cross-sections of (C) and (D) for December 27, 1999, at 0000 UTC. Also indicated: "P," the dark strip of a polar potential vorticity anomaly. (A) Meteosat image; (B) synthetic image derived from ARPEGE analysis; (C) vertical cross-sections of relative humidity (%); (D) vertical cross-sections of relative humidity (%, *pink contours*) and potential vorticity (PVU, *blue contours*).

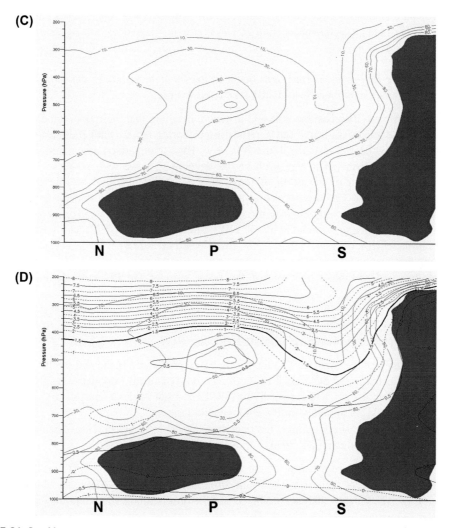

FIGURE 5.21 Cont'd

clear signals for numerical model failure in predicting such a complicated case of rapid cyclogenesis. A mismatch is present in the WV−PV−PWV comparison associated with the zonal jet stream, which is primarily responsible for the cyclogenesis and appears in the satellite WV image in Fig. 5.19A as an undulation (the wave-shaped pattern L) of the high dark/light gradient band. Several features are noteworthy:

- At the very beginning of the rapid cyclogenesis this jet undulation is not analyzed correctly by ARPEGE; the shape of the high-gradient area in the field of 1.5 PVU surface heights does not coincide with the image feature L.

- There is a mismatch between the satellite and the synthetic WV image, associated with the undulation of the jet feature at location L. The synthetic image does not appear as light as the satellite image at the red arrow, and does not appear as dark as the satellite image at the blue arrow.
- The synthetic WV image corresponds better to the PV field than does the satellite image, which exhibits a much more undulated jet stream feature.

This comparison may be considered as a C instance, meaning that the model underestimated the strengthening of the jet streak, which further leads to promoting rapid cyclogenesis.

In addition to the major C instance in the WV−PV−PWV comparison associated with the zonal jet stream, there is a mismatch to the north of the developing wave with respect to the polar jet stream. This jet is evident to the north of location P as a high gradient area in the dynamical tropopause height in Fig. 5.19A as well as northwesterly wind speed greater than 80 kt in Fig. 5.20A. However, the dry zone P (seen by the satellite observations in Fig. 5.21A) corresponds to comparatively high tropopause heights simulated by the NWP model (see Fig. 5.21C) and is associated with light shades in the corresponding synthetic WV image (see Fig. 5.21B). This is a typical C instance, demonstrating that the model has overestimated the ridging in the tropopause heights to the north and, in so doing, has underestimated the role of the polar strip of the PV anomaly in cyclogenesis enforcement.

Therefore, the existence of these two C instances in the WV−PV−PWV comparison is a clear sign of poor prediction of the cyclone development by the operational numerical model. At 0000 UTC on December 27 the model run was in error in the following ways:

- The interaction of the zonal jet with the strong PV anomaly seen in the real satellite WV image was not reflected in the dynamical fields (wind and dynamical tropopause heights).
- Because of the false ridge to the north, the ARPEGE model did not analyze any impact of the northern-latitude PV anomaly on the cyclogenetic process.
- As a result of these shortcomings, the operational model did not predict correctly the strengthening of the zonal jet streak.

These errors led to underestimation of the cyclone deepening. At 0000 UTC on December 28, 1999, as seen in Fig. 5.22A, there is a significant difference between the ARPEGE 24 h forecast and the analysis. Fig. 5.22B shows that, according to surface observations, the operational model underestimated the deepening by 9 hPa; it was also in error with the position of the surface low center (location D).

The main cause of this forecast failure is the underestimation of the interaction between the strong dynamical tropopause anomaly and the zonal jet stream by the operational NWP model. It should be realized, however, that NWP output must be validated not only in close association with the dynamical structure primarily responsible for cyclogenesis but also with respect to any structures that might be able to enforce cyclogenesis, such as the polar intrusion of high PV seen in this case. The operational use of the PV−WV concept can also be of value in highlighting areas of weaker forcing by relatively small-scale dynamical features, which may play a role in the synoptic development as well. A relevant implementation of the WV−PV−PWV comparison may help forecasters to validate numerical model behavior in such complicated cases of cyclogenesis.

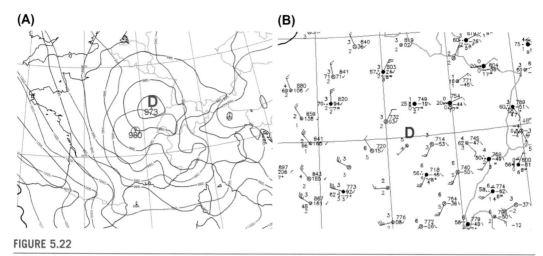

FIGURE 5.22

Comparison among ARPEGE mean sea-level pressure forecast, analysis, and observational data for December 28, 1999, at 0000 UTC. (A) ARPEGE 24 h forecast of mean sea-level pressure (red, hPa) and verifying numerical analysis (blue); (B) surface observational data. Also marked: "D," the position of the surface low center.

5.5 USING POTENTIAL VORTICITY CONCEPTS AND WATER VAPOR IMAGES TO ADJUST NUMERICAL WEATHER PREDICTION INITIAL CONDITIONS AND GET AN ALTERNATIVE MODEL FORECAST

The conserved nature of the PV parameter and the invertibility of PV (as shown in Chapter 1) enable us to build up the flow and temperature structure associated with a given PV anomaly. The technique of PV inversion (Arbogast et al., 2008), whereby the flow and mass fields associated with a particular portion of the PV field are calculated, can be used to adjust NWP initial conditions to get an alternative numerical forecast. There are two ways of using this approach:

- Applying this so-called PV thinking (Hoskins et al., 1985), changes in the initial NWP conditions can be made to deduce which kinds of PV anomalies play important roles in the synoptic development and give us a better understanding of the evolution of tropical and extratropical weather systems.
- If the WV−PV−PWP comparison reveals any signs of numerical model errors, modifications can be appropriately made to the 1.5 PVU surface heights to improve their fit with the satellite WV image and improve the operational NWP forecast.

Though PV is a three-dimensional (3D) field, it is now commonly assumed that upper-level precursors for synoptic development related to upper-level dynamics can be diagnosed using two-dimensional (2D) maps of dynamical tropopause topography/geopotential. A method allowing the forecasters to modify the initial state of the model to perform a new run has been developed at Météo-France. Using a graphical interface on the forecaster's workstation, it is possible to modify (interactively) the 2D field of 1.5 PVU surface height (see Appendix B). In addition, since low-level conditions are important, the low-level temperature or the sea surface temperature can also be adjusted according to the available observational data (which is often necessary in cases of cyclogenesis). After executing the modifications in the regions of interest, we get a new field describing the 1.5 PVU topography (and eventually a new temperature field at 850 hPa or a new sea surface pressure field). The NWP model is then adjusted to fit the modifications at the dynamical tropopause. This modified 3D field is in turn inverted using the PV inversion procedure of Arbogast et al. (2008) to retrieve new wind and temperature fields by means of an iterative inversion process. Finally, at the end of this process, the new initial state is used to perform an alternative forecast by the model.

5.5.1 SENSITIVITY ANALYSES OF UPPER-LEVEL DYNAMICS IN NUMERICAL WEATHER PREDICTION SIMULATIONS

The sensitivity to changes in the initial conditions has been used in several dynamical studies of intense storms (Huo et al., 1999; Romero, 2001; Homar et al., 2002, 2003). For that purpose, the PV inversion method is used to study the impact of selected PV anomalies on the structure of the initial atmospheric fields and on the subsequent dynamical evolution of the simulated circulation systems. This approach has been shown as a powerful tool toward the understanding of important atmospheric aspects related to cyclogenesis, such as baroclinic/barotropic development and convection, and widely used in numerical studies of severe weather cases.

5.5.1.1 Impact of Upper-Level Potential Vorticity Anomalies in Numerical Simulations of Deep Convection in a Cut-off System Over the Western Mediterranean

Forecasting deep moist convection that produces hazardous weather events (called convective storms, which include large hail, damaging wind gusts, heavy rainfall, and in some cases tornadoes) is an essential operational task. Emphasizing the importance of upper-level PV anomalies in forcing convection, Roberts (2000) has performed a climatology of mesoscale PV maxima in the North Atlantic and Western European region, showing that such PV anomalies influence about 60% of the observed thunderstorms. Because of their synoptic scale, the development of upper-level PV anomalies is generally well represented in NWP models. However, even relatively small errors in the upper-level PV field in such models can have a substantial impact on the resulting precipitation (eg, Fehlmann et al., 2000). In Sections 5.5.1.1 and 5.5.1.2 we will consider sensitivity studies of the impact of large upper-level PV anomalies on the initiation and development of deep convection.

As considered in Sections 4.2.3 and 4.3.1, the upper-level blocking circulations may be conductive for strong, persistent development of deep moist severe convection when the low-level convective ingredients are present. Sensitivity analyses are performed here considering a typical case over the Mediterranean on September 28–29, 2012, in a low-level environment of warm and moist air over a complex terrain.

The synoptic situation was characterized by an upper-level cut-off low over the southwestern part of the Iberian Peninsula at 0000 UTC on September 28. The system was progressing eastward and the associated convective development influenced Eastern Spain at 0000 UTC September 29. This is illustrated in Fig. 5.23 where Meteosat IR images with superimposition of the lightning and radar reflectivity are presented. Most of the heavy precipitation that affected the Murcia region on September 28, 2012, was caused by a Mesoscale Convective System (MCS) between 1000 UTC and 1300 UTC. This MCS formed along a convergence line between the warm and moist easterly low-level flow and the strong westerly low-level winds between Southern Spain and North Africa. Daily precipitation amounts reached 240 mm in Andalusia (27 September), 230 mm in Murcia (28 September), and 230 mm in Valencia (29 September).

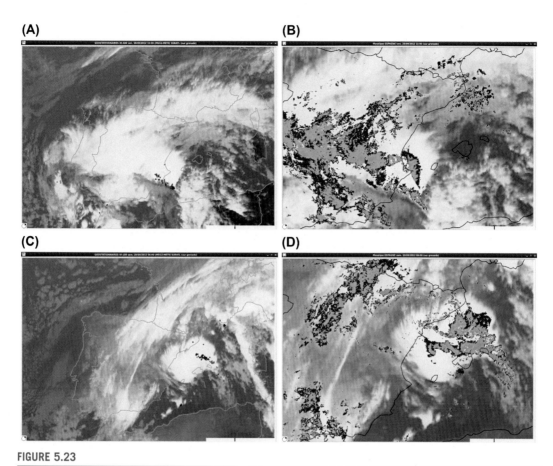

FIGURE 5.23

Meteosat infrared images with superimposition of (A) lightning and (B) radar reflectivity on September 28, 2012, at 1500 UTC; (C) lightning and (D) radar reflectivity on September 29, 2012, at 0600 UTC.

The convective development was conducted by a tide interaction between upper-level and low-level disturbances. This is identified in Fig. 5.24 as the coupling of the approaching upper-level cyclonic PV anomaly (WV image dark band at the red arrow) with a low-level baroclinic zone and related cyclonic cloud structure to the north. This baroclinic zone originates from the subtropical area and it is seen in the WV image as a moisture boundary at the green arrow in Fig. 5.24. The configurations of proximity between strong surface-thermal gradients and upper-level trough disturbances provide extremely favorable environmental conditions, under an appropriate vertical wind shear, to produce the baroclinic growth of disturbances.

In the northeastern flank of the cut-off low, where there is upward forcing, favoring the convection in a deep layer, low-level depression and convergent flow are established, reinforcing the convection and heavy precipitation. The low-level convergence zone, associated with a subsynoptic surface-

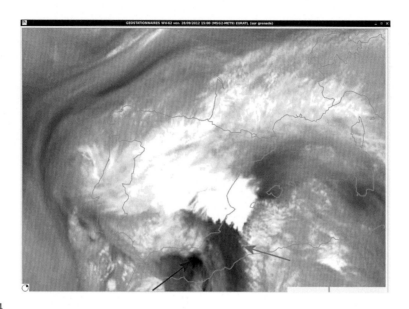

FIGURE 5.24

Meteosat water vapor image on September 28, 2012, at 1500 UTC with an upper-level potential vorticity anomaly (at the *red arrow*) and a low-level baroclinic zone (at the *green arrow*) indicated.

pressure low (seen in the images as a cyclonic cloud structure), progressed northward, reaching the area north of the Balearic Islands at 0000 UTC on September 29 (6 h before the situation in Fig. 5.23C). In order to show how the upper-level PV anomaly contributed to the initiation and development of convection over Spanish coasts, the following questions are addressed:

- What are the structures of the upper-level dynamics affecting the convection in this case?
- How do these upper-level structures interact with low-level features to influence the convection and precipitation pattern?
- How does this case improve our understanding of forcing convection?

The influence of a mid-upper tropospheric cut-off and its corresponding PV anomaly are analyzed here with a PV inversion technique (see Appendix B). The ARPEGE global model is used to show the mechanisms responsible for this convective development. A control simulation starting at 1800 UTC on September 27 and extending out to 48 h is performed. The 3 h cumulated precipitation from the run on September 27 at 1800 UTC is shown in Fig. 5.25 in (A) 21 h forecast and (B) 36 h forecast. Comparing the NWP field in Fig. 5.25A and B with the radar observations in Fig. 5.23B and D shows that the control model run correctly simulated the convective areas: The timing of the MCS northeastern progression is relatively well predicted. Fig. 5.25 also illustrates that the intensive convective precipitations are closely attached to the forward side of the PV anomaly in the highly diffluent part of the blocking system.

A way of assessing the impact of troposphere dynamics in the synoptic development is by calculation of quasi-geostrophic vertical motion due to baroclinicity and nonadiabatic heating following the quasi-geostrophic omega equation (see, eg, Hoskins et al., 1978). The quasi-geostrophic forcing for vertical motion (calculated through the Q vector formulation) is shown in Fig. 5.26 for the

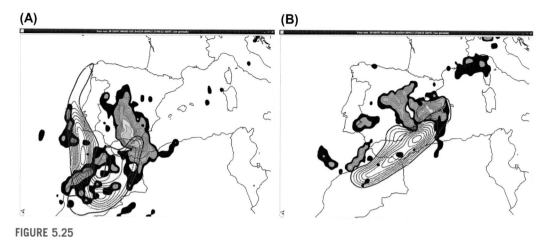

FIGURE 5.25

ARPEGE model fields of 300 hPa potential vorticity (*brown lines*, interval 1 PVU from 2 PVU) and 3 h cumulated precipitation from the run on September 27 at 1800 UTC in (A) 21 h forecast and (B) 36 h forecast.

two control simulations. Comparing Figs. 5.25 and 5.26 reveals that the dynamical tropopause anomalies (cyclonic PV anomalies) in the large upper-level cut-off system are associated with centers of significant dynamical forcing for upward motion in the mid−upper troposphere (700−300 h Pa). These areas of positive forcing influence the convective development over Southeastern Spain (at the red arrow in Fig. 5.26) and move northwards during September 28−29 (not shown). Since convective instability of the Mediterranean air mass during the warm season is also present (not shown), the synoptic environment is highly supportive for the development of deep convection.

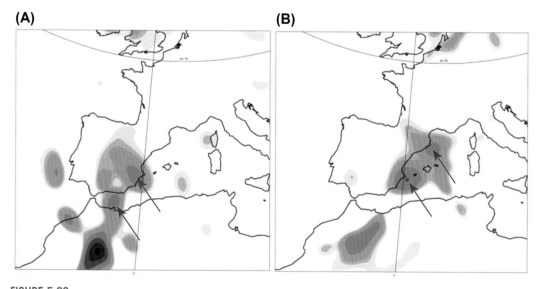

FIGURE 5.26

Mid-upper tropospheric (700−300 hPa) upward quasi-geostrophic forcing (starting at 0.02 Pa s^{-1}, every 0.02 Pa s^{-1}) for the control simulations at (A) 1500 UTC on September 28 and (B) 0600 UTC on September 29.

With the aim of determining the factors responsible for the forcing of the convective development, the effect of the upper-level cut-off is analyzed. This is achieved by performing simulations in which the upper-level cut-off is removed, which is done through the PV inversion procedure. The model initial conditions are artificially modified, and this modified experiment was run with identical boundary conditions and physics options as the control run (Fig. 5.25) as well as for the same 48 h period. Fig. 5.27A and B shows the WV image superimposed by the heights of the 1.5 PVU surface for the initial state of the control (original) run and of the modified experiment, in which the upper-level PV anomaly has simply been eliminated.

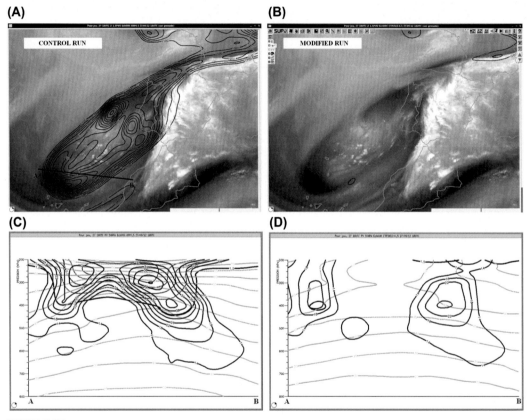

FIGURE 5.27

Geopotential of the 1.5 PVU surface in (A) the control (original) and (B) the modified model initial conditions (interval 100 dam) overlaid by the corresponding water vapor image on September 27, 2012 at 1800 UTC. Vertical cross-section along the line A—B in (A) of potential vorticity (*brown contours*, only ≥1.5 PVU), potential temperature (gray), and absolute vorticity (blue) from the corresponding runs from (C) the original and (D) the modified initial conditions.

5.5.1.1.1 Upper-Level Effects

Fig. 5.27C and D shows vertical cross-sections along the line A—B in Fig. 5.27A of PV anomalies (brown contours), the potential temperature (gray), and the absolute vorticity (blue). The vertical distributions of these parameters simulated by the control (Fig. 5.27C) and modified (Fig. 5.27D) runs illustrate the upper-level effects of the PV modification: High cyclonic absolute vorticity are present at the forward side of the PV anomaly, and cooling under the PV anomaly and warming above are observed in the control initial state (Fig. 5.27A and C). These features are absent in the modified simulation (Fig. 5.27B and D). After removing the PV anomaly from the initial state (Fig. 5.27B), the conditions for dynamical forcing through PV advection associated with the upper-level cut-off low are not present. As a result, favorable conditions for generation and intensification of the cyclone and modifications to the mesoscale flow fields are absent in the modified run (Fig. 5.27D).

The method for incorporation of the PV modifications at the tropopause level includes building up 3D PV corrections. Then, the 3D PV corrections are inverted in order to get wind and temperature corrections. The depth of the corrections in terms of wind and temperature are given by the constraint of no low-level PV modifications just above 850 hPa. This in turn affects the low-level initial conditions. In Fig. 5.28 a comparison of the low-level wind and the total accumulated precipitation among the control run (Fig. 5.28A and B) and the sensitivity experiment (Fig. 5.28C and D) after 21 h and 36 h forecast is displayed.

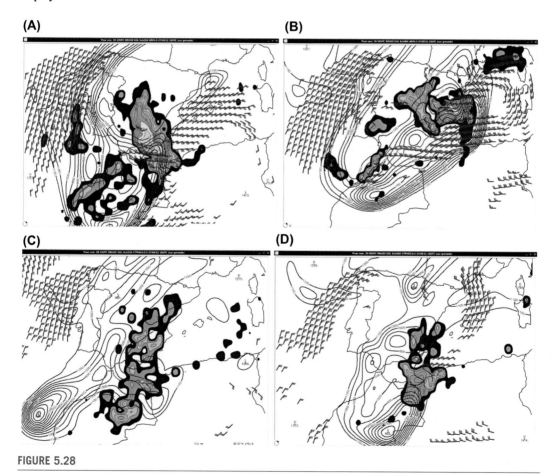

(A)   **(B)**   **(C)**   **(D)**

FIGURE 5.28

Low-level maximum wind at 925 hPa (*red vectors*, threshold at 20 kt), 1.5 PVU surface heights (*brown contours*, every 50 dam from 1100 dam), and 3 h accumulated precipitation (shaded areas). Forecast fields from the control run valid for (A) September 28 at 1500 UTC and (B) September 29 at 0600 UTC as well as from the run of the sensitivity experiment valid for (C) September 28 at 1500 UTC and (D) September 29 at 0600 UTC. Color shades from dark blue to orange indicate precipitations 1, 3, 5, 7, 10, 15, 20, 30, and 50 mm.

5.5.1.1.2 Low-Level Effects

Note that in the modified experiment (Fig. 5.28C and D), the upper-level flow pattern is evolving more slowly around the Iberian Peninsula without the spinning PV anomaly through the Eastern Spanish Mediterranean coast. The resulting surface low-pressure area is not a mobile one and is less intense, remaining near the Algerian coasts (not presented). This scenario resembles many other situations in which the lack of strong and evolving upper-level dynamical forcing implies rather stationary disturbances leeward of the Atlas Mountains that locally enhance the easterly moist flow toward Eastern Spain. Differences in the precipitation distribution among the sensitivity experiments would be mostly rooted in the diversity of dynamical forcing patterns, which ultimately control the cyclone evolution. In the control run at the time of 21 h and 36 h forecast (Fig. 5.28A and B), a vertical interaction between the upper-level PV anomaly and the low-level jet is established as the upper-level cyclonic PV anomaly arrives over a region of low-level baroclinicity. Then the induced circulation will promote warm advection east of it, creating a warm surface anomaly, which in turn will induce its own cyclonic circulation and modify the low-level wind field. This point is illustrated by the evolution of the low-level jet (LLJ), which is depicted through the maximum winds at 925 hPa (red vectors, threshold at 20 kt) in Fig. 5.28, for the control and modified runs. For the modified scenario (Fig. 5.28C and D), the southern part of the westerly LLJ over the Western Mediterranean (seen in the control run) is not present. This implies a confinement of the heavy precipitation about the Southern Mediterranean because of the lack of propagation of the related low-level convergence and moisture supply. For the control experiment (Fig. 5.28A and B), the LLJ enter farther into inland Spain and progress faster toward Southern France in response to the quicker and more extensive surface disturbance.

5.5.1.2 Impact of Upper-Level Potential Vorticity Anomalies in Numerical Simulations of Deep Convection in a Cold Upper-Level Trough Over the Northeast Atlantic and Western Europe

In this section, the PV approach is used to investigate the sensitivity of a convective process, which produced a heavy rain event, to the structure of the upper-level flow. As discussed in Romero et al. (2005), an accurate specification of the troposphere PV distribution can be performed for the successful prediction of such hazardous weather events. The studied severe convective storm developed on May 2–3, 2013, and produced a large amount of precipitations over the eastern part of France, as shown through data from radar observations in Fig. 5.29. This episode developed in the leading diffluent part of a cold mid-level trough, associated with an upper-level PV anomaly over Western Europe. This is seen in Fig. 5.30, which shows the fields of geopotential height and temperature at the 500 hPa isobaric surface as well as PV on the 300 hPa surface (blue contours) 9 h earlier than the time of Fig. 5.29.

The strategy adopted here consists of examining four different ARPEGE consecutive runs. Their forecasts of 3 h accumulated precipitations are shown in Fig. 5.31, as follows:

1. Run (A): The 51 h forecast from May 1 at 0000 UTC in Fig. 5.31A.
2. Run (B): The 39 h forecast from May 1 at 1200 UTC in Fig. 5.31B.
3. Run (C): The 21 h forecast from May 2 at 0600 UTC in Fig. 5.31C.
4. Run (D): The 9 h forecast from May 2 at 1800 UTC in Fig. 5.31D.

Fig. 5.31 shows also superimposed the corresponding different simulations of the PV anomaly as a suitable tracer of upper-level dynamics. We note the following differences in the considered output of the four runs:

- The minimum height of the dynamical tropopause (1.5 PVU surface) for runs (C) and (D) is slightly shifted to the east compared to runs (A) and (B).

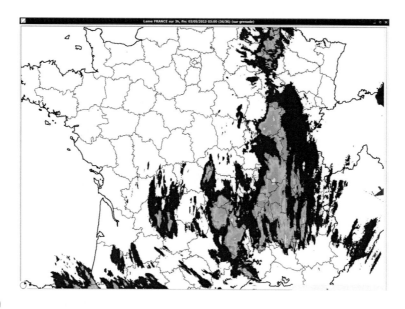

FIGURE 5.29

Accumulated 3 h rainfall at 0300 UTC on May 3, 2013, derived by radar observations, in color shades from violet to yellow indicate precipitations 0.1, 1, 3, 5, 7, 10, 15, and 20 mm.

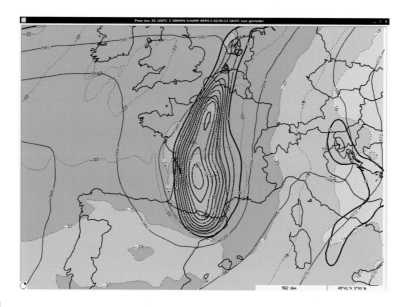

FIGURE 5.30

Geopotential height (*brown contours*) and temperature (shaded) at 500 hPa isobaric surface as well as potential vorticity on the 300 hPa surface (*blue contours*) at 1800 UTC on May 2, 2013.

(A)   **(B)**

(C)   **(D)**

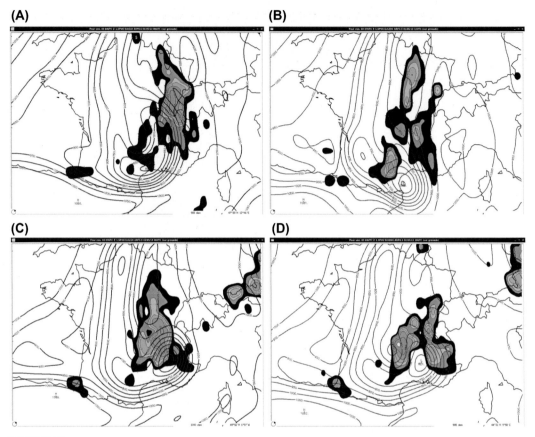

FIGURE 5.31

Forecast fields of 3 h total accumulated rainfall (color-shaded), superimposed on geopotential heights of the 1.5 PVU surface (*brown contours*) on May 3, 2013, at 0300 UTC from different ARPEGE model runs on: (A) May 1 at 0000 UTC, (B) May 1 at 1200 UTC, (C) May 2 at 0600 UTC, and (D) May 2 at 1800 UTC. Color shades from dark blue to yellow indicate precipitations 1, 3, 5, 7, 10, 15, and 20 mm.

- The vorticity distribution patterns (particularly the location of vorticity centers) are quite different.

This in turn leads to the poor skills of the NWP model to capture spatial and quantitative details of the rainfall field. Major differences appear from one run to another, seen in Fig. 5.31 as follows:

- Areas of heavy precipitation are forecasted by runs (A), (C), and (D).
- However, the locations of the heavy precipitation areas are at different positions in the outputs of runs (A), (C), and (D).
- Run (B) seems incapable of simulating an intense convection and predicting rain as heavy as in runs (A), (C), and (D).

Fig. 5.32 shows the 39 h and 21 h forecasts of the convective available potential energy (CAPE; color-shaded area) and 850 hPa relative humidity (blue contours): in Fig. 5.32A from run (B) (forecasted weak rainfall) and in Fig. 5.32B from run (C) (forecasted intensive rainfall). In order to assess the thermodynamic conditions of the environment, vertical cross-sections across the areas of CAPE maximums of θ_w, relative humidity, vertical velocity, and PV are presented in Fig. 5.32C and D. The vertical distribution of PV is given only by the 1.5 PVU contour (green) to depict the dynamical tropopause.

Strong instability, as a favorable condition for deep convection, exists in the forecasts of both run (B) at 1200 UTC on May 1 and run (C) at 0600 UTC on May 2 (high CAPE values). However, among the two runs, the convection produces heavy rain only in run (C) where the region of high instability meets a lowering of the tropopause (Fig. 5.31C). This provides forcing through cyclonic PV advection and allows dry upper-tropospheric and stratospheric air to descend into the middle troposphere. Key aspects in the role of this PV anomaly are the related upward motion ahead of it and downward motion behind it. The combination of advection of low θ_w air above higher θ_w in the region of upward motion ahead of the PV anomaly is conducive for strong thunderstorm development in run (C). Increasing the depth of this convective boundary layer occurred as a result of the lifting induced by the orography and also enhanced through the upper-level PV anomaly. Calculation of quasi-geostrophic forcing for vertical motion shown in Fig. 5.33 reveals important areas of dynamical forcing for upward motion in the mid-upper levels.

In order to analyze the difference between runs (B) and (C) regarding the coupling of the instability and the upper-level forcing, the axes of the cross-sections in Fig. 5.32 are shown in Fig. 5.33. We note the following:

- The cross-sections in Fig. 5.32 are performed across the areas of maximum CAPE predicted in both run (B) and run (C), which predicted in quite a similar way the convective instability.
- However, run (B) and run (C) forecasted different upper-level PV distribution patterns (Fig. 5.31B and C, respectively).
- Since run (B) poorly simulated the upper-level dynamics and the position of the related tropopause folding, it significantly shifted the simulated maximum upper-level forcing from the area of strong CAPE given by the same model run. This takes effect in the following sequences in the NWP model performance:
  - Comparing Fig. 5.32A and B with Fig. 5.33A and B shows that the maximum quasi-geostrophic forcing for run (B) is located to the southwest of the area of strong convective instability.
  - In the output of run (C) these regions of ascent associated with PV anomaly are partially coupled with strong instability (high CAPE values) and, as a result, a strengthening of convection predicted by the numerical simulation.

5.5.1.2.1 Potential Vorticity Modification

To investigate the role that the upper tropospheric disturbance, as represented by its PV field, plays and its impact on the convective parameters, a PV-inversion experiment is conducted by virtually eliminating the PV anomaly. This experiment is run with identical boundary conditions and physics options as the control run. The PV inversion technique is applied to invert the upper-level anomaly at 0600 UTC on May 2, 2013 (the simulation time start), that corresponds to run (C) in Fig. 5.31. Fig. 5.34A shows the fields of PV (blue contours) and wind (red vectors) at the 300 hPa isobaric surface for the control (CTRL) run. In the modified run (Fig. 5.34B), the upper-level cyclonic PV anomaly was removed from the initial state.

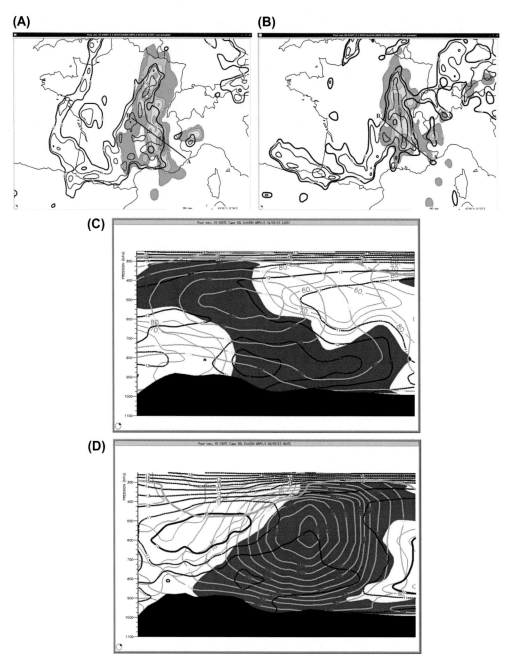

FIGURE 5.32

ARPEGE model (A) 39 h and (B) 21 h forecasts of convective available potential energy (shaded area, interval 250 J/kg, started from 100 J/kg) and 850 hPa relative humidity (*blue contours*, interval 5% from 90%), valid for May 3, 2013, at 0300 UTC; (C) and (D) display vertical cross-sections of θ_w (blue), relative humidity (pink, $\geq$90% shaded), vertical velocity (gray), and potential vorticity (green, only 1.5 PVU contour) along the lines A—B in (A) and (B), respectively.

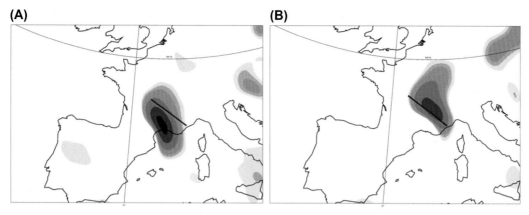

FIGURE 5.33

Mid-upper tropospheric (700-300 hPa) upward quasi-geostrophic forcing (starting at 0.02 Pa s^{-1} every 0.02 Pa s^{-1}) at 0300 UTC on May 3, 2013, from the ARPEGE (A) 39 h forecast and (B) 21 h forecast. The quasi-geostrophic forcing is calculated following the omega equation.

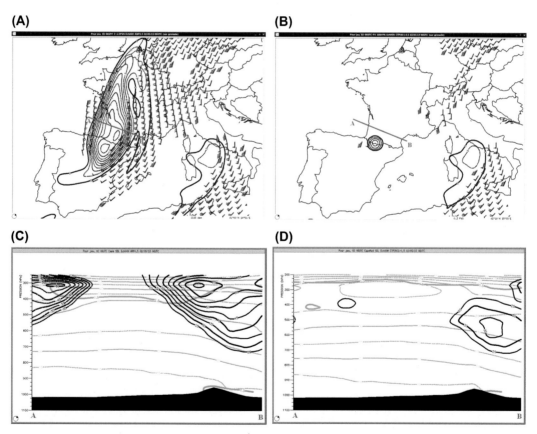

FIGURE 5.34

Potential vorticity (PV) at 300 hPa (*blue lines* starting at 2 PVU every 0.5 PVU) and the vectors of maximum wind (starting at 45 m s^{-1}) at the 300 hPa isobaric surface on May 2, 2013, at 0600 UTC for (A) control (CTRL) run (operational) and (B) modified run. Vertical cross-section along the line A—B shown in upper panels of the PV field (*green line*, only 1.5 PVU contour), potential temperature field (*gray contours*, every 5°C), and wind speed (*black contours*, starting at 25 m s^{-1}) normal to the cross-section plane for (C) CTRL run and (D) modified run.

5.5.1.2.2 Control Run

Fig. 5.35 depicts the evolution of the thermodynamic conditions in the simulation of the control run, by depicting fields of PV anomaly and CAPE (left panels) as well as cross-sections of absolute vorticity and vertical velocity of ascending motions (right panels). On May 3, 2013, a cyclonic vorticity disturbance marked "D" was formed at the surface (Fig. 5.35D), after a significant upper-level vorticity structure, located to the west, had propagated eastward. Fig. 5.35F shows that a tide interaction between upper-level and low-level cyclonic vorticity features has been established:

- The upper-level vorticity advection, associated with the cyclonic PV anomaly, induces a weak circulation at the surface on May 2 at 2100 UTC (Fig. 5.35B). The advection of potential temperature by the induced lower-level circulation can lead to a warm anomaly east of the upper-level anomaly.
- This in turn induces a cyclonic circulation in the upper level that reinforces the original anomaly and leads to amplification of the disturbance on May 3 at 0000 UTC (Fig. 5.35F).
- Further, the upward displacement of isentropic surfaces causes a reduction in static stability beneath and at the forward of the PV anomaly.

The arrival of the upper-level PV anomaly over a low-level baroclinic region and related tropospheric ascent of air ahead of the anomaly are seen in Fig. 5.35 (in the CTRL run). This promotes deep convection that is forecasted by the CTRL run. Fig. 5.36A shows the 24 h forecast of accumulated 6 h precipitations, which exceed 30 mm.

5.5.1.2.3 Modified Run

Removing the upper-level trough from the initial conditions of run (C) produces responses in the CAPE and vertical motion fields at all atmospheric levels (Fig. 5.37). In the modified simulation, the CAPE is reduced beneath and just forward to the east of the experimentally weakened PV anomaly. The dynamical forcing associated with the upper-level PV anomaly is reduced (not shown). This also has produced changes in the vertical motion with a decrease of the ascent forced by the upper-air trough that is seen by comparing Figs. 5.35 and 5.37 (right panels). The precipitation distribution obtained from the modified simulation shows a significant decrease in the accumulated precipitation (Fig. 5.36B) compared to the CTRL run (Fig. 5.36A). This confirms the crucial role played by the upper-level cyclonic PV anomaly on the convective development and related precipitation in this case.

The sensitivity experiments in this section confirm that in addition to the presence of convective ingredients as systematized in Doswell et al. (1996), the development of strong convective storms in midlatitudes involves upper-level forcing as a result of cyclonic PV anomaly advection, which can be tracked by satellite WV imagery. Investigations using the ARPEGE operational forecasts showed that the impact on convection directly beneath and forward from the PV anomaly was in line with the conceptual model presented by Hoskins et al. (1985): The upper-level cold pool destabilized the vertical profile to such an extent that there was high CAPE ready to be released. This finding is also consistent with most of the literature concerning similar convective events (eg, Russell et al., 2008).

5.5.1.3 Applying Potential Vorticity Concepts in Understanding Extratropical Transition of Tropical Cyclones

After a transformation period during which a tropical cyclone (TC) is going into an extratropical environment, the system can intensify as an extratropical cyclone that is sensitive to its interaction with

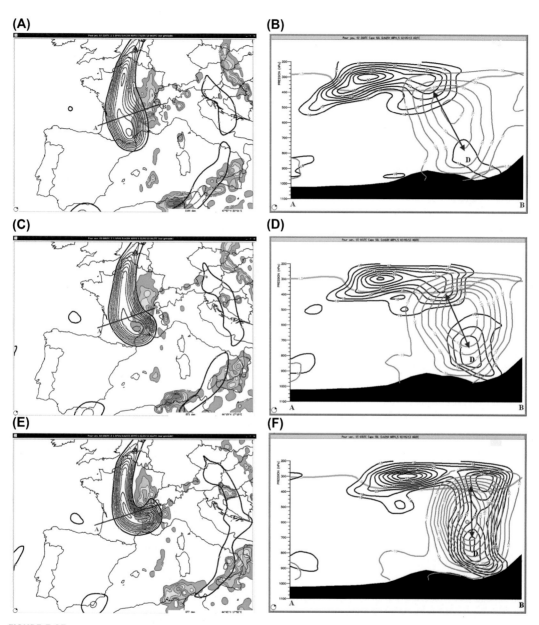

FIGURE 5.35

Time sequence from the control simulation. Left panels are the 300 hPa potential vorticity (PV; *blue contours*) and convective available potential energy (shaded areas, started from 100 J/kg, interval 250 J/kg), for (A) May 2 at 2100 UTC, (C) May 3 at 0000 UTC, and (E) May 3 at 0300 UTC. Right panels display the corresponding vertical cross-section along the lines A—B of absolute vorticity (*blue lines*, starting at $18 \times 10^{-5}\,s^{-1}$), vertical velocity of ascending motions (*orange lines*, starting at 20×10^{-2} Pa s^{-1}), and PV (only 1.5 PVU isoline in green color) for (B) May 2 at 2100 UTC, (D) May 3 at 0000 UTC, and (F) May 3 at 0300 UTC. Also marked "D" a surface cyclonic vorticity disturbance.

(A)　　　　　　　　　　　　　　　　　(B)

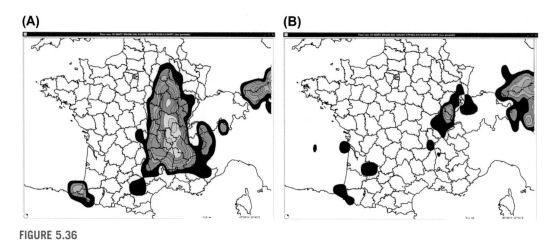

FIGURE 5.36

ARPEGE 24 h forecast of accumulated 6 h precipitations (shaded from 1 to 30 mm) from (A) the control simulation and (B) the modified run valid for 0600 UTC on May 3, 2013.

the midlatitude circulation (see Section 3.6.4). The reintensification occurs in association with the approach of an upper-level PV anomaly from the midlatitudes. In the absence of such a preexisting upper-level dynamical system, the TC would be expected to continue to decay after it moves to the extratropical environment. Jones et al. (2003) provide a review of the knowledge on the processes involved in extratropical transition (ET) and present the PV thinking as an aid to understanding ET. They discussed that a significant contributor to numerical forecast errors during ET is the uncertainty in the initial conditions.

The objective of this section is to illustrate how changing the upper-level environmental conditions at the initial state may influence the intensity of the ET process. This is performed through PV analysis of Typhoon Wipha, which interacted directly with an upper-level trough over the northern Pacific Ocean on October 15, 2013. At that time (as shown in Section 3.6.2), favorable conditions have weakened (the energy flow from the sea surface temperature decreased, increasing vertical wind shear influenced the system, and dry air along the western and southern peripheries wrapped toward the low-level circulation center of the tropical storm). At the same time, a mutual interaction with the mid-latitude baroclinic zone caused the system to begin an ET. A sensitivity study by applying perturbations to the initial conditions is performed to evaluate the impact of the PV anomaly in initial upper-level conditions to the development of the TC at the onset of its ET. The sensitive study starts at the beginning of ET on October 15 at 0000 UTC when the presence of baroclinicity and thus vertical shear is essential for reintensification of the storm as an extratropical system. Fig. 5.38 shows a sequence of WV images of MTSAT from October 15, 2013, at 0600 UTC to October 16, 2013, at 0000 UTC with superimposition of ARPEGE analysis of 1.5 PVU geopotential (brown contours), 300 hPa maximum wind speed (blue isotachs, threshold 75 kt), and 200 hPa divergence (threshold $2\ 10^{-5}\ s^{-1}$, green contours). At the start of the period of reintensification as an extratropical system, the TC is located in a region of progressively enhanced divergence (black arrows in Fig. 5.38) at the exit region of the upstream midlatitude jet (strong gradient of the 1.5 PVU heights). This is associated with strengthening the outflow jet (at the red arrows) just to the north of the tropical system: The maximum

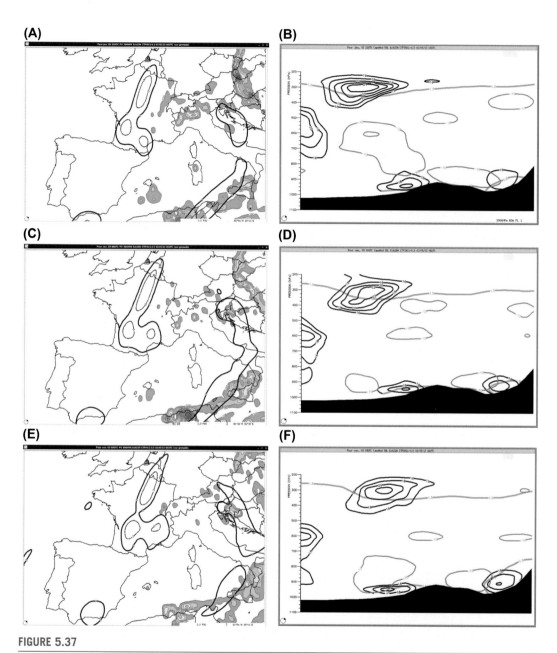

FIGURE 5.37

Time sequence from the potential vorticity modification experiment. The fields in the left panels and the cross-section parameters in the right panels are the same as in Fig. 5.35.

(A)

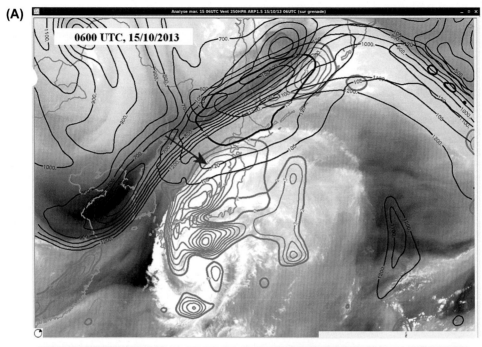

(B)

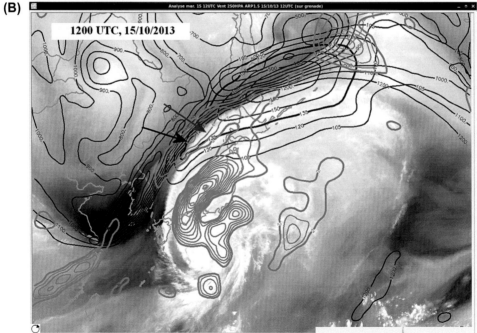

FIGURE 5.38

A sequence of MTSAT water vapor images from October 15, 2013, at 0600 UTC to October 16, 2013, at 0000 UTC with superimposition of ARPEGE analysis of 1.5 PVU geopotential (brown, every 100 dam, only ≤1200 dam), isotachs of 300 hPa maximum wind (blue, every 15 kt, threshold 75 kt), and 200 hPa divergence (green, threshold $2 \times 10^{-5}\,s^{-1}$).

(C)

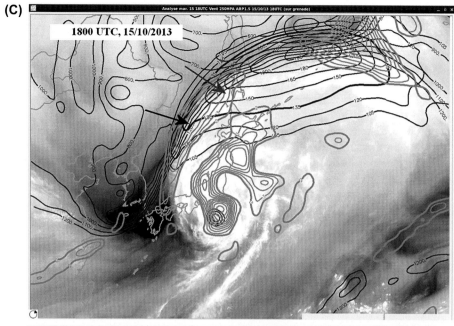

1800 UTC, 15/10/2013

(D)

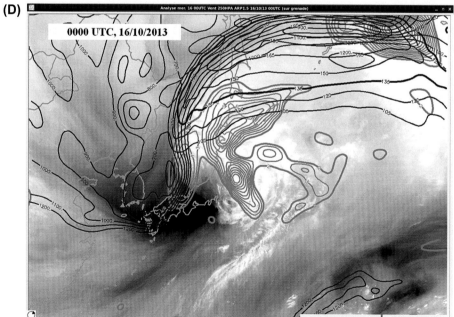

0000 UTC, 16/10/2013

FIGURE 5.38 Cont'd

wind at the jet has increased from 120 kt (Fig. 5.38A) to 160 kt (Fig. 5.38C) for 18 h. WV imagery in Fig. 5.38 confirms the development of strong radial outflow with an enhanced poleward outflow jet stream.

The 6 h sequence of meteorological fields in Fig. 5.39A, C, and E shows that in the same time the PV anomaly (low 1.5 PVU geopotential, brown contours) has approached the TC (deep low in the MSLP field, black contours). This is the period of the most active phase of the mutual interaction between the two dynamical systems leading to strengthening of the outflow jet poleward as seen in the isotachs of 300 hPa wind (shaded area). The corresponding cross-sections in Fig. 5.39B, D, and F show the advection of cyclonic PV (brown contours), associated with the tropopause folding, which progressively extends downward toward the TC, feeding its core with high cyclonic PV.

The evolution of vertical velocity (orange contours in the cross-sections) shows a large fluctuation, as follows:

- Decreasing updraft from $250 \ 10^{-2} \ Pa \ s^{-1}$ at 1200 UTC to $150 \ 10^{-2} \ Pa \ s^{-1}$ at 1800 UTC on October 15 (Fig. 5.39B and D) related to increasing vertical wind shear associated with the approaching trough that weakens the TC system in addition to the negative effect from the cooling sea-surface temperatures in the TC environment.
- Increasing updraft from $150 \ 10^{-2} \ Pa \ s^{-1}$ at 1800 UTC on October 15 to $210 \ 10^{-2} \ Pa \ s^{-1}$ on October 16 at 0000 UTC (Fig. 5.39D and F) related to the presence and strengthening of baroclinicity (as a result of the approaching midlatitude trough, and thus vertical shear becomes an essential factor in reintensification of the TC as an extratropical system in the later stages of ET).

To evaluate the impact of the midlatitude dynamics in the initial upper-level conditions to the development of the extratropical, an experimental initial state is constructed to modify the strength and position of the upper-level PV anomaly and hence its influence on the ET process. Fig. 5.40 displays the 3D consequences of a 2D modification through vertical cross-section. The modification of the initial state includes removal of the active part of the upper-level PV anomaly at the most equatorward side of the trough (brown contours in Fig. 5.40A and B). Fig. 5.40C and D shows vertical cross-sections at the removed PV anomaly of strong wind (black), absolute vorticity (blue), PV anomaly (brown), and isentropic surfaces (green). The modification of the tropopause level (brown thick contour in the cross-sections in Fig. 5.40C and D) is associated with a decrease of the low-stratosphere PV at the most equatorward side of the trough. In effect, an alternative development without interaction of the TC with the upper-level trough is simulated. This kind of modification reflects in the following upper-level dynamical features:

- The static stability is decreased between 500 hPa and 200 hPa (a weaker vertical gradient of potential temperature; green contours in Fig. 5.40D) that is determined by the new level of tropopause.
- Mid-level relative vorticity (blue contours in Fig. 5.40D) is also significantly decreased as a response of the PV modifications.

Applying the PV inversion procedure of Arbogast et al. (2008), a new initial state (EXP1) is obtained and used to perform an alternative forecast with the ARPEGE operational model, after integrating this new initial state in the model run for the next 42 h. The new forecasts of 1.5 PVU surface heights, MSLP,

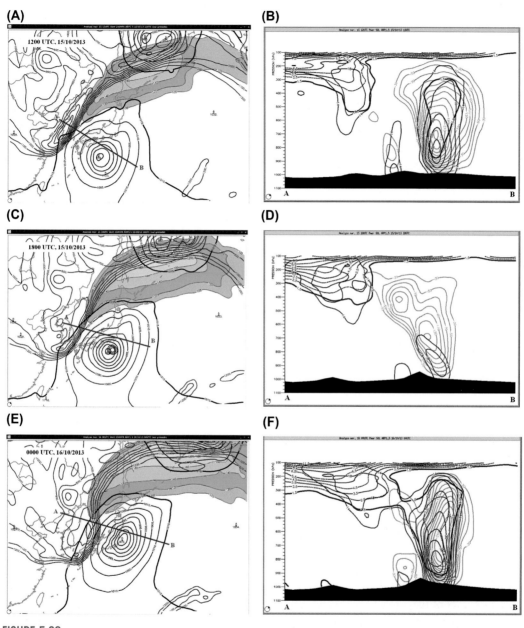

FIGURE 5.39

Extratropical transition of Wipha as a result of interaction with approaching midlatitude potential vorticity anomaly from 0000 UTC October 15, 2013, to 0000 UTC October 16, 2013. In the left panels (A), (C), and (E): mean sea-level pressure field (*black contours*), 1.5 PVU geopotential (*brown contours*), and isotachs of 300 hPa strong wind (every 15 kt, threshold 105 kt, shaded area). In the right panels (B), (D), and (F): the corresponding vertical cross-sections along the *red lines* (A, B) (in the left panels) of 1.5 PVU contour (brown), absolute vorticity (blue in 10^{-5} s^{-1}) and vertical velocity of ascending motions (*orange contour*; interval is 20 10^{-2} Pa s^{-1} above 30 10^{-2} Pa s^{-1}).

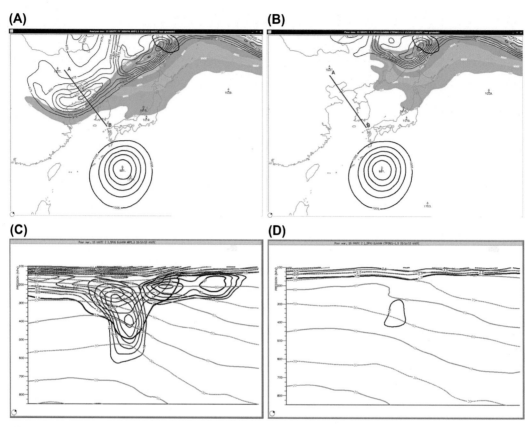

FIGURE 5.40

ARPEGE model analysis at 0000 UTC on October 15, 2013, of 1.5 PVU surface heights (brown, only ≤975 dam), 300 hPa wind (≥80 kt, shaded area), mean sea-level pressure (*black contours*) for the runs from (A) nonmodified (CTRL) and (B) modified (EXP1) initial states. Vertical cross-sections along the axis displayed in (A) and (B) from (C) the CTRL and (D) after the tropopause modification (EXP1) of absolute vorticity in blue (≥16 10^{-5} s^{-1}), potential temperature (green), potential vorticity (≥1.5 PVU, in brown), and wind speed normal to the cross-section plane (≥75 kt, black).

strong winds, and relative humidity at 300 hPa as well as 200 hPa divergence for the next 30 h starting from October 15, 2013, 0000 UTC are compared to the forecasts from the CTRL run in Fig. 5.41.

The removal of the upper-level PV anomaly at the most equatorward side of the trough results in a removal of the related westerly jet stream. There is a relatively weaker and more zonal upper-level jet north and northeast of the TC center. During the first 12 h (Fig. 5.41A and B), the intensity and track of TC are not much different in CTRL and EXP1 forecasts. The dry air observed in CTRL and EXP1 extending southward to the west of the storm can play a negative role in the TC development, but the presence of warm moist air to the southeast of the TC (not shown) may lessen this detrimental impact on TC

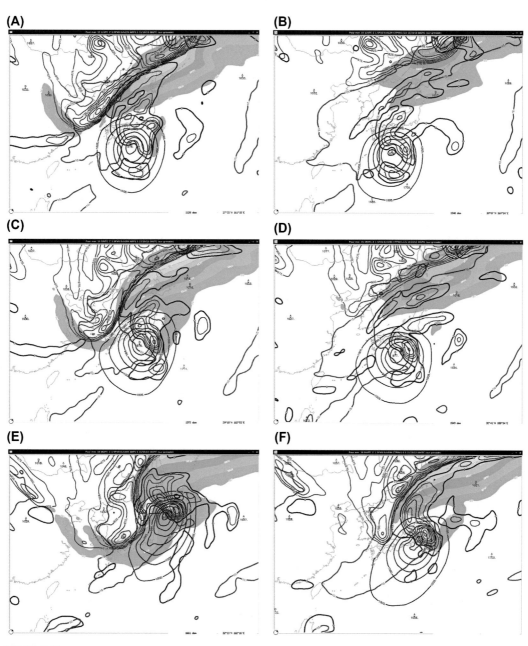

FIGURE 5.41

ARPEGE model forecasts of 1.5 PVU surface heights (brown, ≤975 dam), 300 hPa strong winds (≥80 kt, shaded area), mean sea-level pressure (black), relative humidity at 300 hPa (*red contour*, only 10%), and 200 hPa divergence (blue) starting from October 15, 2013, at 0000 UTC. The runs from nonmodified analysis (CTRL, left panel) for October 15 are shown in (A) for 1200 UTC and (C) for 1800 UTC and for October 16 are shown in (E) for 0600 UTC. The runs from the modified initial state (EXP1, right panels) for October 15 are shown in (B) for 1200 UTC and (D) for 1800 UTC and for October 16 are shown in (F) for 0600 UTC.

intensity and thus keep the TC intensifying. At this time, the upper-level jet and the associated strong westerly flow are located far to the north and northwest of the TC circulation. Because the upper trough is far from the center, vertical wind shear is weak. The divergence is strong and covers a large area.

The modification of the initial state by removing the active part of the upper-level PV anomaly takes effect in the 18 h forecast.

5.5.1.3.1 Control Run

At 1800 UTC in the CTRL forecast (Fig. 5.41C), the upper trough on the western side of the TC approaches, and the upper-tropospheric jet strengthens, the vertical wind shear continues to rise, and the divergent wind becomes stronger. The upper-level westerly trough brings dry air into the western side of the TC and could impact the system in several ways such as reducing CAPE and suppressing the convective development within the TC. But the fusion of warm, moist air from the south/southeast limits these effects. The strong trough approaching the storm center from the north remains at a distance, presumably due to the strong opposing divergent flow near the center. The upper-level environment can be interpreted as favorable for development because the tropical cyclone center tends to fall within the left exit of the approaching westerly jet and lies in the right entrance region of the northerly jet, a region favored for divergence and upward motion (Fig. 5.41A and C). TC intensification occurred because the stronger upper-level divergence (and implied upward motion) mitigated the detrimental impact of increasing vertical wind shear in the control run (the vertical wind shear decreased with time as the superposition occurred, and the deepening begins approximately when the PV anomaly reaches the storm center).

5.5.1.3.2 Modified Run

A comparison with the perturbed experiment (right panels in Fig. 5.41) indicates that the differences in the structure and orientation of the PV anomaly (Fig. 5.41F) can produce environmental conditions so that the storm weakens rather than strengthens. The weakening occurs even though the upper-level divergence is about equally strong and the TC remains in the favorable right entrance region of the outflow jet.

After 30 h, differences between the control and perturbed forecasts increased in magnitude. Fig. 5.42 compares the same fields as in Fig. 5.41 of the CTRL to the EXP1 runs in (A) and (B) from the 36 h forecast valid for 1200 UTC as well as in (C) and (D) from the 42 h forecast valid for 1800 UTC on October 16. In these perturbed forecasts (Fig. 5.42B and D), the cyclone depths are decreased with 12 hPa and 20 hPa than in the control forecasts respectively (Fig. 5.42A and C). The sensitivity experiment reveals that by removing the active part of the upper-level cyclonic PV anomaly from the upper-level environment, the TC moves more slowly northward due to a weaker southwesterly flow.

In the control forecast, the upper-level PV anomaly covers a large area of the cyclone and the leading part of the tropopause folding (high gradient of the 1.5 PVU surface geopotential) approaches the surface low center (at the pink arrow in Fig. 5.42A). The mutual interaction between these upper- and low-level dynamical features contributes to the extratropical intensification. In the perturbed forecast (Fig. 5.42B), the upper-level trough does not cover as extensive a region as in the CTRL forecast. The TC remnants are located too far to the east of this area, and therefore the PV anomaly does not contribute to the extratropical baroclinic development. There is no phasing between the TC and the upper-level trough, and this case dissipates. The peak deepening (seen in the MSLP) is strongly dependent on the relative positioning of the TC and midlatitude trough. Reintensification as an extratropical cyclone was found to occur when the combination of the upper-level dynamics and the thermodynamic processes acted in tandem to create a region that was favorable for extratropical cyclone development. Such an environment often develops when the upper-level TC outflow enhances the upper-level wind speed at the equatorward entrance region of a downstream jet streak and the remnant TC circulation interacts with the lower-tropospheric baroclinic zone.

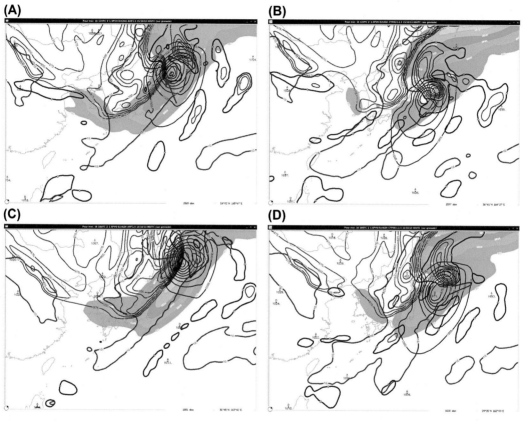

FIGURE 5.42

ARPEGE model forecasts of 1.5 PVU surface heights (brown, ≤ 975 dam), 300 hPa wind (≥80 kt, shaded area), mean sea-level pressure (black), relative humidity at 300 hPa (*red contour*, only 10%), and 200 hPa divergence (blue) starting from October 15, 2013, at 0000 UTC. Control compared to the perturbed forecast in (A) and (B) for the 36 h forecast valid for October 16 at 1200 UTC as well as in (C) and (D) for the 42 h forecast valid for 1800 UTC.

The sensitivity experiment confirms that the interaction with a cyclonic upper-level PV anomaly is a critical factor in the ET process. Simulating correctly the phasing between the tropical cyclone and midlatitude trough may be the single most important factor in accurately predicting future intensification trends of a transitioning cyclone in NWP models. If the interaction between a TC and the midlatitude PV anomalies at the upper levels is misrepresented in a forecast model, there will be an impact on developments downstream and thus a reduction in the accuracy of medium-range forecasts. Any uncertainties in the initial upper-level PV distribution may amplify and result in significant NWP errors in the forecast of ET development of the storm.

5.5.2 IMPROVING NUMERICAL FORECASTS BY POTENTIAL VORTICITY INVERSION ADJUSTMENTS

Errors in numerical weather forecasts can be attributed to two causes: deficiencies in the modeling system and inaccurate initial conditions (Zhu and Thorpe, 2006). The scope of this chapter concerns the understanding of the character of initial condition uncertainties at the upper level and related growth of forecast spread. During a forecast some of these initial errors can amplify and result in significant forecast failures. Moreover, the representation of the upper-level dynamics and thermodynamic context of the atmosphere by numerical algorithms introduces further uncertainties associated, for instance, with truncation errors, with uncertainty of parameters describing subgrid-scale processes such as cumulus convection in a global model. Therefore, a key issue of scientific research is the improvement of this initial state. Initial conditions are produced by assimilating recent observations together with an earlier short-range forecast, which serves as background information. The model assimilation will reject observations that deviate too much from the guess field. Due to the availability of sparse observation data or if erroneous observations have previously been accepted, the guess-field may contain errors and new good observations may be rejected. So, a lack of observations and approximations in the data assimilation procedures inevitably lead to imperfect initial conditions and subsequently to forecast errors, especially when initial errors are located in so-called sensitive regions. This is still a problem encountered in NWP model initialization despite the increasing quality in assimilation techniques.

5.5.2.1 An Example of Effects on Forecasting Cyclogenesis With Strong Surface Winds

This section presents a case study example of an experiment to improve the NWP initial state and forecast of cyclogenesis in strong Atlantic zonal flow associated with the formation of a sting jet considered in Section 3.5.4 (see Arbogast et al., 2012).

As a first step of the procedure, the field of 1.5 PVU surface heights derived at the 6 h forecast by the ARPEGE model control run valid for 1200 UTC on January 23, 2009, is superimposed on the corresponding Meteosat WV image in Fig. 5.43.

The comparison shows several significant differences:

- The process of cyclogenesis over the Atlantic is underestimated by the model according to the satellite WV image: The wavy feature of the dynamical tropopause is too much weaker with respect to an already forming cloud-head system in the image.
- The dark area in the satellite image along the leading edge of the dry slot (at the red arrow), related to downward vertical motion, is not well reflected by the NWP model simulation: The highest brightness temperatures measured by the WV channel, which appear most dark on the image, do not correspond to the tropopause folding (the lowest heights of the 1.5 PVU surface).

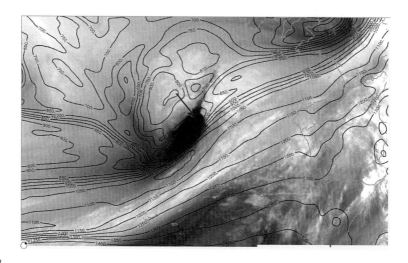

FIGURE 5.43

Comparison between 6 h forecast and satellite observations valid at 1200 UTC on January 23, 2009: geopotential of 1.5 PVU surface (*brown contours*) superimposed on a Meteosat 6.2 μm image.

As a second step, the satellite WV image on January 23, 2009, at 1200 UTC (Fig. 5.43) can be compared with the corresponding synthetic WV image (Fig. 5.44) derived from the 6 h forecast from the ARPEGE model with a grid of 0.5 × 0.5 degrees. This WV−PWV comparison confirms that the upper-level features seen in the imagery related to the PV anomaly (the dark area at the tropopause folding and the cloud system ahead of it) are too weak in the model-simulated image. The following

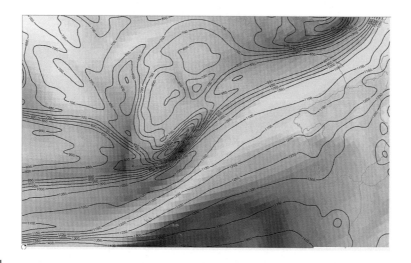

FIGURE 5.44

Comparison between 6 h forecast of 1.5 PVU surface geopotential (*brown contours*) and the corresponding synthetic water vapor image valid at 1200 UTC on January 23, 2009.

differences in the synthetic product regarding the corresponding features in the satellite observations are visible:

- The dark zone of the dry intrusion on the synthetic WV image is not pronounced enough compared to the reality.
- Much less lightening is produced in the synthetic image at the area of moist ascent (more distinct cloud head in the satellite image).

Therefore, the objective approach (WV—PWV comparisons) and the subjective approach (comparison between PV field and WV image in the first step) give rise to the same conclusion.

5.5.2.1.1 PV Modification

The experiment is performed to improve the poor upper-level analysis by manual modifications in order to better fit the gradient zone of the PV field to the moisture boundaries and the dry intrusion indicated by the satellite observations.

Fig. 5.45 shows the modified initial field of the dynamical tropopause height superimposed on the WV image. The modification of the dynamical tropopause mainly consists of reducing the level of the tropopause to increase the amplitude of the precursor in the Atlantic at the latitudes of the Portugal coast and increasing the level of tropopause at the latitudes of the Brittany coast. The lower boundary of the inversion domain is taken as the top of the planetary boundary layer (model level close to 850 hPa). The boundary conditions are chosen as Dirichlet boundary conditions, that is, the wind (or geopotential) on the boundary is fixed.

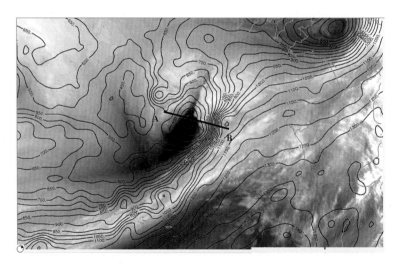

FIGURE 5.45

Modified initial tropopause height field superimposed on the water vapor image from Meteosat on January 23 at 1200 UTC.

Fig. 5.46 shows vertical cross-sections of vorticity maximum (blue contours), PV (red contour, only the 1.5 PVU surface), wind speed orthogonal to the cross-section plane (black contours), and isentropic surfaces (green contours) before and after modification of the dynamical tropopause and

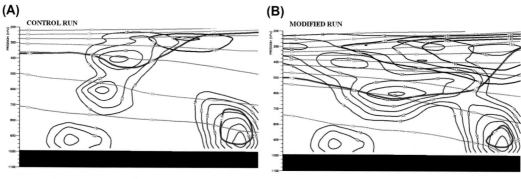

FIGURE 5.46

Vertical cross-sections along the axis displayed in Figure 5.45 of cyclonic vorticity (*blue contours*), potential vorticity (*red contour*, only 1.5 PVU surface), wind speed normal to the cross-section plane (*black contours*), and isentropic surfaces (*green contours*) (A) before the modification and (B) after the modification of the dynamical tropopause and inversion at 1200 UTC on January 23, 2009.

inversion. The cross-section is performed along the axis displayed in Fig. 5.45 within the dry slot on the WV image that is the upper-level precursor of the deepening Cyclone Klaus (see Section 3.5.4). The inversion of the modified PV distribution (Fig. 5.46B) takes effect that the tropopause folding is enhanced and the static stability is increased at the 500 hPa level, which is now above the dynamical tropopause. The cross-sections reveal that the strengths of the mid-level and low-level dynamical features are significantly increased as a response to the PV modifications:

- In the area of the upper-level dry slot a lowering of the tropopause level leads to an increase of the upper-level vorticity maximum (blue contours).
- Consistently, the system increases the depth of the low-stratosphere PV (red contour). The PV inversion, in turn, builds the modified wind and temperature fields in the surroundings.
- The increase of the upper-troposphere PV (Fig. 5.46B) leads to a redistribution of the isentropic surfaces (green contours) and an increase of the relative vorticity maxima.
- The wind component orthogonal to the cross-section is then significantly increased (black contours).

5.5.2.1.2 Improved Forecast

The operational global ARPEGE model is run with the modified initial condition. The forecast is verified against ARPEGE analysis of 10 m wind gusts and MSLP over the southwestern part of France. Fig. 5.47E and F shows the 24 h and 30 h modified ARPEGE forecasts valid at 0600 and 1200 UTC on January 24. The added value of the modified run is assessed by comparing these forecasts with the corresponding analyses (Fig. 5.47A and B) and the forecasts (Fig. 5.47C and D) from the operational model run at 0600 and 1200 UTC. The new simulation is quite satisfactory. The forecast skill of the modified run improves the 0600 UTC run forecast:

- The MSLP forecast is significantly improved. Central pressure dropped to 970 hPa at 0600 UTC on January 24, 3 hPa higher than the analyzed pressure (Fig. 5.47A and B) and 10 hPa deeper than in the operational (CTRL) run (Fig. 5.47C and D). At 1200 UTC on January 24, the central minimum pressure is 975 hPa, about 3 hPa higher than observed.

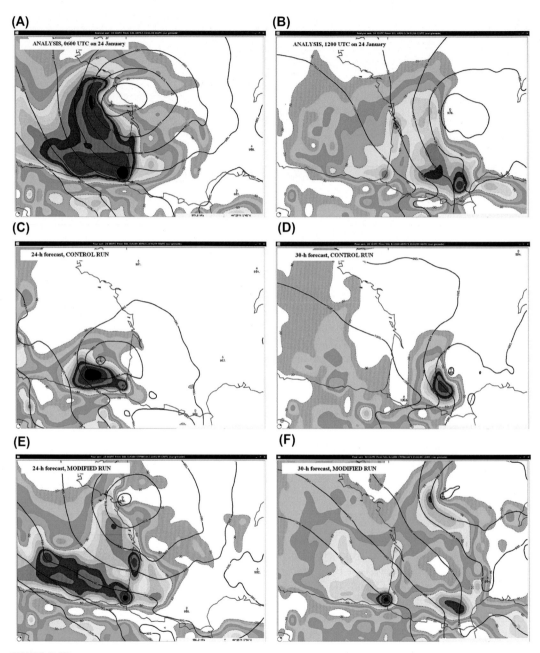

FIGURE 5.47

Comparison of the mean sea-level pressure (*black contours*) and 10 m wind gusts (shaded area from 30 to 80 kt): (A) and (B) from ARPEGE analysis, (B) and (C) from the operational model run, and (E) and (F) from the modified run for 24 h forecast and 30 h forecast, respectively, valid for January 24, 2009, at 0600 UTC and 1200 UTC, respectively.

- The cyclone moves slightly faster than in the operational (CTRL) run, and the track is slightly to the north compared to the analyzed position. The main areas of the strongest gusts are much better simulated, concentrated around the Pyrenees. However, the impacts of the storm are felt over a wider area including important urban sites in France (Bordeaux, Narbonne, Perpignan) as shown in the analysis and in the modified run.

These results confirm that human modifications add some correct observational information to the system simulation. One way in which human expertise in weather forecast can be expressed consists of monitoring NWP models by comparing PV fields to the satellite observations or satellite image brightness temperatures compared directly to those computed from predicted fields by the numerical model. A further step would be to translate the qualitative judgment in terms of PV modifications. The rerun of the model with new initial conditions after the PV inversion enables us to adjust the physical relationship between features that are present in satellite images and hence to improve the model simulation of the ensuing development.

5.5.2.2 An Example of Effects on Forecasting Upper-Level Forcing of Convection

As broadly discussed and illustrated in Chapter 4, the development of strong convective systems over the middle latitudes and subtropical areas usually is associated with clear upper-level synoptic forcing, in addition to a conditional instability of the atmosphere. Following the upper-level flow evolution by WV imagery offers the means to check on the validity of the NWP model simulations of the development of strong convection (see Section 5.4.4) and help to improve the operational numerical model forecasts by PV modifications.

Using the WV—PV relationship to adjust the initial state of NWP simulations first requires a subjective identification of the areas where PV modifications should be performed. As shown in Section 5.4.5, the use of the PV—WV concept can be of value in highlighting areas of forcing by dynamical features of various scales and positions, which may play a role in the synoptic development. On the other hand, an objective quantification of the significant PV—WV disagreement that needs PV modification based solely on WV images seems to be impossible. This section illustrates a way to avoid as much as possible the subjectivity in applying the WV—PV method by using the singular vector (SV) analysis to determine objectively the sensitive area for PV modification. The SV analysis has been a subject of numerous studies on NWP and atmospheric predictability and has found applications in improving the NWP initial state (eg, Buizza and Palmer, 1995; Buizza and Montani, 1999; Zhu and Thorpe, 2006).

The approach will be illustrated by the case study example shown in Fig. 5.48 where the observed radar reflectivity is superimposed on the Meteosat IR image and overlaid by ARPEGE analyses of geopotential of the 1.5 PVU surface on October 4, 2013. This convective system initiates, grows, and produces intensive precipitations in the leading part of the upper-level PV anomaly (minimum or trough in the geopotential of the 1.5 PVU surface). A sensitivity analysis system using an SV method with a moist total energy (TE) metric at both the initial and final times is used to identify the sensitivity area where changes in the initial modeled condition would have the greatest impact on the forecast in the verification domain (Buizza and Palmer, 1995).

The sensitivity analysis was performed using a forecast field of the operational global ARPEGE model with a lead time of 24 h. For this daily sensitivity analysis, a fixed target area was specified. The case has been evaluated on October 4, 2013 (based on the analysis at 0000 UTC), so the targeting time is

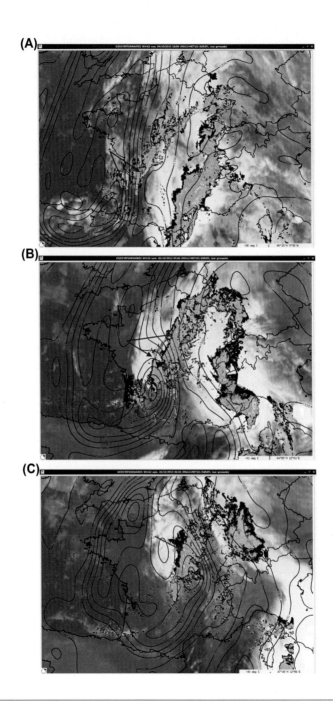

FIGURE 5.48

Radar reflectivity superimposed on the Meteosat infrared image overlaid by ARPEGE analyses of geopotential heights of the 1.5 PVU surface (brown) on (A) October 4, 2013, at 1800 UTC as well as on October 5 at (B) 0000 UTC and (C) 0600 UTC.

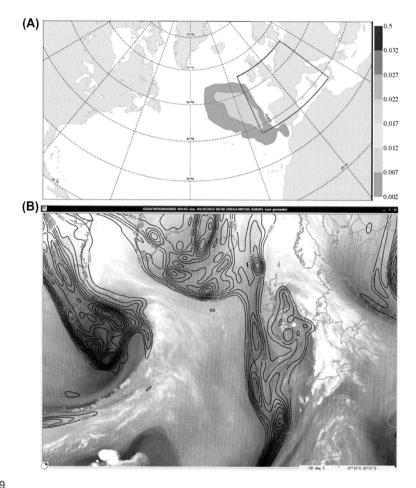

FIGURE 5.49

(A) Targeting singular vector analysis based on the Météo-France method. Vertically integrated total energy metric (color-shaded) represents the sensitive area regarding the impact of upper-level temperature and wind fields (from 200 to 500 hPa levels) on the 24 h forecast within the target area (*red box*) from the model run on October 4, 2013, at 0000 UTC; (B) the water vapor image at 0000 UTC on October 4, 2013, with the 1.5 PVU surface heights superimposed (≤1000 dam), from the ARPEGE control run.

on October 5, 2013, at 0000 UTC. In order to detect dynamically active sections of an SV, the independent and linear evolution of perturbations initially confined in the lower-troposphere is considered here. Fig. 5.49A shows sensitivity guidance for the target region (red box), along with the vertically integrated total energy (between 200 and 500 hPa) representation of the leading SV normalized by the maximum value of TE in the global model with a 24 h lead time. The growth of an SV can be described in terms of PV concept. The portion of the SV located above 500 hPa was found to play a major role in the interaction processes between the perturbation and the basic-state fields and to be responsible for the development of the perturbation (Montani and Thorpe, 2002). It follows that the perturbation of PV tends to increase with time if it is located within a region of a strong basic-state gradient of PV. Such gradients are associated with the upper-level jet or a disturbance at a lower level where thermal gradients will promote growth of low-level PV perturbation. The lower-tropospheric 24 h singular vector appears to be located in just such a region of strong upper-level PV gradient.

Based on these considerations, we may conclude that for the target region in Fig. 5.49A (over the red box) the color-shaded area across the Atlantic is a sensitive area with the possibility of an upper-level forcing and any mismatch in the WV−PV comparison over that area would have an impact on the quality of the operational forecast.

Fig. 5.49B shows the WV image from 0000 UTC on October 4, 2013, overlaid by the field of the 1.5 PVU surface heights. The upper-level diagnosis confirms the presence of a dynamical tropopause anomaly and associated dry intrusion across the Atlantic to the west of the target area. This dynamical feature seen in the WV image and in the upper-level PV field in Fig. 5.49A collocates with the eastern part of the sensitive area (color-shaded in Fig. 5.49A) derived by the SV method. Therefore, modifications of PV anomaly at the mid- and upper levels at these positions could be beneficial to the numerical simulation. Fig. 5.50 offers necessary information to assess the validity of the model performance and indicate relevant structures, which need "dynamical corrections," through a WV−PV−PWV comparison of the 6 h forecast fields from the operational model valid for 0600 UTC on October 4, 2013.

It is noted that the PV distribution around the sensitivity area over the Eastern Atlantic poorly captures the evolution of the WV features.

- The field of 1.5 PVU heights fits the pseudo WV image (derived by the same NWP model), but there are differences when compared to the satellite WV image (Fig. 5.50A).
- Comparison between the satellite WV and synthetic WV images shows that the ARPEGE model did not simulate well the moist ascent at the location of the blue arrow and the gradient zone in the 1.5 PVU surface heights between the blue and the black arrows.
- At the same time, the satellite WV image reveals a different appearance of the dry descent areas (red arrows): The darker dry areas in the satellite WV images are not correlated with the synthetic WV image, where the NWP model has simulated a moist (light) area.
- The dry feature of PV anomaly labeled "A1" is not well located regarding the observed darkest zone on the satellite WV image.

These disagreements between the 1.5 PVU heights and the observed dynamics seen in the satellite WV image can be classified as C instances of WV−PV−PWV comparison (according to the scheme in Fig. 5.5). Therefore, the upper-level initial state of the model can be modified in the direction given by the observed upper-level dynamical features in the Meteosat WV image.

Fig. 5.51 shows the new state of the dynamical tropopause height on October 4 at 0600 UTC superimposed onto the corresponding satellite WV image. Several modifications have been made according to the evolution observed from the real WV image:

- The area of tropopause folding (PV anomaly A1) was moved over the dry slots observed in the satellite WV image and the gradient zone was strengthened around the anomaly.
- The configuration of the 1.5 PVU gradient zone on the northeast side of this anomaly was modified (red arrow) to better fit the development of the humidity wave observed on the satellite WV image to the east.
- The ridge associated with the wave in the Atlantic northern flow (black arrow) was also modified.

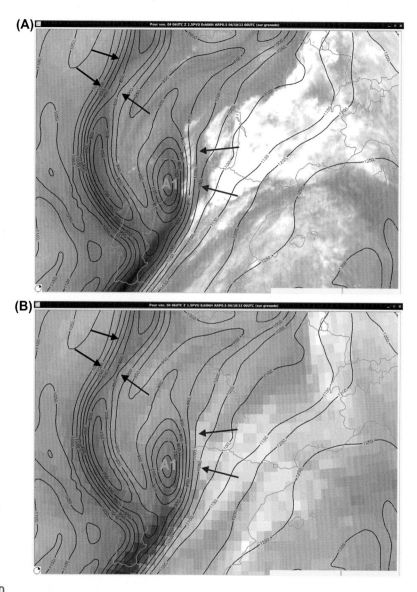

FIGURE 5.50

(A) Meteosat water vapor image and (B) synthetic water vapor image, superimposed on ARPEGE 6 h forecast of 1.5 PVU surface heights (interval 50 dam) valid for October 4 at 0600 UTC.

No modifications were made to the low levels.

The modified PV anomalies were then inverted (in terms of geopotential heights, wind, and temperature fields) and incorporated into the numerical output following the PV inversion method presented by Arbogast et al. (2008). A new run of the ARPEGE model was launched starting from this

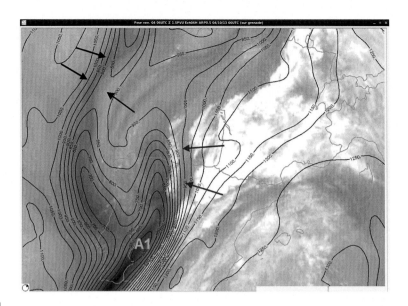

FIGURE 5.51

The new initial state of the geopotential of the 1.5 PVU surface (brown, interval 50 dam) on October 4, 2013, at 0600 UTC after handmade modifications and potential vorticity inversion, superimposed on the corresponding Meteosat water vapor image.

new initial state on October 4, 2013, at 0600 UTC. Fig. 5.52A and B shows the 30 h forecast of the geopotential heights of the 1.5 PVU surface valid for October 5, 2013, at 0600 UTC from the ARPEGE CTRL run and from the modified ARPEGE run. They clearly demonstrate the positive impact of applying PV corrections at the initial time. The PV correction in the sensitive area has taken effect on

(A)   **(B)**

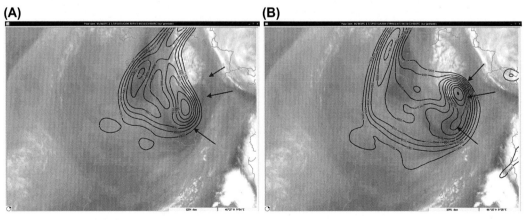

FIGURE 5.52

Meteosat water vapor image, overlaid by the ARPEGE field of 1.5 PVU surface heights (≤1000 dam), valid for October 5, 2013, at 0600 UTC: (A) 30 h forecast from the operational control run on October 4, 2013, at 0000 UTC and (B) 24 h forecast from the modified run based on the new initial state on October 4, 2013, at 0600 UTC.

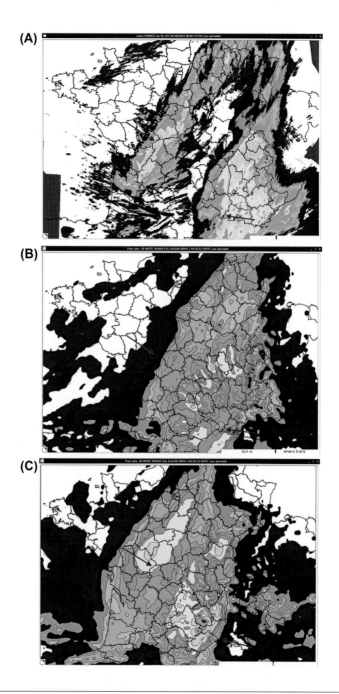

FIGURE 5.53

The 6 h total amount of rainfall valid for October 5, 2013, at 0000 UTC from (A) radar observations, compared with the ARPEGE fields of 24 h forecast from (B) the operational control run on October 4, 2013, at 0000 UTC and (C) from the modified run based on the new initial state on October 4, 2013, at 0000 UTC. Color shades from violet to red indicate precipitations 0.1, 1, 3, 5, 7, 10, 15, 20, 30, 50, and 70 mm.

the evolution of upper-level flow: The modification propagated toward the east/northeast and induced after 30 h of simulation a stronger correlation between the position of high PV areas and dark zones on WV imagery (red arrows).

The gain provided by the initial corrections is associated with a substantial improvement in the precipitation forecast. Fig. 5.53 shows the 6 h total amount of rainfall valid for October 5, 2013, at 0000 UTC from radar observations, compared with the ARPEGE fields of the 24 h forecast from the operational CTRL run and from the modified run based on the new initial state on October 4, 2013, at 0000 UTC.

Comparing with the radar observations in Fig. 5.53A, it is seen that the total rainfall values forecasted by the new run are stronger (Fig. 5.53C) in the South of France in better accordance with the maximum observed values (at the bottom right red arrows). The precipitating structure observed to the northwest (upper-left red arrows) is also better simulated by the new run: It extends over Northeast France and fits the observations better. A comparison with Fig. 5.52B shows a strong correlation between the position of the precipitating area and high PV values. It appears that an accurate forecast of the intensity and location of the precipitation is strongly dependent on both the position and the value of the upper-level PV anomaly.

The improvement we achieved with the modification of the initial state of the ARPEGE model is important, and we can expect that a limited-area mesoscale model based on such a global run should give a rather accurate forecast. This experiment clearly shows the great impact of the upper-level synoptic-scale dynamics on such convective events at midlatitudes. It appears that local corrections of the upper-level PV field (or of the topography of the dynamical tropopause) guided by satellite WV channel imagery may lead to a substantial improvement of the forecast in terms of accumulated precipitation. After the precipitation fields obtained from the modified model show significant improvement, a limited-area mesoscale model based on such a global run should give a rather accurate forecast.

5.6 SUMMARY OF THE CONCLUSIONS
5.6.1 COMPARISON BETWEEN WATER VAPOR IMAGERY AND DYNAMICAL FIELDS AS AN OPERATIONAL TOOL

Comparing WV imagery with model-derived PV fields (PV—WV comparison) provides a basis of methods for validating NWP output. The dynamically active regions in the upper-tropospheric circulation are associated with processes that demonstrate a relationship between WV imagery and the PV distribution at these altitudes (PV—WV relationship; Sections 5.1.1 and 5.1.2).

- The PV—WV relationship is complex (Section 5.1.3): The same darkening process on the imagery denotes larger (respectively, smaller) PV anomalies in the higher (respectively, lower) latitudes. A poor relationship may be associated with blocking circulation systems.
- The comparison between PV fields and WV imagery is useful in dynamically active regions and in cases of a meaningful PV—WV relationship.

The approach of combined use of PV fields and WV imagery may be considered from two general points of view (see Section 5.1.1):

- In a dry (dark) feature PV—WV comparison, the PV—WV relationship is examined over dark-gray shade features in the imagery, which are associated with dynamical tropopause anomalies.

- In a moist (light) feature PV—WV comparison, the PV—WV relationship is analyzed using light patterns on the imagery. The purpose of a moist comparison is to establish any mismatches at the moist side of the jet associated with high geopotential of the dynamical tropopause and light image gray shades.

Synthetic (pseudo) WV images are also derived for use in weather forecasting. These are brightness temperature fields (for the WV channel) of the atmosphere simulated by a numerical model (Section 5.2).

The use of synthetic WV images may enrich the efficiency of the PV—WV comparison as an operational tool by allowing comparison of model-generated PV fields with two different data sources (Section 5.3).

- Usually, the quickest way to check NWP analyses and very short period forecasts is to compare real (derived by satellite) and synthetic PWV imagery.
- However, PV fields should also be compared with the imagery, because when a real error is detected, this comparison may provide knowledge of the error in the PV distribution as well as of a potential adjustment of NWP initial fields by PV modifications.
- In addition, synthetic WV images are useful when compared not only with satellite WV images but also with PV fields to indicate whether any mismatches correspond to real NWP model errors (Section 5.4).
- Vertical cross-sections may be used to gain knowledge about the vertical structure of the PV distribution, humidity, and vertical motions that may facilitate the interpretation of WV imagery (Sections 5.3 and 5.4).

5.6.2 COMPARING SATELLITE AND SYNTHETIC WATER VAPOR IMAGERY WITH POTENTIAL VORTICITY FIELDS TO VALIDATE NUMERICAL WEATHER PREDICTION OUTPUT

Applying the approach presented in Section 5.3.1 for the comparison between the three kinds of data sources, we can distinguish five possible cases of agreement or disagreement, each indicating a different kind of NWP validity or inaccuracy. We can classify these cases into three main groups:

A. Agreement between the WV image and PV field as well as between the PV field and the synthetic image (WV $\equiv$ PV $\equiv$ PWV): No sign of NWP errors (Section 5.3.2).
B. Mismatches between the synthetic image and the PV field (PWV $\neq$ PV): NWP errors are possible, but the specific nature of the errors is unclear (Section 5.3.2).
Three subinstances are possible:
  B1. Agreement between the WV image and the PV field (WV $\equiv$ PV).
  B2. Mismatches between the WV image and the PV field as well as between the WV image and the synthetic image (WV $\neq$ PV and WV $\neq$ PWV).
  B3. Agreement between the WV image and the synthetic image (WV $\equiv$ PWV).
C. Mismatches between the WV image and the PV field (WV $\neq$ PV) and agreement between the PV field and the synthetic image (PV $\equiv$ PWV): Clear signs of NWP errors (Section 5.4).

5.6.3 USING THE TECHNIQUE OF POTENTIAL VORTICITY INVERSION TO ADJUST NUMERICAL WEATHER PREDICTION INITIAL CONDITIONS TO BETTER UNDERSTAND THE SYNOPTIC DEVELOPMENT

The conserved nature of the PV parameter and the invertibility of PV (as shown in Chapter 1) enable us to build up the flow and temperature structure associated with a given PV anomaly. The technique of PV inversion, whereby the flow and mass fields associated with a particular portion of the PV field are calculated, can be used to adjust NWP initial conditions to get an alternative numerical forecast. Applying the PV thinking, changes in the initial NWP conditions can be made to deduce which PV features play important roles in the process and give us a better understanding of the evolution of both tropical and extratropical weather systems (Section 5.5.1).

5.6.4 USING SATELLITE AND SYNTHETIC WATER VAPOR IMAGES AND POTENTIAL VORTICITY CONCEPTS TO ADJUST NUMERICAL WEATHER PREDICTION INITIAL CONDITIONS AND TO GET AN ALTERNATIVE NUMERICAL FORECAST

If any significant disagreement of satellite WV images with the model-generated PV fields and synthetic WV images is detected, a local modification of PV may be applied to adjust initial conditions in the operational NWP model (Section 5.5.2).

- PV modifications are appropriate to be made for the 1.5 PVU surface height to improve its fit to the satellite WV image and to get a new initial state by applying a PV inversion procedure.
- In addition, adjustment of the low-level conditions can improve the fit of the low-level temperature field to the observational data.
- Finally, an alternative forecast may be computed by rerunning the NWP model using this new initial state, and this could thereby improve the operational numerical forecast.

Conclusion

As shown and discussed in Part II of this book, water vapor (WV) imagery provides an opportunity to make real-time observations of the upper-level circulation of the atmosphere. This is very important because upper-level circulation represents the main tropospheric motions and governs the low-level flow. Hence, this is essential from an operational weather forecasting perspective since perturbations of upper-level circulation represents the upper-troposphere dynamics, which plays an essential role in the development of midlatitude high-impact weather systems. Animating a sequence of WV images helps to emphasize the evolution of the upper-level moisture flow and therefore enables us to survey the evolution of the upper-level dynamics. Comparing the analyzed and forecast dynamical fields with the real dynamics in the troposphere as observed in the satellite images offers a means to control the behavior of the numerical models. For convective storm nowcasting, a combined use of complementary observation techniques and products is considered to be very beneficial and ensures the optimal use of the satellite data in an operational context.

Potential vorticity (PV) fields and related concepts have proven to be valuable operational tools in the dynamic evaluation of model analysis and very short-range forecasts; these are especially efficient in cases where cyclogenesis or upper-level forcing of convection in midlatitudes as well as extratropical transition of a tropical cyclone are likely to occur. Comparing upper-level PV fields and WV imagery can help to detect errors in the operational numerical weather prediction (NWP) models and to alert forecasters to the model shortcomings. Cross-sections must also be used to get a more complete and accurate view of the model behavior. In addition, to obtain a more direct comparison with real (satellite) WV imagery, synthetic WV images (calculated from the numeric model analyses and forecasts fields) are now available. Various examples of the use of these products were presented in Chapter 5.

One of the most important tasks of the forecasters is to validate operational NWP models. In particular, since many NWP forecast errors are due to problems in the numeric analysis, correctly assessing the initial state of the model can be crucial for the prognosis of severe weather events. For this reason, the ability to take into consideration all evidence from the model output along with observational and satellite data together and to assess correctly the rapid synoptic development proves to be an extremely difficult operational task. This guide delivers an important message to forecasters concerning the feasibility of easily applying such a complex approach by comparing satellite and synthetic WV imagery with PV fields. As shown in Sections 5.4 and 5.5, this concept may significantly help to facilitate understanding the synoptic evolution and hence to improve the forecast in various typical cases of cyclogenesis and convective developments in which the NWP models might be in error.

Currently, a powerful solution technique is available that uses methods to invert the PV fields into winds and temperatures that can then be used to correct the numeric analysis and to rerun the NWP model. This approach can successfully be applied in sensitivity analyses of upper-level dynamics through NWP simulations to deduce which dynamical features play important roles in the process and give us a better understanding of the evolution of both tropical and extratropical weather systems.

Despite improvements in operational NWP analysis schemes and the significant increase of available remotely sensed data, there remain situations when, even at very short ranges, NWP products are misleading because errors are present in the initial conditions and these errors become amplified in the subsequent forecast. Although modern four-dimensional variational systems of NWP data assimilation have been implemented, they work point by point, with no specific method allowing preservation of crucial structures such as, for instance, strong humidity gradients. Using satellite imagery in an operational forecasting environment provides an accurate global view of the organization of moisture and cloud systems and gives information about their structures. By monitoring the NWP products—in particular, applying a satellite imagery approach—forecasters can assess the numeric information, detect failures in the initial state of the model, and improve their forecasts. Assuming an error in NWP output can be detected, the problem then becomes one of how to improve the forecast. For nowcasting and very short-range forecasts, the forecaster has to extrapolate the evolution of the detected systems by comparing against a conceptual model. As shown in Section 5.5.2, local corrections of the upper-level PV field (or of the topography of the dynamical tropopause) guided by WV channel imagery may lead to a substantial improvement of the initial conditions, thereby improving the forecast in an alternative run. The approach described in this book is not inconsistent with numeric analysis schemes; it is a complementary solution. Forecasters need efficient and easily applied tools and techniques to perform their tasks to provide the best possible forecast, and this is particularly essential when severe weather is expected. The authors hope that the material in this book on the interpretation of satellite WV imagery and PV fields over midlatitudes and subtropical and tropical areas as well as the methodology presented in Chapter 5 can help in this context.

RADIATION MEASUREMENTS IN WATER VAPOR ABSORPTION BAND

A

A.1 GEOSTATIONARY METEOROLOGICAL SATELLITES AND THEIR WATER VAPOR CHANNELS

Current geostationary meteorological satellites operated by the main space agencies provide imagery in various channels of the strong 6.3 μm absorption band of water vapor (WV). Information for some of them is presented in Table A.1. The Russian Electro-L N1 satellite (launched in 2011 for Indian Ocean coverage) is not mentioned since the WV channel of its instrument is not functional because of excessive noise, as reported at the 42nd Meeting of the Coordination Group for Meteorological Satellites (CGMS), May 22–23, 2014, in Guangzhou, China.

The definitions of the WV channels of the instruments on board the geostationary satellites that are currently in operation are given in Table A.1.

The Meteosat First Generation (Meteosat-7) instrument, which still performs only the Indian Ocean Data Coverage Service, the 6.3 μm channel band is 5.7–7.1 μm. Since 2003, the European Organization for the Exploitation of Meteorological Satellites (EUMETSAT) has launched four Meteosat Second Generation (MSG) satellites (Meteosat-8 to Meteosat-11). The MSG geostationary satellites brought out in space the instrument Spinning Enhanced Visible and InfraRed Imager (SEVIRI), which performs measurements in two channels centered at 6.2 and 7.3 μm wavelengths

301

Table A.1 Water Vapor Channels on Current Geostationary Satellites

Sector	Longitude	Name (Mission)	Operator	Center (µm)	Bandwidth (µm)
East Atlantic	0°E	Meteosat Second Generation (MSG) (full scan)	EUMETSAT	6.25 7.35	5.35−7.15 6.85−7.85
East Atlantic	9.5°E	MSG (rapid scan)	EUMETSAT	6.25 7.35	5.35−7.15 6.85−7.85
Indian Ocean	57.3°E	Meteosat-7	EUMETSAT	6.4	5.7−7.1
West Atlantic	75°/105°W	GOES East	NOAA	6.55	5.8−7.3
East Pacific	135°W	GOES West	NOAA	6.55	5.8−7.3
Indian Ocean	55°E	INSAT	ISRO	6.8	5.7−7.1
West Pacific	145°E	MTSAT	JMA	6.75	6.5−7.0
	140.7°E	Himawari	JMA	6.2 6.9 7.3	6.06−6.43 6.89−7.01 7.26−7.4
West Pacific	105°E	FY-2E	CMA	6.9	6.2−7.6

within the WV absorption band that are sensitive to the radiation in the spectral intervals 5.35−7.15 µm and 6.85−7.85 µm, respectively.

GOES Imager on board the geostationary satellites of National Oceanic and Atmospheric Administration (NOAA) generates WV images in 6.55 µm channel (5.8−7.3 µm). The FY-2E satellite of China Meteorological Administration (CMA) performs measurements in WV absorption band 6.2−7.6 µm. For MTSAT satellite of Japan Meteorological Agency (JMA), the WV channel band is 6.5−7.0 µm (centered at 6.75 µm), while for the Indian National Satellite (INSAT) of Indian Space Research Organization (ISRO) it is 5.7−7.1 µm (centered at 6.8 µm).

The new generation JMA geostationary satellite Himawari-8 became operational for a meteorological mission in 2015. The satellite has a new payload called Advanced Himawari Imager (AHI) with three WV channels of relatively narrow bands centered at 6.2, 6.9, and 7.3 µm. The strength of these three WV channels is that they provide information about moisture at three different levels in the layer between the lower-middle troposphere to the upper troposphere.

A.2 RADIATIVE TRANSFER THEORY FOR THE WATER VAPOR CHANNELS

The meteorological satellites measure radiation emitted by objects such as cloud-top elements and the Earth's surface in various channels covering different bands of the infrared (IR) spectrum. For the WV channels, some portion of the upcoming radiation will be absorbed by WV and then reradiated by the atmosphere. The absorption and reradiation depend on the vapor pressure in various ways for different channels. Accordingly, the satellite may measure quite different intensity of radiation, which is referred to as "radiance" at the corresponding channel.

Normally, the radiative transfer theory for the WV channels is developed with the assumption that the scattering of the long-wave radiation may be neglected because of the strong absorption within the spectral response region of the satellite instrument. The radiance I_ν emitted at the top of a non-scattering atmosphere at zenith angle θ and wave number ν is given by the radiative transfer equation:

$$I_\nu = (I_0)_\nu \tau(\nu,\theta,p_0) + \int_{p_0}^{0} B\{\nu,T(p)\} \frac{d\tau(\nu,\theta,p)}{dp} dp, \qquad [A.1]$$

where $(I_0)_\nu$ is the radiation from the surface (land, sea, or cloud top) and the integration is performed from the surface at pressure $p = p_0$ to space ($p = 0$). $B\{\nu,T(p)\}$ is the Plank function corresponding to the atmospheric temperature $T(p)$ at pressure p and is given by:

$$B\{\nu,T(p)\} = \frac{c_1\nu^3}{\exp\{c_2\nu/T(p)\} - 1}, \qquad [A.2]$$

where $c_1 = 1.19104 \times 10^{-5}$ mW m^{-2} sr^{-1}(cm^{-1})$^{-4}$ and $c_2 = 1.43877$ K(cm^{-1})$^{-1}$ are constants. $\tau(\nu,\theta,p)$ is the transmittance of an atmospheric path at zenith angle θ from the surface to space that depends on the vertical humidity profile by the expression:

$$\tau(\nu,\theta,p) = \exp\left\{ -\sec\theta \int_p^0 k(\nu,p)c(p)\rho(p)dp \right\}, \qquad [A.3]$$

where $k(\nu,p)$ is the absorption coefficient, $c(p)$ is the mass mixing ratio profile, and $\rho(p)$ is the atmospheric density profile. The transmittance, at a given level, describes the ratio of the radiation arriving from below this level and penetrates to the satellite. In other words, it describes how much absorption will be accomplished on the radiation reaching that level by the all WV above this level (see Weldon and Holmes, 1991).

Eq. [A.1] presents the monochromatic radiance at a single wave number ν expressed in [W m^{-2} sr^{-1}] units. Since the satellite radiometer exhibits a specific resolution, it measures radiation $I_{\Delta\nu}$, which is a value of the IR radiation averaged in the spectral band $\Delta\nu = \nu_1 - \nu_2$:

$$\begin{cases} I_{\Delta\nu} = \int_{\nu_1}^{\nu_2} A(\nu)I_\nu d\nu, \\[2em] \int_{\nu_1}^{\nu_2} A(\nu)d\nu = 1, \end{cases} \qquad [A.4]$$

where $A(\nu)$ is the spectral response function of the satellite radiometer and (ν_1,ν_2) are its limiting wave numbers. I_ν is the monochromatic radiance given by Eq. [A.1].

Thus, using the integral over the spectral response in [A.4], the spectral radiance measured by the satellite per unit wavelength (instead of per unit frequency) is derived in [W m^{-2} sr^{-1} cm^{-1}] that are simply called the radiance unit.

A.3 WATER VAPOR CHANNELS OF METEOSAT SECOND GENERATION SATELLITES AND THEIR SPECTRAL RESPONSE

Fig. A.1 illustrates the spectral response of MSG SEVIRI radiometer in the view of space-based inter-calibration system approach (Hewison et al., 2013) derived through measurements by IR Atmospheric Sounding Interferometer (IASI), which provides high-quality hyperspectral reference observations from its low Earth orbit (LEO) MetOp platform. It is adapted from Fig. 3 of Hewison et al. (2013) showing example radiance spectra measured by IASI (black), convolved with the Spectral Response Functions of SEVIRI channels 5–11 from right to left (colored-shaded areas) as well as approximate positions of H_2O absorption lines.

Fig. A.1 shows the spectral response of the two MSG WV channels, centered at 6.2 μm (5.35–7.15 μm) in orange and 7.3 μm (6.85–7.85 μm) in yellow. The WV absorption within the band 8.3–9.1 μm (pink) is much weaker and the 8.7 μm channel may be considered as an Infra Red (IR) channel in the WV absorption band. The integration in Eq. [A.4] takes effect in eliminating the fine-scale structure of the spectrum seen in Fig. A.1.

In the spectral region of the two MSG WV channels centered at 6.2 and 7.3 μm wavelengths (1380–1834 cm^{-1} and 1272–1449 cm^{-1}, respectively) the dominant absorbing gas is WV. Although there are spectral lines of other absorbing gases like CH_4 and N_2O (see Fischer et al., 1981; Tjemkes

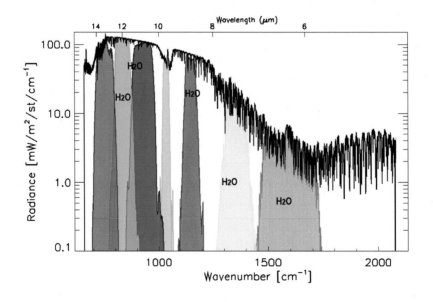

FIGURE A.1

Example of radiance spectra measured by Infrared Atmospheric Sounding Interferometer (black), convolved with the Spectral Response Functions of Spinning Enhanced Visible and InfraRed Imager channels 5–11 from right to left (colored-shaded areas) as well as approximate indications for H_2O absorption lines.

Adapted from Fig. 3 of Hewison, T.J., Wu, X., Yu, F., Tahara, Y., Hu, X., Kim, D., Koenig, M., 2013. GSICS inter-calibration of infrared channels of geostationary imagers using Metop/IASI. IEEE Trans. Geosci. Remote Sens. 51 (3), http://dx.doi.org/10.1109/TGRS.2013.2238544 and kindly provided by Tim Hewison (EUMETSAT).

and Schmetz, 2002), their influence on the radiance signal can be neglected for the investigated wide spectral intervals.

The radiation characteristics of the MSG channels will be illustrated in Section A.3, based on calculation, kindly performed for the purposes of this book by Stephen Tjemkes of EUMETSAT. To calculate the propagation of the radiance through the atmosphere the accurate radiative transfer code LBLRTM has been used. Its current version is v12.2. The associated line parameter set is the AER v3.2, which is based on the latest High-resolution Transmission database (HITRAN). The LBLRTM code is maintained by Atmospheric and Environmental Research (AER) in Boston, USA (http://rtweb.aer.com/lblrtm.html). LBLRTM is one of the few high-resolution line-by models, which means it solves the monochromatic radiative transfer in Eq. [A.1] (see also Tjemkes and Schmetz, 1997). The calculations are performed for vertical profiles of temperature and humidity from models of "standard atmospheres," shown in Fig. A.2.

- tropical (red), which is the warmest one among the others and most humid especially in the lower troposphere
- midlatitude summer (green)
- midlatitude winter (dark blue), the coldest one and most dry among the others
- subarctic summer (light blue)

Above the 250 hPa level the temperature profiles for the standard atmospheres in Fig. A.2 are complicated but this does not influence contribution functions because the specific humidity is practically zero at these altitudes.

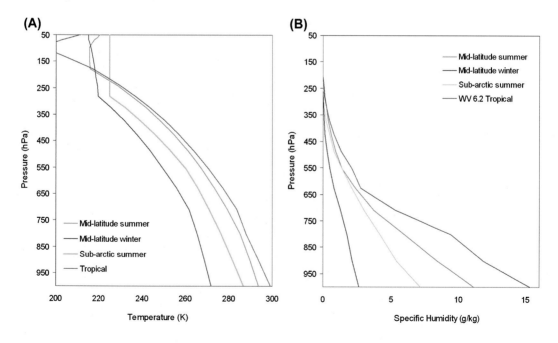

FIGURE A.2

(A) Temperature and (B) water vapor profiles of standard atmospheres: tropical (red), midlatitude summer (green), midlatitude winter (dark blue), and subarctic summer (light blue).

A.3.1 TRANSMITTANCE

The transmittance profile [A.3] depends on the absorption properties of the atmosphere regarding the radiation in the specific channel and the vertical distribution of WV, which is the dominant absorbing gas. Fig. A.3 shows the transmittances for 6.2, 7.3, and IR 8.7 µm channels of SEVIRI instrument on MSG for the different standard atmospheres, for which the humidity profile is quite different (Fig. A.2B). A different nadir angle is chosen for each atmospheric profile in order to give a more realistic idea of the absorption of the radiation sensed by the geostationary satellites. In Fig. A.3 the calculations for tropical, midlatitude summer, and subarctic summer models are performed for zenith angles 0°, 51.78°, and 62.68°, respectively, corresponding to latitudes 0°, 45°, and 55°N at the 0° meridian.

As seen in Fig. A.1, the radiation in IR 8.7 µm is slightly absorbed by moist air, while the WV 6.2 µm and WV 7.3 µm channels of MSG are within the band of strong absorption by WV around 6.3 µm wavelength. The significant absorption prevents a great part of the radiation in the WV channels emitted from the surface escaping into space. This is clearly demonstrated by the transmission function, which is zero at low levels for the two WV channels. For the IR 8.7 µm channel a large fraction of radiation emitted by the surface can escape into space. This is reflected by the profile of the transmission function, which has a value around 0.8 at the surface, which means about 80% of the radiation in the IR 8.7 µm channel emitted from the surface will reach the satellite.

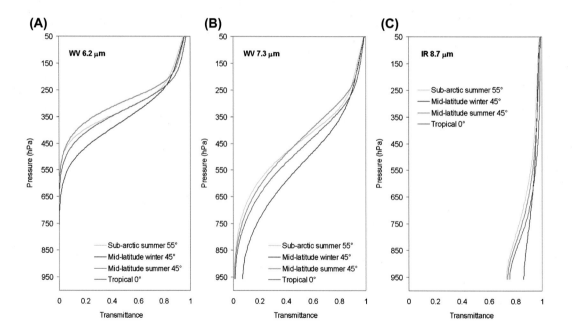

FIGURE A.3

Transmittance for (A) water vapor (WV) 6.2 µm, (B) WV 7.3 µm, and (C) infrared 8.7 µm channels of Spinning Enhanced Visible and InfraRed Imager on Meteosat Second Generation for four different standard atmospheres: tropical (red), midlatitude summer (green), midlatitude winter (dark blue), and subarctic summer (light blue).

Fig. A.3 shows that atmospheric profiles with increased temperature and humidity (tropical, red; and midlatitude summer, green) decrease the transmittance of the radiation to space much more than for the winter standard atmosphere (dark blue). For all channels, the transmittance is maximum along the entire profile for the midlatitude winter atmosphere because the low temperatures decrease the specific humidity and the absorption is reduced. For WV 6.2 μm, this effect will cause moisture at low-level down to 700 hPa level altitude to be detectable, while for all three summer standard atmospheres the transmittance becomes zero below 600 hPa level.

Meteosat and other geostationary satellites are located at 0° latitude, and the radiance measured by the satellite depends on the nadir angle of the moisture layer sensed (see Fig. A.4). The radiometer scans the Earth disc viewing different locations under different angles that also affects the resolution of the satellite instrument. Due to the fact that the horizontal resolution decreases with increasing distance from the subsatellite point, the part of the image to be quantitatively processed is usually restricted to a circular area inside 55°E to 55°W and 55°S to 55°N. That is why, for the subarctic atmosphere in Fig. A.3, there is the chosen nadir angle that corresponds to the 55° latitude and the limit for quantitative application of the information from geostationary satellites.

Fig. A.3 shows that although the tropical profiles are warmer and more humid the transmittance (in red) is larger than for the midlatitude and subarctic atmospheres during the summer. This is a result of differences in the radiation path of the earth's scanning at different nadir angles from a geostationary satellite. This effect is illustrated considering the transmittances of the midlatitude summer (Fig. A.5A) and midlatitude winter (Fig. A.5B) atmospheres for the radiation in 7.3 μm WV channel.

As shown in Fig. A.5 varying the nadir angle, and hence the slant path through the atmosphere, affects the transmittance and hence the radiance. With increases of zenith angle the transmittance decreases because of the increasing the path of radiation length. Also seen in Fig. A.5B is that during the winter the transmittance for the radiation in 7.3 μm WV channel is significantly larger than zero and a fraction of radiation emitted by the surface can escape into space. This is especially pronounced at the lower latitudes due to the shorter radiation path length and allows high altitude mountain terrain to be visible in the 7.3 μm channel images in cases of dry mid- and upper-troposphere (see Chapter 2, Fig. 2.17). This effect will be further illustrated considering the contribution of the WV absorption and reradiation at a different level than the radiance at the top of the atmosphere.

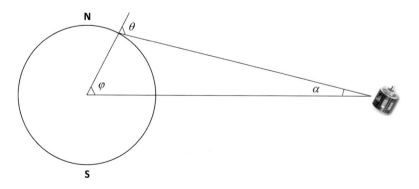

FIGURE A.4

Definition of the zenith angle θ; Meteosat nadir angle α; φ is the latitude.

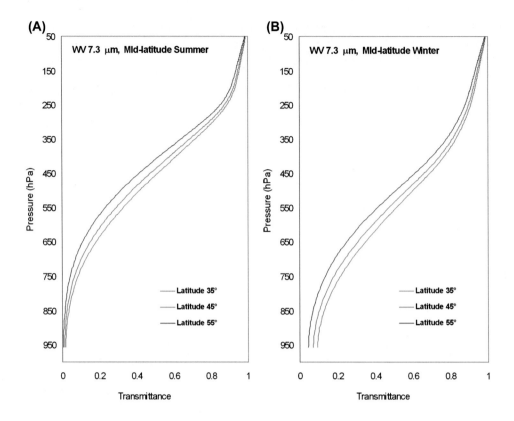

FIGURE A.5

Transmittances for the radiation in Meteosat Second Generation 7.3 μm water vapor channel of
(A) midlatitude summer and (B) midlatitude winter atmospheres. The curves for zenith angles 40.62° (red),
51.78° (green), and 62.68° (blue) are presented that correspond to 35°, 45°, and 55° latitude, respectively.

A.3.2 WEIGHTING FUNCTION

In order to analyze the transmittance ability of the atmosphere with regard to the radiance at any IR
channel, the weighting function is defined by using the derivative of the transmittance profile [A.3]
with respect to height (or pressure level). The weighting function illustrates the relative vertical
contribution of radiation by moisture of the atmosphere and it is presented by Eq. [A.5] (see Fischer
et al., 1981):

$$W(\log p) = \int_{v_1}^{v_2} A(v) \frac{d\tau(v, \theta, \log p)}{d \log p} dv. \qquad [A.5]$$

Weighting functions are important diagnostic parameters to analyze the radiation measured by the
satellite. They indicate at which layers the radiance can originate due to the absorption properties of

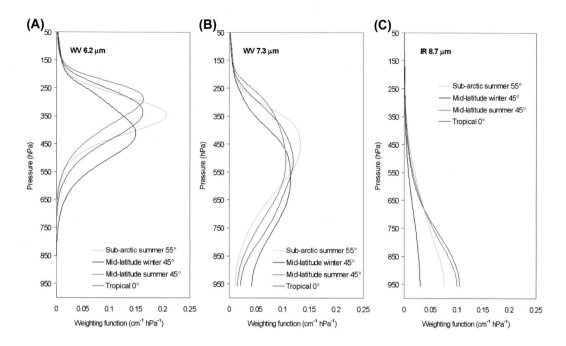

FIGURE A.6

Weighting functions for (A) water vapor (WV) 6.2 μm, (B) WV 7.3 μm, and (C) infrared 8.7 μm channels of Spinning Enhanced Visible and InfraRed Imager on Meteosat Second Generation for four different standard atmospheres: tropical (red), midlatitude summer (green), midlatitude winter (dark blue), and subarctic summer (light blue).

the atmosphere in cloud-free conditions for specific channels. The weighting functions, as a function of the logarithm of the pressure for the different standard atmospheric models, are shown in Fig. A.6. The plots demonstrate that the three MSG channels have been selected such that they provide good information on clouds, the Earth's surface, and WV. Fig. A.6 shows that radiation at the top of the atmosphere in the WV channels originates from levels around the middle troposphere, which is indicated by the local maximums in the curves at these levels. The WV 6.2 μm weighting functions peak at lower pressure values (around 350 hPa level) than the WV 7.3 μm weighting functions (around 500 hPa level) due to the stronger WV absorption of the 6.2 μm radiation. The weighting functions for IR 8.7 μm channel show peaks near the surface, indicating sensitivity to low-level moisture. Fig. A.6 also shows that in different atmospheric profiles of temperature and humidity, the peaks of the weighting functions appear at different levels, hence the radiance, measured in the WV absorption channels originates from different altitudes.

The weighting functions for WV channels are highly dependent on the humidity profile (through the transmittance) in the layer of the channel's sensitivity. The fact that the weighting function for the 6.2 μm channel (Fig. A.6A) in the midlatitude summer model lies much higher than for the midlatitude winter atmosphere is due to the corresponding much higher WV density (Fig. A.2B) of the summer profile. Increasing the viewing angle also plays a certain role to shift the weighting function curves at upper levels due to increasing radiation path length and hence increasing the total content of moisture, which absorbs. This effect causes the level of weighting function maximum in subarctic atmosphere and nadir angle 7.68° to be the same (350 hPa) as in the tropical atmosphere and nadir angle 0°, although the specific humidity in the tropical model is higher.

The influence of the humidity profile on the weighting function can be seen when considering the 7.3 μm channel, whose sensitivity covers almost the entire troposphere. The curves in Fig. A.6B for midlatitude (green) and subarctic summer (light blue) models illustrate the following radiation characteristics:

- In the midlatitude-summer atmosphere, greater parts of the radiation emitted by moisture originate from the upper (above 300 hPa) and lower-middle (below 600 hPa) layers comparing with the corresponding weights in the subarctic model. This is due to the increased specific humidity in the midlatitude summer profile (Fig. A.2B, green) at these layers in comparison with the subarctic model values (Fig. A.2B, light blue).
- Accordingly, the relative vertical weight of radiation emitted by moisture in the layer 600–300 hPa is much larger for the subarctic model (Fig. A.6B, light blue) than for the midlatitude summer profile.

Therefore, if the humidity of a layer is reduced, the contribution from the layer decreases.

A.3.3 CONTRIBUTION FUNCTION

The channels in the WV absorption band exhibit specific sensitivity not only to the moisture; the radiance depends also on the temperature profile in the path of radiation to the satellite, and thus provides information for a wide range of meteorological applications. With the air temperature sounding fixed, the radiation intensities measured by the satellite in different channels are composed by the radiation arriving from different origins. Therefore, for interpreting satellite measurements in the WV absorption band an altitude association of the observable phenomena in corresponding images is required. Although the weighting function can be used to characterize the weights for each level to the radiation measured by a space-borne instrument, it is not sufficient to diagnose the contribution of each level to the radiance at the top of the atmosphere. This is because it accounts for only the transmittance, and the amount of emitted radiation of each level is not taken into account in the weighting function [A.5].

Due to the strong temperature dependence of the Plank function [A.2] within the considered spectral region, the whole integrand of the radiative transfer Eq. [A.1] must be analyzed (Fischer et al., 1981). It is the product of the weighting function and the Plank function that is called contribution function:

$$C(\log p) = \int_{\nu_1}^{\nu_2} A(\nu)B\{\nu, T(p)\} \frac{d\tau(\nu, \theta, \log p)}{d \log p} d\nu, \qquad \text{[A.6]}$$

Thus, taking into account both the transmittance and the energetic level of the emitted radiation, the contribution function [A.6] describes the contribution of different atmospheric layers to the radiance measured at the satellite. It can be defined as the fraction of radiation at a particular level, which will leave the top of the atmosphere (Tjemkes and Schmetz, 2002). The contribution function follows this as the gradient of the cumulative contribution function with pressure. It thus can be used to analyze the relative contribution by moisture and temperature of the atmosphere to the radiation measured by the satellite. The contribution function has the radiance units (W m^{-2} sr^{-1} cm^{-1}) per hPa, while in the end the total radiance is calculated as the integral of this function over height (pressure). By virtue of its definition, the contribution function resembles the weighting function to a large extent. In certain situations, however, there is a significant difference between them.

Fig. A.7 shows comparisons of the two functions, normalized with respect to their maximum value, for 6.2 and 7.3 μm channels in tropical atmosphere at the nadir (0° latitude). The contribution function lies below the weighting function everywhere. This is because of the effect that only a specific portion of the thermal radiation coming from below will be absorbed by tropospheric moisture and then reemitted at the temperature of the top of the moist layer (to be considered as "portioning effect" in Section A.4). In the upper troposphere, a large amount of the radiation arriving from below passes through this layer, since this moisture has a low mixing ratio and a low density. Because of this effect, the contribution function is lower than the weighting function in colder conditions above the level of maximum contribution and higher than the weighting function in warmer conditions below the level of peak response. The maximum contribution comes from different levels for the different channels. This is determined by the reversal of roles between the radiation passing through the moist layer from below and radiation emitted by the moist layer itself, considered by Weldon and Holmes (1991) as "cross-over effect" (to be illustrated in Section A.4).

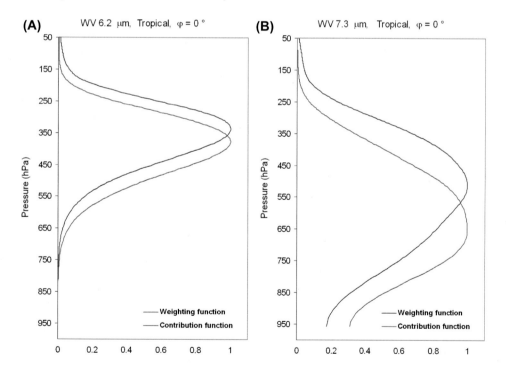

FIGURE A.7

Normalized weighting function (blue) and contribution function (red) for (A) 6.2 μm and (B) 7.3 μm channels of Spinning Enhanced Visible and InfraRed Imager on Meteosat Second Generation for tropical atmosphere profile at nadir.

Fig. A.7 also illustrates that the differences between the weighting and contribution functions depend on the radiation characteristics of the channel and on the moisture profile. While the contribution function for 6.2 µm channel exhibits a peak near 300 hPa level, the 7.3 µm channel has a "blunt" maximum over a large layer between 700 and 550 hPa levels. With this regard, the following conclusions are noteworthy:

- The reason for this behavior is that the weighting function of 7.3 µm channel shows a decrease of its maximum due to decreased moisture around 600 hPa level in the tropical atmosphere (as seen in Fig. A.2B).
- In the same time the relatively increased specific humidity around 750 hPa (Fig. A.2B) appears as a bulge in the curve of the weighting function around this level.
- These features of the tropical atmospheric model produce a large difference between the profiles of weighting and contribution functions of 7.3 µm channel. The contribution function accounts for the amounts of the emitted radiation that contribute to the radiance through the portioning effect and produces a blunt maximum in the lower-middle troposphere.
- For 6.2 µm channel the contribution function resembles the weighting function since there are no such anomalies of humidity profile in the layer of sensitivity of this channel (500−250 hPa), where the specific moisture decreases monotonically with height (Fig. A.2B).

The behaviors of the weighting and contribution functions for standard atmosphere profiles, considered above, concern the radiation effects depending on the simultaneous variation of temperature and humidity distribution. In order to separate the influence of these two variables, calculations for very dry and wet atmosphere without changing the temperature profile are performed, by using simple scaling of the specific humidity multiplied by 0.5 and 1.5 and compared with the standard case (scaling factor 1.0). Fig. A.8 shows the weighting and contribution functions for 6.2 µm channel in midlatitude winter atmosphere, considering wet, standard, and dry atmospheric models of humidity with temperature profiles unchanged. It is seen that the maximum of the contribution function lies always below the peak of the weighting function, due to the dominant role of the higher temperature of the mid-level air through the cross-over effect (see Section A.4).

Comparing the contribution functions in Fig. A.8B, one finds that the spread of the contribution functions for dry, standard, and wet profiles with respect to altitude is distinctly smaller in the upper troposphere (above the peak contribution) than that in the middle troposphere. Since the increase of the Plank function with growing temperature is steeper at higher temperature the differences are larger below the level of a peak response where the wet atmosphere absorbed much radiation at a higher pressure and hence higher mixing ratio and then reradiated at higher energetic level than the absorption and reradiation in the upper troposphere. During the winter (Fig. A.5B), the contribution function lies considerably below the weighting function at the most upper level (about 250 hPa). The bulge in the curve of the weighting function in the upper part of the troposphere for an extremely wet profile (150% specific humidity) is because the weighting function accounts for the contribution from moisture and the increased moisture content increases the anomaly. Because of the portioning effect there is not such a bulge in the curve of the corresponding contribution function due to the opposite effect of the lower Plank function values from the low temperatures in the middle and upper troposphere.

Fig. A.8B shows considerable altitude shifts of peak contribution in both directions with respect to the standard atmosphere: the drier the atmosphere, the lower the altitude of the peak contribution. In addition, because of the portioning effect, the contribution function depends on the WV profile and not

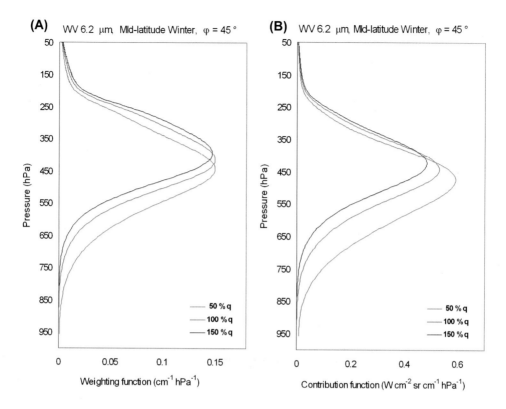

FIGURE A.8

Comparisons between (A) weighting function and (B) contribution function for 6.2 μm channels of Spinning Enhanced Visible and InfraRed Imager on Meteosat Second Generation for midlatitude summer atmosphere temperature profile and wet (blue), standard (green), and dry (red) atmospheric profile by using simple scaling of the standard values of specific humidity multiplied by 0.5, 1.0, and 1.5.

only on the total moisture content in the vertical column through the transmittance [A.3]. Contribution functions for 6.2 and 7.3 μm channels of SEVIRI on MSG for different atmospheric profiles of temperature and humidity are shown in Fig. A.9.

It is seen that the contribution functions of MSG WV channels are peaking at different altitudes: These are in the upper troposphere for the 6.2 μm radiation and in the lower-middle troposphere for the 7.3 μm channel. The effect of stratospheric WV and temperature on the radiance is negligible because of very low WV density above 250 hPa level for all standard atmospheres considered (see Fig. A.2B). Therefore, changing water content at a given level influences in a different way the radiance in each one of the two channels thus allowing for a sort of "sounding" capability of MSG WV imagery to observe moisture regimes in different layers of the troposphere. The absorption is strong enough to block a great part of radiation of the two MSG WV channels from escaping into space, such that most of the radiation, which leaves the atmosphere, originates from within the atmosphere itself. Thus, a combination of channels provides useful data for air mass analyses, eg, to infer information on atmospheric instability.

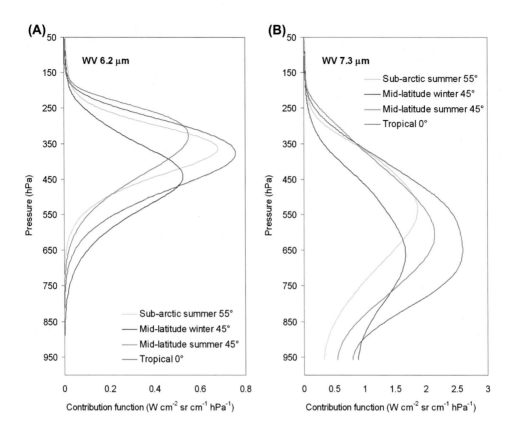

FIGURE A.9

Contribution functions for (A) 6.2 μm and (B) 7.3 μm channels of Spinning Enhanced Visible and InfraRed Imager on Meteosat Second Generation for four different standard atmospheres: tropical (red), midlatitude summer (green), midlatitude winter (dark blue), and subarctic summer (light blue).

The following radiation effects on the radiance in WV channels of MSG can be illustrated in Fig. A.9, considering variations of the contribution functions for the considered atmospheric models:

- For the two channels, WV 6.2 μm and WV 7.3 μm, the levels of contribution function maximums decrease with decreasing of WV density. As a result, for the midlatitude winter atmosphere (dark blue), which is the driest one along the entire troposphere, they are closer to the Earth surface than for midlatitude summer (green) and tropical (red) models.
- The fact that the contribution function of 6.2 μm channel (Fig. A.5A) in the upper troposphere for midlatitude summer (green) lies higher than for tropical atmosphere (red) is mainly due to the corresponding higher WV density of midlatitude summer profile (slightly higher specific humidity at upper levels above 350 hPa level; see Fig. A.2B). This is partially also a result from the warmer profile of the tropical atmosphere: As shown in Fischer et al. (1981), in the upper-troposphere the contribution function (for WV 6.3 μm channel) of a warm atmosphere

lies at lower pressures, while at lower levels the normalized contribution functions of a warmer and a colder profile do not show significant differences.

- On the contrary (see Fig. A.9A), in the middle troposphere the contribution function of 6.2 μm channel for midlatitude summer (green) lies lower than for tropical atmosphere (red) because the WV below 350 hPa level is much denser for a tropical than for a midlatitude summer atmosphere (see Fig. A.2B).
- The contribution function of WV 7.3 μm channel (see Fig. A.9B) for tropical atmosphere (red) has higher values than for midlatitude summer (green) in a deep layer from 350 hPa level down to the low troposphere, due to its profile of higher WV density and higher temperature in the layer (see Fig. A.2).

Comparing Fig. A.9A and A.9B shows that radiation in 7.3 μm band is able to penetrate the WV to a great extent than the 6.2 μm radiation. If the low-level air is cold and the ground is cold, this causes little difference in the radiance measured by the two channels. However, if the low-level air and the ground are warm, the 7.3 μm radiation arrives to the satellite from warmer air, producing a much warmer brightness temperature than by the 6.2 μm radiation. Since the absorption by WV in 6.2 μm channel is higher, smaller amounts of moisture, commonly at higher altitudes, are detectable by 6.2 μm than by 7.3 μm radiation measurements. Because of this, low-level features are totally obscured on 6.2 μm imagery.

A.4 ALTITUDE ASSOCIATION OF THE CONTRIBUTION TO THE METEOSAT WATER VAPOR CHANNEL RADIANCE: EFFECT OF PORTIONING AND CROSSOVER EFFECT

A.4.1 RADIANCES FROM MOISTURE STRATIFIED IN SINGLE LAYERS

With a temperature profile fixed, in real cases the moisture could often be stratified in single layers at different altitudes. With the aim to quantitatively evaluate the response of the absorption/air mass channels of SEVIRI radiometer 6.2, 7.3, and 8.7 μm to stratification of humidity in the troposphere, calculations by using RTTOV8 radiation model (Saunders and Brunel, 2005) are performed. A standard temperature profile with troposphere profiles of moisture stratified in individual layers are handled by the model to learn about the response of these three channels to various cases of vertical distribution of humidity. The calculations are performed at 35°, 45°, and 55° latitude and correspond to zenith angles 40.62°, 51.78°, and 62.68°, respectively. Table A.2 shows the distribution of the layers and temperature profile for the standard atmosphere used. Eighteen different moisture profiles are considered at a given standard vertical distribution of the air temperature. For each profile, one of the 18 layers is defined as a moist layer by setting relative humidity to a constant value of 97%, 60%, 30%, and 30%. All other 17 layers of the profile are considered as dry layers of "background moisture": the minimum values of humidity defined by the RTTOV8 model with the fixed temperature sounding. The background moisture in the dry layers varies from 0.12×10^{-2} g per kg to 0.67×10^{-1} g per kg mixing ratio, and from 0.17% to 2.15% relative humidity, accordingly. Moist layers are defined by setting higher moisture at the two border RTTOV8 levels with a gradually decrease in humidity, eg, from 97% to the background moisture along the entire adjacent dry layers.

Table A.2 Description of RTTOV8 Levels Used for Monthly Mean Standard Temperature Profile (ECMWF Reanalysis Midlatitude 45°N) and Definition of the Moist Layers

RTTOV Level No.	Definition	Level Temperature (°C)	Moist Layer No.	Bottom–Top Pressure of Layer (hPa)	RTTOV Level No.	Definition	Level Temperature (°C)	Moist Layer No.	Bottom–Top Pressure of Layer (hPa)
22	222.94 hPa	−55.0			32	610.60 hPa	−8.5	9	610.6–565.5
23	253.71 hPa	−52.0	18	253.7–222.9	33	656.43 hPa	−5.2	8	656.4–610.6
24	286.60 hPa	−47.2	17	286.6–253.7	34	702.73 hPa	−2.1	7	702.7–656.4
25	321.50 hPa	−41.9	16	321.5–286.6	35	749.12 hPa	0.6	6	749.1–702.7
26	358.28 hPa	−36.0	15	358.3–321.5	36	795.09 hPa	3.0	5	795.1–749.1
27	396.81 hPa	−31.0	14	396.8–358.3	37	839.95 hPa	5.5	4	839.9–795.1
28	436.95 hPa	−25.8	13	436.9–396.8	38	882.82 hPa	7.6	3	882.8–839.90
29	478.54 hPa	−20.7	12	478.5–436.9	40	957.44 hPa	11.1	2	957.4–882.8
30	521.46 hPa	−16.3	11	521.5–478.5	43	Surface (2 m)	13.7	1	1000.0–957.4
31	565.54 hPa	−12.2	10	565.5–521.5					

For the model simulations, a mean temperature profile, derived by the European Center for Medium-Range Weather Forecasting (ECMWF) 40-year reanalysis is used at a set of 19 isobaric levels. These levels are defined by their pressures as are for the levels from No 22 to No 43 of the RTTOV8 radiation code. Thus the calculations are performed for 18 moist layers at different altitude locations and at 45°N latitude, which corresponds to zenith angle 51.78°, respectively. Descriptions of the RTTOV8 levels, which are used as well as definitions of the moist layers, are presented in Table A.2. Standard atmospheric pressure of 1000 hPa is accepted for the surface.

For each of the 18 model profiles defined, the radiance, brightness temperature, and transmittance from the bottom of the layer to space are calculated for the MSG WV channels.

A.4.2 BRIGHTNESS TEMPERATURE

The Brightness temperature is a measure of the intensity of radiation thermally emitted by an object, given in units of temperature since there is a correlation between the intensity of the radiation emitted and physical temperature of the radiating body that is given by the Stefan–Boltzmann Eq. [A.7].

$$I = \varepsilon \sigma T_e^4 \qquad [A.7]$$

here ε is the IR emissivity; $\sigma = 5.6696 \times 10^{-8}$ Wm^{-2}deg^{-4} is the Stefan–Boltzman constant; and T_e is the effective radiating temperature.

For split-window IR channels of MSG (10.8 and 12 μm), the brightness temperature equals the surface blackbody temperature of the object, which radiates. Because of the absorption by WV, the brightness temperature derived in WV channels (especially 6.2 μm) may be totally different from the physical temperature of the object depending on the vertical distribution of humidity. In general the radiation from WV that reaches the satellite does not arrive from a single surface or level, but from some layer of finite depth. WV—in typical concentrations—is semitransparent for the radiation in 6.2 and 7.3 μm channels, except for the low-level. Therefore, the brightness temperature measured by the satellite in the WV channels is a "net" temperature of some layer of moisture, not the temperature of any single surface or level.

A.4.3 RADIANCE

For the measurements in the IR channels, the *radiance* is related to the *brightness temperature* through the Plank function. The following analytic relation between the radiances (R) and the equivalent brightness temperatures T_b for the Meteosat (SEVIRI) channels is adopted:

$$R(\nu_c) = C_1 \nu_c^3 / \{\exp[C_2 \nu_c / (\alpha T_b + \beta)] - 1\} \qquad [A.8]$$

with: $C_1 = 1.19104 \times 10^{-5}$ mW m^{-2} sr^{-1}(cm^{-1})$^{-4}$; $C_2 = 1.43877$ K(cm^{-1})$^{-1}$; ν_c the central wave number of the channel, presented in Table A.2; and α, β coefficients; see Table A.3.

A.4.4 EFFECT OF PORTIONING AND CROSSOVER EFFECT

The unique property of the weighting and contribution functions of the WV channels to peak around the middle troposphere (Figs. A.6 and A.7, respectively) is due to the effect that an amount of the radiation coming from Earth's surface passes through the tropospheric moist layers and another

Table A.3 Central Wave Numbers and Coefficients Used in RTTOV for Plank Function Calculations

Parameter for Meteosat Second Generation Channel	Water Vapor (WV) 6.2 μm	WV 7.3 μm	Infrared 8.7 μm
ν_c	1588.790140	1359.929538	1148.276868
α	0.9942812723	0.9986568065	0.9995245584
β	1.706687325	0.3196454542	0.09399094187

portion is absorbed and reemitted at low energy by a colder air. This specific radiation effect, produced by the WV absorption was defined as the crossover effect by Weldon and Holmes (1991) for the WV channels of GOES satellites.

With the air temperature sounding fixed, WV channel radiance, measured by the satellite, is produced by radiation intensities arriving from different origins. Below a specific level in the lower troposphere, the atmosphere becomes relatively opaque to the radiation. This threshold level changes with the WV channel and varies with the humidity and temperature. If some moisture is present in layers of finite depth above the threshold level, significant amounts of radiation pass through the layers from some warmer origin located below. The other portion is absorbed and reemitted at lower energy by colder air. Therefore, depending on the mixing ratio and the density of the WV in the layer, the radiation at the satellite is partly originated from the moist layer and partly from below. This effect is referred to as the "portioning effect" by Georgiev et al. (2007), equally "crossover effect" by Weldon and Holmes (1991). The "portioning effect" influences the radiation measured by different WV channels in different ways and allows these channels to be sensitive to the moisture distribution at different altitudes. Fig. A.10 illustrates the portioning effect for the two MSG WV channels, where the radiances, derived by RTTOV8 simulations, are presented in radiance units (W m^{-2} sr^{-1} cm^{-1}).

In order to interpret the portioning effect, the following three radiation curves are considered:

- Total radiance (black curve) is the net radiation intensity from which the brightness temperature is derived.
- "Contribution from the moist layer" (blue) is that portion contributed by the WV of the considered moist layer (as well as by the background moisture above).
- "Contribution from below" (red) is that portion of the total radiance contributed by the radiation that passes through the considered moist layer (coming from the earth surface as well as from the background moisture below the layer). In an idealized case of absolutely dry air below the moist layer, this portion represents the contribution of Earth's surface to the total radiance.

The total radiance is the clear-air radiance $L_{clr}(n_1)$ derived directly by RTTOV8 at the bottom of the moist layer for each one of the 18 profiles, representative for the 18 moist layers at different altitude locations.

The contribution from the moist layer (R_{layer}) to the radiance is derived from the RTTOV output for cloudy radiances, referred to as "overcast" radiance $L_{overcast}$, which represents cloud-affected radiances assuming black, opaque clouds. In order to eliminate the cloud effect in the overcast radiance and to obtain only the atmospheric contribution, the overcast radiance obtained for the bottom level n_1

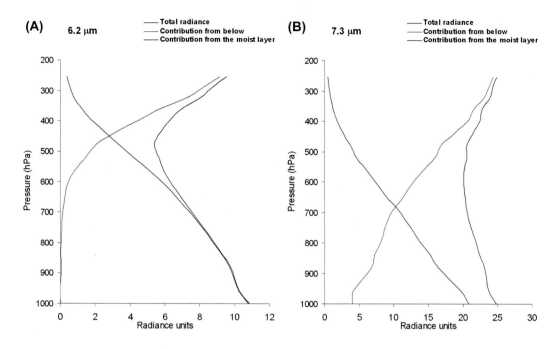

FIGURE A.10

Portioning and crossover effects for the two water vapor channels (A) 6.2 μm and (B) 7.3 μm illustrated by setting the relative humidity in the moist layers to 30%.

of the moist layer is diminished by the cloud-top radiance. The cloud-top radiance is derived from the radiation $R(n_1)$ of the bottom level multiplied by the transmittance $\sigma(n_1)$ of that level to space. Thus the portion contributed by the WV of the considered moist layer (and the background moisture above) is assigned to the contribution of the moist layer, obtained by expression [A.9].

$$R_{\text{layer}} = L_{\text{overcast}}(n_1) - R(n_1)\sigma(n_1) \qquad [\text{A.9}]$$

The radiation $R(n_1)$ at the bottom of the layer is calculated using the expression [A.8], assuming that the temperature of the bottom of the layer is T_b.

Finally, in the contribution from below the moist layer R_{below} is obtained as a difference between the clear-air radiance $L_{\text{clr}}(n_1)$ derived by RTTOV8 at the bottom of the layer and the contribution from the moist layer obtained according to expression [A.9] as follows:

$$R_{\text{below}} = L_{\text{clr}}(n_1) - R_{\text{layer}} \qquad [\text{A.10}]$$

At each level, the radiation of a specific channel is influenced by two opposite mechanisms as follows:

- Absorption of the radiation by the moist layer and re-radiation at a lower energetic level.
- Transparency of the layer for radiation coming from below the layer that moderates the cooling of the radiance due to absorption in the moist layer.

Due to the specific portioning effect of these opposite mechanisms, the "cooling effect" on the radiance exhibits its maximum at different levels for different absorption channels. These are the levels, around which a moist layer produces the coldest brightness temperature. The concept of radiance portioning reveals the following important characteristics of the radiances in 6.2 and 7.3 μm channels, seen in Fig. A.10:

- Layers at low levels often absorb all or most of the radiation coming from below and produce warm brightness temperatures because the moist layer itself is warm. The contribution from below is small, mostly absorbed by the low-level moisture. The moist layer contribution to the total radiance is high and dominates the contribution from below.
- Layers of moisture at high altitudes produce relatively warm brightness temperatures. The water content of such a layer is very small and much of the radiation from lower warmer sources penetrates through the layer. For that reason, the moist layer contribution is very small and the most part of the measured radiation is emitted by the lower-level moisture as well as by Earth's surface, then passes through the moist layer.
- For 6.2 and 7.3 μm WV channels the curves of the two portions of the radiances cross each other at the middle troposphere, and for that reason the effect of portioning was referred to as "crossover" effect by Weldon and Holmes (1991).
- Moist layers at middle altitudes produce the coldest brightness temperatures measured by the WV channels due to the crossover effect between radiation penetrating from below and radiation from the moist layer.

Fig. A.10 shows that the 6.2 and 7.3 μm channels exhibit similar overall crossover characteristics, but there are significant differences in their crossover altitudes. Since the absorption by WV in 6.2 μm channel is higher, the radiances measured by this channel are lower, and, accordingly, its brightness temperatures are colder than those for 7.3 μm channel. The crossover level for the 6.2 μm radiance is located at a higher altitude than this for 7.3 μm channel. This allows the radiances measured in the two channels to be used for assessing total moisture content, diagnosing dynamical processes at two different tropospheric layers, and distinguishing related thermodynamic structures (see Santurette and Georgiev, 2007; Georgiev and Santurette 2009).

Since the transmittance depends on the humidity in the moist layers, the crossover altitude and the level of maximum contribution depend on the vertical humidity profile. The influence of the relative humidity on the portioning effect and the crossover altitude was evaluated by performing the calculations for very low (10%) and high (97%) relative humidity of the moist layers. The results are presented in Fig. A.11 and show that the crossover altitudes for the two WV channels appear at lower levels with decreasing relative humidity of the moist layers from 97% to 10%.

The portioning effect for the IR 8.7 μm channel is presented in Fig. A.12. In cases of low humidity to nearly saturated WV in the moist layers (30−97% relative humidity), there is no crossover effect for the IR 8.7 μm radiance, since the contribution from the moist layer is very small due to very low absorption by WV and by the radiation in this channel.

It should be mentioned that the moist layers definition in Table A.2 is far from realistic, and hence all calculations in the RTTOV8 simulated radiances concerns highly idealized troposphere profiles of single moist layers and very dry troposphere above and below this moist layer. At the same time, for obtaining the RTTOV8 regression model coefficient, the training sample consists of about 50 real

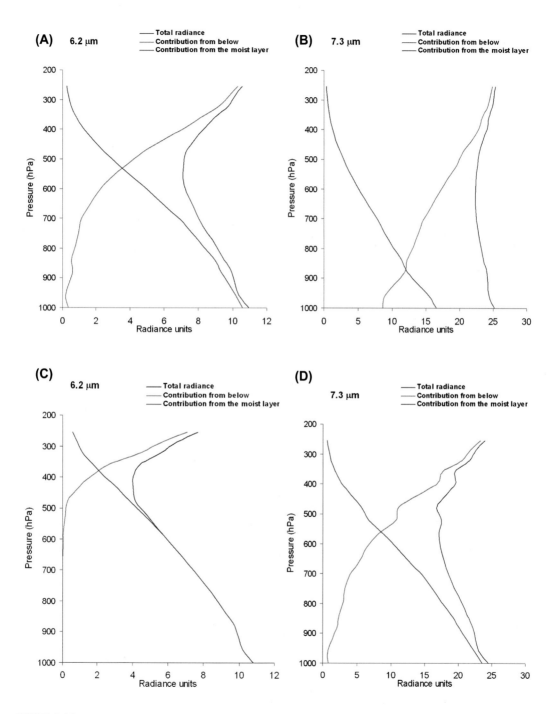

FIGURE A.11

Portioning and crossover effects for the two water vapor channels moist layer in different moisture regimes: relative humidity 10% for (A) 6.2 μm and (B) 7.3 μm; relative humidity 97% for (C) 6.2 μm and (D) 7.3 μm.

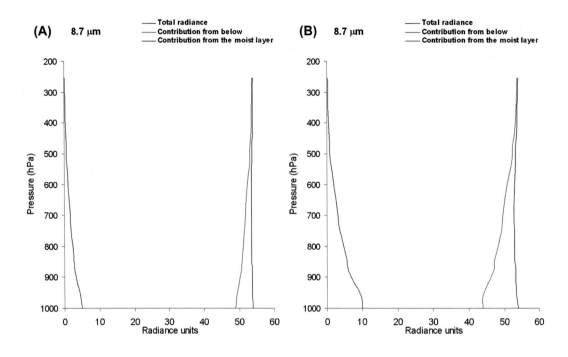

FIGURE A.12

Portioning effect of the infrared 8.7 μm channel for relative humidity in the moist layer (A) 30% and (B) 97%.

profiles. This problem reflects the levels of peak in absorption, and we are not sure how precise are the assessments of the crossover altitudes.

However, the portioning effect is as fundamental to the processes of radiation measurements as the basic concept of crossover effect is (see Weldon and Holmes, 1991). For that reason, Figs. A.10 and A.11 provide a basis of quantitative comparisons of the characteristics of 6.2 and 7.3 μm radiation with different vertical distributions of WV. They clearly show the differences between the two WV channels listed below.

- The coldest 6.2 μm brightness temperature is most likely to be produced by moisture located near 350 hPa for 97% relative humidity and near 450 hPa for 30% relative humidity.
- For 7.3 μm channel, the coldest brightness temperature appears near 550 hPa level for a moist layer of 97% relative humidity, and between 750 hPa for 30% relative humidity.

The minimum of radiance and hence the coldest brightness temperature appear around the levels of maximum contribution to the radiance for each channel (see also Fig. A.9).

POTENTIAL VORTICITY MODIFICATION TECHNIQUE AND POTENTIAL VORTICITY INVERSION TO CORRECT THE INITIAL STATE OF THE NUMERICAL MODEL

B

Although numerical weather prediction (NWP) models become increasingly accurate, there are still cases in which extreme weather conditions (thunderstorms, windstorms, snow, and heavy precipitation) are predicted at too late a stage. In most cases, the quality of the forecast is limited by the accuracy of the initial weather analysis that may suffer from a lack of observations or from an erroneous rejection of extreme values in the data-assimilation system. Forecasters can detect errors in the numerical weather analysis by comparing with observations. A misfit between the model analysis (or 3—6 h forecast) and observations leads to the subjective modification of some of the features of upper-level fields (potential vorticity [PV] field) and/or surface fields (temperature at 850 hPa isobaric surface or mean sea-level pressure). The analysis (or 3—6 h forecast) is then modified using the wind and temperature corrections provided by the inversion of the quasi-geostrophic PV correction and an alternative forecast is produced from this new state.

The PV is a useful meteorological parameter to study the dynamical structures at the synoptic scale, thanks to the conservative and invertibility principles (Hoskins et al., 1985). Both principles make the PV a suitable tracer of upper-level dynamics, which plays an important role in midlatitude synoptic developments. Upper-level positive PV anomalies can be interpreted as upper-level disturbances penetrating into the upper troposphere linked to the undulation of the tropopause.

The water vapor (WV) imagery in 6.2 μm (6.5/6.7 μm) channel of geostationary satellites is an important source of information in areas where conventional observations are rare, and it is an important tool for synoptic-scale analyses as it mostly reveals the water vapor content in the middle and upper troposphere. As shown in Chapter 3, dark and light patterns on WV images can be related to the structure of PV around the tropopause.

As Santurette and Joly (2002) showed, only the level of dynamical tropopause, generally expressed by the 1.5 PVU surface (1 PVU = 10^{-6} m^2 s^{-1} K kg^{-1}) over midlatitudes, is needed to diagnose upper-level dynamics if balance in the atmosphere and a monotonic vertical PV gradient are assumed. Taking advantage of the relationship between the WV images and the PV distribution close to the dynamical tropopause, some modifications can be carried out to the analyses of the PV field to improve the match with the satellite image. These modifications can be efficiently applied to the dynamical tropopause surface defined by the height of the 1.5 PVU surface. However, there is not a one-to-one relation between PV and WV images, and it is extremely difficult to find quantitatively a correct PV modification because of the large differences from case to case. It appeared that the skill of a forecaster in terms of pattern recognition and the meteorological experience must be exploited for successful application of this approach.

A way to improve the skills of the method is by comparing the imagery, derived by satellite measurements in the specific channel to synthetic imagery from a numerical model forecast (introduced in Section 5.2). Such a comparison is successfully applied as a diagnostic tool to evaluate forecast using synthetic images in 6.2 μm channel. Synthetic satellite imagery is generated using the fast radiative transfer model for TIROS Operational Vertical Sounder (RTTOV), originally developed at European Center for Medium Range Weather Forecast (ECMWF) in the early 1990s (Eyre, 1991). Input data used are vertical profiles of temperature and humidity for each grid point of an NWP model output data bank. The result of the calculation is a field of brightness temperature on the same grid as the numerical model grid.

The process of initializing the PV structures based on the information from satellite and synthetic images involves the following points:

- Compare the WV images from satellite observations with the synthetic imagery in the WV channel at the simulation initial time (or 3 h and 6 h forecast).
- Compare the dynamical tropopause (1.5 PVU surface height) with the WV brightness temperature distribution in areas with mismatch between the synthetic and satellite WV images (as broadly discussed in Section 5.3).
- Reduce the mismatch by modifying the topography of the dynamical tropopause accordingly.
- For each horizontal grid point of the numerical model, the new vertical height of the dynamical tropopause defines a PV correction, which is actually the difference between the original PV value at that height and 1.5 PVU. A 1DVAR method based on known forecast error statistics is then applied to build a vertical profile of PV modification over the corresponding grid point.
- Invert the modified PV field given the mass—wind balance condition derived by Charney (1955) following the methodology presented by Arbogast et al. (2008).
- Perform the corresponding numerical run using this improved initial state.

For illustration, Fig. B.1 shows the comparison of the synthetic and Meteosat WV images in channel 6.2 μm overlaid by the geopotential of the 1.5 PVU surface for August 6, 2014 (the NWP products, calculated using the ARPEGE forecast). The synthetic and satellite images appear in phase. The dry slots are well correlated with minimums of the dynamical tropopause height. Strong gradients in height of the dynamical tropopause follow the dark/light contrast in the satellite WV image. In this case, the conclusion is that the simulation of the upper-level dynamics by the NWP model is correct and therefore modifications of the geopotential of the 1.5 PVU surface are not necessary.

In case of disagreement between satellite observations and the operational NWP model output, one can decide how to modify the unperturbed PV field to reduce the mismatch, using the relationship between WV images and PV distribution near the dynamical tropopause. Fig. B.2 shows such a case on September 26, 2012. The locations of the areas, which need modifications, are indicated by red arrows: The PV gradient could be decreased (at the arrow to the north) and the curvature to be strengthened (two southern arrows) to match the curvature of the PV feature, which is present in the real atmosphere and seen on the satellite WV image.

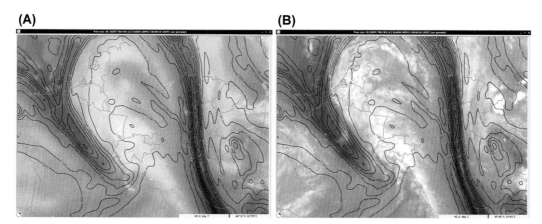

FIGURE B.1

Overlay for August 6, 2014, at 1500 UTC of the geopotential on the 1.5 PVU surface (3 h forecast, contour interval 50 dam) and (A) synthetic satellite image; (B) meteosat water vapor imagery.

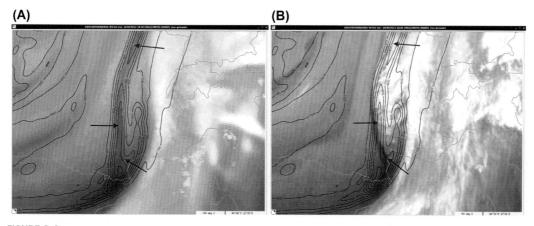

FIGURE B.2

As with Fig. B.1 except for the September 26, 2012, at 1800 UTC (6 h forecast). The arrows indicate locations where the potential vorticity modifications can be applied.

A method that enables forecasters to modify the initial state of the model to perform a new run has been developed at MétéoFrance. A graphical interface on the forecaster's workstation allows one to move a structure of the field, to increase or decrease a minimum, and also to refine the pattern by drawing new contours. Using this interactive tool, it is possible to modify interactively the two-dimensional (2D) 1.5 PVU height field over regions where a disagreement between satellite and synthetic WV imagery is observed. The 2D field of 1.5 PVU heights can be modified using predefined modifications: the addition of a source/sink, the strengthening/weakening of a structure, or the displacement of a structure. All modifications act on a selected area (free parameter). The modification is strongest in the center of the area and falls with increasing distance from the center to guarantee continuity. Modifications in the model initial fields can be carried out for the 1.5 PVU heights and also for the mean sea-level pressure or the temperature at 850 hPa at low level.

A one-dimensional optimal analysis method is then applied at each vertical profile in order to provide three-dimensional (3D) fields of PV modifications. The modified 3D PV field is, in turn, inverted, using the PV inversion procedure of Arbogast et al. (2008).

A forecaster's duty consists of defining a height displacement δz of the 1.5 PVU surface height for each vertical level corresponding in turn to a δPV that must be applied to the two closest model levels. To keep the PV vertical gradient unchanged, we apply the same δPV at the four model levels above and the four levels below. This information is incorporated into the model state using an optimal analysis based on the best linear unbiased estimate where the modifications are treated as an observation vector, and the initial model state P_{old} as the background. The new state P_{new} is written as:

$$P_{new} = P_{old} + BH^{T}\left(HBH^{T} + R\right)^{-1}\delta P$$

where R is the observation error matrix, B is the background error covariance matrix, and δP is the vector of eight PV modifications. The background error vertical covariances are estimated using a one-month sample matrix. As in most data assimilation systems, R is taken to be diagonal, which is a crude assumption within this context. We then assume that $HBH^{T} \gg R$; say, the modification must exceed the unmodified analysis P_{old}, as we wish that P_{new} would fit the modifications. Therefore, it becomes

$$P_{new} = P_{old} + BH^{T}\left(HBH^{T}\right)^{-1}\delta P$$

Each observation, as has been defined previously, impacts the new vertical PV distribution following background covariances (rows or columns of B) weighted by the term within parentheses, which is the background error variance at the observation location—acting only as a weight.

References

Agustí-Panareda, A., Thorncroft, C.D., Craig, G.C., Gray, S.L., 2004. The extratropical transition of Hurricane Irene (1999): a potential vorticity perspective. Quart. J. Roy. Meteor. Soc. 130, 1047–1074.

Agustí-Panareda, A., Gray, S.L., Craig, G.C., Thorncroft, C.D., 2005. The extratropical transition of tropical cyclone Lili (1996) and its crucial contribution to a moderate extratropical development. Mon. Wea. Rev. 133, 1562–1573.

Al Badi, H., Al Maskari, J., Roesli, H.-P., 2009. Two Cases of Severe Convection along the Red Sea Coast of Saudi Arabia (25 November and 22 December 2009). EUMETSAT Image Gallery Casa Studies. EUMETSAT, Darmstadt. http://oiswww.eumetsat.org/WEBOPS/iotm/iotm/20091222_convection/20091222_convection.html.

Arakawa, A., Schubert, W.H., 1974. Interaction of a cumulus cloud ensemble with the large-scale environment, part I. J. Atmos. Sci. 31, 674–701.

Arbogast, P., Maynard, K., Crépin, F., 2008. Ertel potential vorticity inversion using a digital filter initialization method. Quart. Jour. Roy. Meteor. Soc. 134, 1287–1296.

Arbogast, P., Maynard, K., Piriou, C., 2012. About the reliability of manual model PV corrections to improve forecasts. Wea. Forecasting 27, 1554–1567.

Argence, S., Lambert, D., Richard, E., Chaboureau, J.-P., Arbogast, P., Maynard, K., 2009. Improving the numerical prediction of a cyclone in the Mediterranean by local potential vorticity modifications. Quart. J. Roy. Meteor. Soc. 135, 865–879.

Bader, M.J., Forbes, G.S., Grant, J.R., Lilley, R.B.E., Waters, A.J. (Eds.), 1995. Images in Weather Forecasting. A Practical Guide for Interpreting Satellite and Radar Imagery. University Press, Cambridge, 499 pp.

Bao, J.-W., Michelson, S.A., Neiman, P.J., Ralph, F.M., Wilczak, J.M., 2006. Interpretation of enhanced integrated water vapor bands associated with extratropical cyclones: their formation and connection to tropical moisture. Mon. Wea. Rev. 134, 1063–1080.

Berrisford, P., Hoskins, B.J., Tyrlis, E., 2007. Blocking and Rossby wave breaking on the dynamical tropopause in the southern hemi-sphere. J. Atmos. Sci. 64, 2881–2898.

Bluestein, H.B., 1993. Synoptic-Dynamic Meteorology in Midlatitudes. In: Observations and Theory of Weather Systems, vol. II. Oxford University Press, 594 pp.

Bosart, L.F., Lin, S.C., 1984. A diagnostic analysis of the Presidents' day storm of February 1979. Mon. Wea. Rev. 112, 2148–2177.

Bosart, L.F., Bracken, W.E., Molinari, J., Velden, C.S., Black, P.G., 2000. Environmental influences on the rapid intensification of Hurricane Opal (1995) over the Gulf of Mexico. Mon. Wea. Rev. 128, 322–352.

Brennan, M.J., Lackmann, J.M., 2005. The influence of incipient latent heat release on the precipitation distribution of the 24–25 January 2000 US East Coast cyclone. Mon. Wea. Rev. 133, 1913–1936.

Browning, K.A., 1997. The dry intrusion perspective of extra-tropical cyclone development. Meteorol. Appl. 4, 317–324.

Browning, K.A., 2004. The sting at the end of the tail: damaging winds associated with extratropical cyclones. Quart. Jour. Roy. Meteor. Soc. 130, 375–399.

Browning, K.A., 2007. The convective storm initiation project. Bull. Amer. Meteor. Soc. 88, 1939–1955.

Buizza, R., Montani, A., 1999. Targeting observations using singular vectors. J. Atmos. Sci. 56, 2965–2985.

Buizza, R., Palmer, T.N., 1995. The singular vector structure of the global atmospheric global circulation. J. Atmos. Sci. 52, 1434–1456.

Carlson, T.N., 1980. Airflow through midlatitude cyclones and the comma cloud pattern. Mon. Wea. Rev. 108, 1498–1509.

Carroll, E.B., 1997a. Use of dynamical concepts in weather forecasting. Meteorol. Appl. 4, 345–352. © UK Crown copyright, published by Met Office.

Carroll, E.B., 1997b. Poorly forecast trough disruption shown in water vapour images. Meteorol. Appl. 4, 229–234. © UK Crown copyright, published by Met Office.

Chaigne, E., Arbogast, P., 2000. Multiple PV inversions in two FASTEX cyclones. Quart. J. Roy. Meteor. Soc. 126, 1711–1784.

Chang-Seng, D.S., Jury, M.R., 2010. Tropical cyclones in the SW Indian Ocean. Part 2: structure and impacts at the event scale. Meteorol. Atmos. Phys. 106, 163–178.

Charney, J., 1955. The use of the primitive equations of motion in numerical prediction. Tellus 7, 22–26.

Clark, P.A., Browning, K.A., Wang, C., 2005. The sting at the end of the tail: model diagnostics of fine-scale three-dimensional structure of the cloud head. Quart. J. Roy. Meteor. Soc. 131, 2263–2292.

Conte, D., Miglietta, M.M., Levizzani, V., 2011. Analysis of instability indices during the development of a Mediterranean tropical-like cyclone using MSG-SEVIRI products and the LAPS model. Atmos. Res. 101, 264–279.

COMET® Program, 2009. Dynamic Feature Identification. Deformation Zone Analysis. University Corporation for Atmospheric Research, Boulder, CO. https://www.meted.ucar.edu/norlat/sat_features/dz_analysis/. Copyright 2007.

COMET® Program, 2015. Assessing NWP with Water Vapour Imagery. University Corporation for Atmospheric Research, Boulder, CO. https://www.meted.ucar.edu/norlat/sat_features/anwv/navmenu.php. Copyright 2015.

COMET® Program, 2016. Forecasting Sensible Weather from Water Vapour Imagery. University Corporation for Atmospheric Research, Boulder, CO. https://www.meted.ucar.edu/training_module.php?id=1124. Copyright 2016.

Courtier, P., Freydier, C., Geleyn, J.-F., Rabier, F., Rochas, M., 1991. The ARPEGE project at Météo-France. In: Proceedings ECMWF Seminar, vol. II. ECMWF, Reading, United Kingdom, pp. 193–231.

Demirtas, M., Thorpe, A.J., 1999. Sensitivity of short-range forecasts to local potential vorticity modifications. Mon. Wea. Rev. 127, 922–939.

Doswell C.A., III, 1987. The distinction between large-scale and mesoscale contribution to severe convection: a case study example. Wea. Forecasting 2, 3–16.

Doswell C.A., III, Brooks, H.E., Maddox, R.A., 1996. Flash flood forecasting: an ingredients-based methodology. Wea. Forecasting 11, 560–581.

Doswell, C.A., Bosart, L.F., 2001. Extratropical synoptic-scale processes and severe convection. Ch. 2. In: Doswell C.A., III (Ed.), Severe Convective Storms, Meteorological Monographs, pp. 27–69.

Durkee, J.D., Mote, T.L., 2010. A climatology of warm-season mesoscale convective complexes in subtropical South America. Int. J. Climatol. 30, 418–431.

Emanuel, K.A., 2007. Quasi-equilibrium dynamics of the tropical atmosphere. In: Schneider, T., Sobel, A.H. (Eds.), The Global Circulation of the Atmosphere. Princeton University Press, pp. 186–218.

EUMeTrain/EUMETSAT, 2012. SatNanu. Manual of Synoptic Satellite Meteorology. © Copyright EUMeTrain. ZAMG, Vienna. http://www.eumetrain.org/satmanu/index.html.

EUMETSAT, 2005. Upper Level Divergence Product Algorithm Description. EUM/MET/REP/05/0163. EUMETSAT, Germany, 16 pp.

EUMETSAT, 2012. The Conversion from Effective Radiances to Equivalent Brightness Temperatures. EUM/MET/TEN/11/0569. EUMETSAT, Darmstadt. http://www.eumetsat.int/website/home/Data/Products/Calibration/MSGCalibration/index.html?lang=EN.

Ellrod, G.P., 1990. A water vapor image feature related to severe thunderstorms. Natl. Wea. Dig. 15 (4), 21–29.

Eyre, J.R., 1991. A Fast Radiative Transfer Model for Satellite Sounding Systems. ECMWF Research Dept. Tech. Memo. ECMWF, Reading, 176 pp.

Fehlmann, R., Quadri, C., Davies, H.C., 2000. An Alpine rainstorm: sensitivity to the mesoscale upper level structure. Wea. Forecasting 15, 4–28.

Fischer, H., Eigenwillig, N., Müller, H., 1981. Information content of METEOSAT and Nimbus/THIR water vapor channel data: altitude association of observed phenomena. J. Appl. Meteorol. 20, 1344–1352.

Fujita, T.T., 1978. Manual of Downburst Identification for Project NIMROD. SMRP Research Paper 156. University of Chicago, 104 pp.

Funatsu, B.M., Waugh, D.W., 2008. Connections between potential vorticity intrusions and convection in the eastern tropical Pacific. J. Atmos. Sci. 65, 987–1002.

Georgiev, C.G., 1999. Quantitative relationship between Meteosat WV data and positive potential vorticity anomalies: a case study over the Mediterranean. Meteorol. Appl. 6, 97–109.

Georgiev, C.G., Martín, F., 2001. Use of potential vorticity fields, Meteosat water vapour imagery and pseudo water vapour images for evaluating numerical model behaviour. Meteorol. Appl. 8, 57–69.

Georgiev, C.G., 2003. Use of data from Meteosat water vapour channel and surface observations for studying pre-convective environment of a tornado-producing storm. Atmos. Res. 67–98, 231–246.

Georgiev, C.G., Santurette, P., Dupont, F., Brunel, P., 2007. Quantitative evaluation of 6.2 µm, 7.3 µm, 8.7 µm Meteosat channels response to tropospheric moisture distribution. In: Proceedings Joint 2007 EUMETSAT Meteorological Satellite Conference and the 15th AMS Satellite Meteorology & Oceanography Conference, Amsterdam, The Netherlands, 24–28 September 2007, ISBN:92-9110-079-X, ISSN:1011-3932, Darmstadt.

Georgiev, C.G., Santurette, P., 2009. Mid-level jet in intense convective environment as seen in the 7.3 µm satellite imagery. Atmos. Res. 93, 277–285.

Georgiev, C.G., Kozinarova, G., 2009. Usefulness of satellite water vapour imagery in forecasting strong convection: a flash-flood case study. Atmos. Res. 93, 295–303.

Georgiev, C.G., Santurette, P., 2010. Quality of MPEF DIVergence product as a tool for very short range forecasting of convection. In: Proceedings of 2010 EUMETSAT Meteorological Satellite Conference (Córdoba 20 – 24 September 2010). ISSN: 1011-3932.

Georgiev, C.G., 2013. Information content of MPEF DIVergence product in diagnosing the environment of deep convection. Atmos. Res. 123, 337–353.

Ghosh, A., Loharb, D., Dasc, J., 2008. Initiation of Nor'wester in relation to mid-upper and low-level water vapor patterns on METEOSAT-5 images. Atmos. Res. 87 (2), 116–135.

Griffiths, M., Thorpe, A.J., Browning, K.A., 1998. Convective destabilization by a tropopause fold diagnosed using potential vorticity inversion. JCMM Intern. Rep. 96 pp.

Grimes, A., Mercer, A.E., 2015. Synoptic Scale Precursors to Tropical Cyclone Rapid Intensification in the Atlantic Basin. Adv. Meteorology, 2015, Art. ID 814043, 16 pp. http://dx.doi.org/10.1155/2015/814043.

Hamilton, D.W., Lin, Y.-L., Weglarz, R.P., Kaplan, M.L., 1998. Jetlet formation from diabatic forcing with applications to the 1994 Palm Sunday tornado outbreak. Mon. Wea. Rev. 126, 2061–2089.

Hanley, D., Molinari, J., Keyser, D., 2001. A compositive study of the interactions between tropical cyclones and upper-tropospheric troughs. Mon. Wea. Rev. 129, 2570–2584.

Hello, G., Arbogast, P., 2004. Two different methods to correct the initial conditions, potential vorticity modifications and perturbations with sensitivities: an application to the 27 December storm over south of France. Meteor. Appl. 11, 41–57.

Hewison, T.J., Wu, X., Yu, F., Tahara, Y., Hu, X., Kim, D., Koenig, M., 2013. GSICs inter-calibration of infrared channels of geostationary imagers using Metop/IASI. IEEE Trans. Geosci. Remote Sens 51 (3). http://dx.doi.org/10.1109/TGRS.2013.2238544.

Highwood, E.J., Hoskins, B., 1998. The tropical tropopause. Quart. Jour. Roy. Meteor. Soc. 124, 1579–1604.

Hofer, L., Schipper, J., Georgiev, C., 2011. MPEF Divergence, 12 September 2010. Available at EUMeTrain: http://www.eumetrain.org/data/1/181/contrib.htm.

Homar, V., Romero, R., Ramis, C., Alonso, S., 2002. Numerical study of the October 2000 torrential precipitation event over eastern Spain: analysis of the synoptic-scale stationarity. Ann. Geophys. 20, 2047–2066.

Homar, V., Romero, R., Stensrud, D., Ramis, C., Alonso, S., 2003. Numerical diagnosis of a small, quasi-tropical cyclone over the western Mediterranean: dynamical vs. boundary factors. Quart. Jour. Roy. Meteor. Soc. 129, 1469–1490.

Hoskins, B.J., Draghici, I., Davies, H.C., 1978. A new look at the ω-equation. Quart. Jour. Roy. Meteor. Soc. 104, 31–38.

Hoskins, B.J., McIntyre, M.E., Robertson, A.W., 1985. On the use and significance of isentropic potential vorticity maps. Quart. Jour. Roy. Meteor. Soc. 111, 877–946.

Hoskins, B., 1997. A potential vorticity view of synoptic development. Meteorol. Appl. 4, 325–334.

Huo, Z., Zhang, D.-L., Gyakum, J.R., 1999. Interaction of potential vorticity anomalies in extratropical cyclogenesis. Part II: sensitivity to initial perturbations. Mon. Wea. Rev. 127, 2563–2575.

Insel, N., Poulsen, C.J., Ehlers, T.A., 2010. Influence of the Andes Mountains on South American moisture transport, convection, and precipitation. Clim. Dyn. 35, 1477–1492.

Iwabe, C.M.N., da Rocha, R.P., 2009. An event of stratospheric air intrusion and its associated secondary surface cyclogenesis over the South Atlantic Ocean. J. Geophys. Res. 114, D09101. http://dx.doi.org/10.1029/2008JD011119.

Jones, S.C., Harr, P.A., Abraham, J., Bosart, L.F., Bowyer, P.J., Hanley, D.E., Hanstrum, B.N., Lalaurette, F., Sinclair, M.R., Smith, R.K., Thorncroft, C., 2003. The extratropical transition of tropical cyclones: forecast challenges, current understanding and future directions. Wea. Forecasting 18, 1052–1092.

Knippertz, P., Martin, J.E., 2007. A Pacific moisture conveyor belt and its relationship to a significant precipitation event in the semiarid Southwestern United States. Wea. Forecasting 22, 125–144.

Köenig, M., de Coning, E., 2009. The MSG global instability indices product and its use as a nowcasting tool. Wea. Forecasting 24, 272–285.

Kraucunas, I., Hartmann, D.L., 2005. Equatorial superrotation and the factors controlling the zonal mean zonal winds in the tropical upper troposphere. J. Atmos. Sci. 62, 371–389.

Krennert, T., Zwatz-Meise, V., 2003. Initiation of convective cells in relation to water vapour boundaries in satellite images. Atmos. Res. 6768, 353–366.

Leroux, M.-D.E., Plu, M., Barbary, D., Roux, F., Arbogast, P., 2013. Dynamical and physical processes leading to tropical cyclone intensification under upper-level trough forcing. J. Atmos. Sci. 70, 2547–2565.

Lin, J.-L., Qian, T., Shinoda, T., Li, S., 2015. Is the tropical atmosphere in convective quasi-equilibrium? J. Clim. 28, 4357–4372.

Mansfield, D.A., 1996. The use of potential vorticity as an operational forecast tool. Meteorol. Appl. 3, 195–210.

Malardel, S., 2008. Fondamentaux de météorologie, à l'école du temps, 2è édition. Cépadues, Météo-France, Toulouse. (2.85428.631.6). 709 pp.

Martín, F., Elizaga, F., Riosalido, R., 1999. The mushroom configuration in water vapour imagery and operational applications. Meteorol. Appl. 6, 143–154. © UK Crown copyright, published by Met Office.

McCann, D.W., 1983. The enhanced-V: a satellite observable severe storm signature. Mon. Wea. Rev. 111, 887–894.

Merrill, R., 1988. Environmental influences on Hurricane intensification. J. Atmos. Sci. 45, 1678–1687.

Michel, Y., Bouttier, F., 2006. Automated tracking of dry intrusions on satellite water vapour imagery and model output. Quart. J. Roy. Meteor. Soc. 132 (620), 2257–2276.

Molinari, J., Vollaro, D., Alsheimer, F., Willoughby, H.E., 1998. Potential vorticity analysis of tropical cyclone intensification. J. Atmos. Sci. 55, 2632–2644.

Montani, A., Thorpe, A.J., 2002. Mechanisms leading to singular-vector growth for Fastex cyclones. Quart. J. Roy. Meteor. Soc. 128, 131–148.

Morwenna, G., Thorpe, A.J., Browning, K.A., 2000. Convective destabilization by a tropopause fold diagnosed using potential-vorticity inversion. Quart. J. Roy. Meteor. Soc. 126, 125–144.

NASA, 2010. Stunning NASA Infrared Imagery of Hurricane Igor Reveals a 170 Degree Temperature Difference! Hurricane and Tropical Storms. Latest storm images and data from NASA. http://www.nasa.gov/mission_pages/hurricanes/archives/2010/h2010_Igor.html.

NASA, 2013. Wipha (Eastern/Western Pacific). Satellite Sees Extra-tropical Typhoon Wipha Affecting Alaska. http://www.nasa.gov/content/goddard/wipha-easternwestern-pacific/#.VKhJqsnmnTQ.

NWCSAF, 2015. Satellite Application Facility on Support to Nowcasting/and Very Short-range Forecasting. http://www.nwcsaf.org/HD/MainNS.jsp.

Olander, T.L., Velden, C.S., 2009. Tropical cyclone convection and intensity analysis using differenced infrared and water vapor imagery. Wea. Forecasting 24, 1558—1572.

Pagé, C., Fillion, L., Zwack, P., 2007. Diagnosing summertime mesoscale vertical motion: Implications for atmospheric data assimilation. Mon. Wea. Rev. 135, 2076—2209.

Pankiewicz, G.S., Swarbrick, S.J., Watkin, S.C., 1999. Automatic estimation of potential vorticity from Meteosat water vapour imagery to adjust initial fields in NWP. In: The 1999 Meteorological Satellite Data Users' Conference, EUM P 26. ISSN: 1011-3932. EUMETSAT, Lighthouse Multimedia, Darmstadt, pp. 387—394.

Pasch, R.J., Kimberlain, T.B., 2011. Tropical Cyclone Report Hurricane Igor (AL112010), 8—21 September 2010. NOAA National Hurricane Season. http://www.nhc.noaa.gov/data/tcr/AL112010_Igor.pdf.

Prasad, K., 2006. Environmental and Synoptic Conditions Associated with Nor'westers and Tornadoes in Bangladesh—an Appraisal Based on Numerical Weather Prediction (NWP) Guidance Products. SMRC—No. 14. SAARC Meteorol. Res. Centre, Dhaka, Bangladesh. http://www.saarc-smrc.org/Report%20No14.pdf.

Pedder, M.A., 1997. The omega equation: Q-G interpretations of simple circulation features. Meteorol. Appl. 4, 335—344.

Plu, M., Arbogast, P., Joly, A., 2008. A wavelet representation of synoptic-scale coherent structures. J. Atmos. Sci. 65, 3116—3138.

Rabin, R., Corfidi, S.F., Brunner, J.C., Hain, C.E., 2004. Detecting winds aloft from water vapour satellite imagery in the vicinity of storms. Weather 59, 251—257.

Ralph, F.M., Neiman, P.J., Wick, G.A., 2004. Satellite and CALJET aircraft observations of atmospheric rivers over the eastern North Pacific Ocean during the winter of 1997/98. Mon. Wea. Rev. 132, 1721—1745.

Ralph, F.M., Neiman, P.J., Rotunno, R., 2005. Dropsonde observations in low-level jets over the northeastern Pacific Ocean from CALJET-1998 and PACJET-2001: mean vertical-profile and atmospheric-river characteristics. Mon. Wea. Rev. 133, 889—910.

Rappin, E.D., Morgan, M.C., Tripoli, G.J., 2011. The impact of outflow environment on tropical cyclone intensification and structure. J. Atmos. Sci. 68, 177—194.

Roberts, N.M., 2000. The relationships between water vapour imagery and thunderstorms. JCMM Intern. Rep 110, 40 pp. Available from Joint Centre for Mesoscale Meteorology, Dept. of Meteorology, University of Reading.

Romero, R., 2001. Sensitivity of a heavy rain producing western Mediterranean cyclone to embedded potential vorticity anomalies. Quart. Jour. Roy. Meteor. Soc. 127, 2559—2597.

Romero, R., Martin, A., Homar, V., Alonso, S., Ramis, C., 2005. Predictability of prototype flash flood events in the western Mediterranean under uncertainties of the precursor upper-level disturbance: the HYDROPTIMET case studies. Nat. Hazards Earth Syst. Sci. 5, 505—525.

Russell, A., Vaughan, G., Norton, E.G., Morcrette, C.J., Browning, K.A., Blyth, A.M., 2008. Convective inhibition beneath an upper-level PV anomaly. Quart. Jour. Roy. Meteor. Soc. 134, 371—383.

Santurette, P., Joly, A., 2002. ANASYG/PRESYG, Météo-France's new graphical summary of the synoptic situation. Meteorol. Appl. 9, 129—154.

Santurette, P., Arbogast, P., 2002. Water vapour image: a tool allowing the forecaster to monitor and improve the model initial state. In: Proceedings of the 2002 Meteorological Satellite Data Users' Conference, Dublin, 2—6 September 2002. EUM P 36, ISSN 1011-3932, EUMETSAT, 193—199.

Santurette, P., Georgiev, C.G., 2005. Weather Analysis and Forecasting: Applying Satellite Water Vapor Imagery and Potential Vorticity Analysis. Academic Press, Elsevier Inc., 179 pp.

Santurette, P., Georgiev, C.G., 2007. Water vapour imagery analysis in 7.3 µ/6.2 µ for diagnosing thermo-dynamic context of intense convection. In: Proceedings Joint 2007 EUMETSAT Meteorological Satellite Conference and the 15th AMS Satellite Meteorology & Oceanography Conference, Amsterdam, The Netherlands, 24—28 September 2007, ISBN 92-9110-079-X, 1011—3932, Darmstadt.

Saunders, R.W., Brunel, P., 2005. RTTOV_8_7 Users Guide. NWPSAF-MO-ud-008. EUMETSAT, Darmstadt.

Schmetz, J., Turpeinen, O.M., 1988. Estimation of the upper tropospheric relative humidity field from METEOSAT water vapor image data. J. Appl. Meteor. 27, 889—899.

Schmetz, J., Borde, R., Holmlund, K., König, M., 2005. Upper tropospheric divergence in tropical convective systems from Meteosat-8. Geophys. Res. Lett. 32, L24804.

Schultz, D., Sienkiewicz, J., 2013. Using Frontogenesis to identify sting jets in extratropical cyclones. Wea. Forecasting 28, 603–613.

Scofield, R.A., 1985. Satellite convective categories associated with heavy precipitation. In: Preprint, Sixth Conference on Hydrometeorology, October 29, 1985. American Meteorological Society, Indianapolis, pp. 42–51.

Shapiro, M.A., 1981. Frontogenesis and ageostrophic forced secondary circulations in the vicinity of jet stream-frontal zone systems. J. Atmos. Sci. 38, 954–973.

Shapiro, M.A., 1982. Mesoscale Weather Systems of the Central United States. NOAA Tech. Rep. CIRES, Boulder, CO, 78 pp.

Shapiro, M.A., Keyser, D., 1990. Fronts, jet streams and the tropopause. In: Newton, C.W., Holopainen, E.O. (Eds.), Extratropical Cyclones, the Erik Palmen Memorial Volume. American Meteorological Society, pp. 167–191.

Shapiro, L.J., Willoughby, H.E., 1982. The response of balanced hurricanes to local sources of heat and momentum. J. Atmos. Sci. 39 (2), 378–394.

Simmons, A.J., Hollingsworth, A., 2002. Some aspects of the improvement in skill of numerical weather prediction. Quart. Jour. Roy. Meteor. Soc. 128, 647–677.

Smith, R.K., 1980. Tropical cyclone eye dynamics. J. Atmos. Sci. 37, 1227–1232.

Sodemann, H., Stohl, A., 2013. Moisture origin and meridional transport in atmospheric rivers and their association with multiple cyclones. Mon. Wea. Rev. 141, 2850–2867.

Stoyanova, J.S., Georgiev, C.G., 2013. SVAT modelling in support to flood risk assessment in Bulgaria. Atmos. Res. 123, 384–399.

Swarbrick, S.J., 2001. Applying the relationship between PV fields and water vapour imagery to adjust initial conditions in NWP. Meteorol. Appl. 8, 221–228.

Thiao, W., Scofield, R., Robinson, J., 1993. The Relationship Between Water Vapor Plumes and Extreme Rainfall Events During the Summer Season. NOAA Tech. Memor. NOAA Tech. Memor. NESDIS 67. Sat. Appl. Lab., Washington, 69 pp.

Tjemkes, S., Schmetz, J., 1997. Synthetic satellite radiances using the radiance sampling method. J. Geophys. Res. 102 (D2), 1807–1818.

Tjemkes, S.A., Schmetz, J., 2002. Radiative Transfer Simulations for the Thermal Channels of METEOSAT Second Generation. EUMETSAT, Darmstadt. http://www.eumetsat.int/website/wcm/idc/idcplg?IdcService= GET_FILE&dDocName=PDF_TM01_MSG-RADIO-TRANS-SIM&RevisionSelectionMethod=LatestReleased &Rendition=Web.

Uccellini, L.W., Johnson, D.R., 1979. The coupling of upper- and lower-tropospheric jet streaks and implications for the development of severe convective storms. Mon. Wea. Rev. 107, 682–703.

Velden, C.S., Olander, T.L., Zehr, R.M., 1998. Development of an objective scheme to estimate tropical cyclone intensity from digital geostationary satellite infrared imagery. Wea. Forecasting 13, 172–186.

Weldon, R.B., Holmes, S.J., 1991. Water Vapor Imagery: Interpretation and Applications to Weather Analysis and Forecasting. NOAA Technical Report. NESDIS 57. NOAA, US Department of Commerce, Washington D.C., 213 pp.

Wikipedia, 2013. Typhoon Wipha. http://en.wikipedia.org/wiki/File:Wipha_2013_track.png, http://en.wikipedia. org/wiki/Typhoon_Wipha_%282013%29.

Yasunari, T., Miwa, T., 2006. Convective cloud systems over the Tibetan Plateau and their impact on meso-scale disturbances in the Meiyu/Baiu Frontal zone. J. Meteor. Soc. Jpn. 84, 783–803.

Zhu, Y., Newell, R.E., 1998. A proposed algorithm for moisture fluxes from atmospheric rivers. Mon. Wea. Rev. 126, 725–735.

Zhu, H., Thorpe, A., 2006. The predictability of extra-tropical cyclones: the influence of the initial condition and model uncertainties. J. Atmos. Sci. 63, 1483–1497.

Glossary

2D	Two-dimensional
3D	Three-dimensional
4D	Four-dimensional
1DVAR	One-dimensional variational data assimilation
AMVs	Atmospheric motion vectors
ARPEGE	MétéoFrance operational numerical weather forecast model (Action de Recherche Petite Echelle-Grande Echelle). Stretched global model; T 798 stretching factor C 2.4; equivalent grid resolution varies from 30 to 15 km from North America to Western Europe (10 km over France), 35 km over subtropical North Pacific, and 60 km over New Zealand; 70 vertical levels. ARPEGE was updated in April 2015 (105 Levels, T 1198, C 2.2, equivalent grid resolution 7.5 km over France, 18 km over North America, 25 km over subtropical North Pacific, and 36 km over New Zealand)
ARs	Atmospheric rivers
CAPE	Convective available potential energy
CIN	Convective inhibition
COMET	Education and training program for the environmental sciences
CTP	Cloud top pressure
DIV (product)	Divergence (in the upper-troposphere derived by satellite data)
ECMWF	European Center for Medium-range Weather Forecasts
EUMETSAT	The European Organization for the Exploitation of Meteorological Satellites
ET	Extratropical transition
GII	Global instability indices
GOES	Geostationary Operational Environmental Satellites of NOAA
Himawari	(New generation) Geostationary satellite of JMA
HRV	High-resolution visible (the Meteosat imaging channel operating in this part of the spectrum)
INSAT	Geostationary meteorological satellite of Indian Space Research Organization
IR	Infrared (the satellite imaging channel operating in this part of the spectrum)
iso-θ	Constant potential temperature surfaces
ITCZ	Inter tropical convergence zone
JMA	Japan Meteorological Agency
LFC	Level of free convection
LHR	Latent heat release
LLJ	Low-level jet
MCB	Moisture conveyor belt
MCS	Mesoscale Convective System
Meteosat	The EUMETSAT Geostationary Meteorological Satellite
MLJ	Mid-level jet
MPEF	Meteosat Product Extraction Facilities of EUMETSAT
MSG	Meteosat Second Generation
MSLP	Mean sea-level pressure
MTSAT	Geostationary satellite of JMA
NCEP	National Centers for Environmental Prediction
NESDIS	National Environmental Satellite Data and Information Service
NOAA	National Oceanic and Atmospheric Administration (of the United States)
NWC SAF	Nowcasting Satellite Application Facility
NWP	Numerical weather prediction

PV	Potential vorticity
PVU	Potential vorticity unit ($1\,PVU = 10^{-6}\,m^2\,s^{-1}\,K\,kg^{-1}$)
PWV	Pseudo water vapor image (synthetic water vapor image derived from atmospheric model)
RTTOV	TIROS Operational Vertical Sounder
SCF	Surface cold front
SEVIRI	Spinning Enhanced Visible and InfraRed Imager
SST	Sea-surface temperature
SV	Singular vector
TIROS	Television Infrared Observation Satellite Program of NOAA
TE	Total energy
TC	Tropical cyclone
θ	Potential temperature
θ_w	Wet-bulb potential temperature
UCF	Upper cold front
UTC	United coordinated time
VIS	VISible (the Meteosat imaging channel operating in this part of the spectrum)
WCB	Warm conveyor belt
WMO	World Meteorological Organization
WV	Water vapor (the satellite imaging channel operating in this part of the spectrum)

Index

335

vorticity (*Continued*)
definition of, 3
potential. *See* potential vorticity (PV)

W

warm advection, 15, 86, 91, 100, 183, 185, 210, 266
warm anomaly, 15, 76—77, 103—104, 154, 181, 183, 202,
  252—254
Warm Conveyor Belt (WCB), 88, 205
water vapor absorption band, radiation measurements in,
  301—322, 302t, 304f—309f, 311f, 313f—314f, 316t,
  318t, 319f, 321f—322f
water vapor imagery. *See also* satellite water vapor imagery,
  interpretation problem of; water vapor imagery
  features associated with synoptic thermodynamic
  structures
  mismatch between PV fields and, 241—258
    cyclogenesis with upper-level precursor in strong zonal
      flow, 244—247, 245f—246f
    cyclone development within cut-off low system, 241—244,
      242f—244f
    moist ascent at cut-off upper-level flow, 247—250,
      248f—249f
    upper-level influence on deep convection within cut-off
      low system, 250—254, 251f, 253f
  operational use of relationship between PV fields and,
    224—228, 296—297
    dry intrusion—PV anomalies relationship, complexity of,
      227—228
    information content of vorticity fields related to water
      vapor imagery, 226—227
    nature and usefulness of, 224—226, 225f
    PV fields—WV imagery—synthetic WV images comparison,
      230—241, 231f, 297
    instances of, 233—241, 233f, 236f—237f, 239f—240f
    numerical weather prediction output, validating, 231—233,
      231f
water vapor imagery features associated with synoptic
  thermodynamic structures, 55—156
  blocking regime, 91—97
    in which easterlies result from anticyclogenesis, 93—95,
      94f, 96f
    in which easterlies result from cyclogenesis, 93—97,
      94f, 98f
  cyclogenesis with upper-level precursors, 103—115
    cyclone deepening, WV imagery dry slot as precursor of,
      113—115, 117f—118f
    cyclone development in Western North Atlantic, 104—110,
      105f, 107f, 109f, 111f
    explosive cyclogenesis in Southern West Pacific, 110—113,
      112f, 114f, 116f
    Hurricane Igor, WV imagery analysis of, 132f, 134—136

intensification, 134—136, 137f
weakening, 136, 137f
intensification of tropical cyclones on anticyclonic shear side
  of jet streams, 138f, 139—142, 140f—141f
interaction effects with midlatitude upper-level troughs on
  tropical cyclones intensity, 142—143
light and dark imagery features, interpretation of, 57—68
  dry (dark) bands/spots, 60, 61f—62f
  dry intrusions, 60—67, 61f—64f, 66f—67f
  jet stream moisture boundaries, 68, 69f
  medium-gray to light-gray features, 58f, 59
  nearly white to white features, 58—59, 58f
middle troposphere wind field, 68—91
  dry air mass over Northeast Africa, 90—91, 92f
  jet stream interaction with tropopause dynamic
    anomaly, 74
  moisture movement, 87—90, 87f, 89f
  synoptic context, 83—86, 84f—85f
radiation measurements in WV channels, operational use of,
  56—57
reintensification, 136—138, 137f—138f
split cold front, in WV channels imagery, 122—129,
  124f—128f
sting jet and related surface wind gusts, WV imagery for
  identifying, 116—122, 118f—122f
superposition of WV imagery and dynamical fields, 155
tropical cyclones interaction with upper-level dynamic
  structures, 129—152, 132f
  tropical cyclone track satellite data from National Oceanic
    and Atmospheric Administration (NOAA) NESDIS,
    132—133, 132f
  tropical cyclones role in extratropical development associated
    with upstream upper-level positive potential vorticity
    anomaly, 143—152, 145f—147f, 149f—151f
  Typhoon Wipha, WV imagery analysis of, 132f,
    133—134, 135f
    intensification, 133—134, 135f
upper troposphere wind field, 68—91
  deep convection in midlatitudes, 77—78, 79f
  deep convection in tropical areas, 80—83, 81f—82f, 83t
  jet stream interaction with tropopause dynamic anomaly,
    74, 75f
  patterns of, 70—74, 71f, 73f
  synoptic scale upper-level perturbations, 76—77, 78f
weighting function, 312
wet-bulb potential temperature (θ_w), 84, 121f, 128f, 160,
  162—163, 174, 178f, 179—180, 207—210, 228, 269
wind maximum, 12—13, 85—86, 88, 90, 183, 185, 206—207
wind vectors, 68, 74, 77, 83—84, 88, 91, 95—97, 99, 104,
  118—119, 133, 144, 167, 176—178, 197, 208—210,
  216, 244—245, 254—255
WV bands, 21